Adoption and Optimization of Embedded and Real-Time Communication Systems

Seppo Virtanen
University of Turku, Finland

Managing Director: Lindsay Johnston
Editorial Director: Joel Gamon
Book Production Manager: Jennifer Yoder
Publishing Systems Analyst: Adrienne Freeland
Development Editor: Heather Probst
Assistant Acquisitions Editor: Kayla Wolfe
Typesetter: Christina Henning
Cover Design: Jason Mull

Published in the United States of America by
Information Science Reference (an imprint of IGI Global)
701 E. Chocolate Avenue
Hershey PA 17033
Tel: 717-533-8845
Fax: 717-533-8661
E-mail: cust@igi-global.com
Web site: http://www.igi-global.com

Library of Congress Cataloging-in-Publication Data

Adoption and optimization of embedded and real-time communication systems / Seppo Virtanen, editor.
p. cm.
Includes bibliographical references and index.
Summary: "This book presents innovative research on the integration of embedded systems, real-time systems and the developments toward multimedia technology"--Provided by publisher.
ISBN 978-1-4666-2776-5 (hardcover) -- ISBN 978-1-4666-2777-2 (ebook) -- ISBN 978-1-4666-2778-9 (print & perpetual access) 1. Embedded computer systems--Research. 2. Real-time programming--Research. 3. Multimedia communications--Research. I. Virtanen, Seppo, 1972-
TK7895.E42A36 2013
006.2'2--dc23

2012032564

British Cataloguing in Publication Data
A Cataloguing in Publication record for this book is available from the British Library.

Editorial Advisory Board

List of Reviewers

Table of Contents

Detailed Table of Contents

Section 1
Designing and Engineering Embedded and Sensor Networks

During past decades, Ethernet progressively became the most widely used Local Area Network (LAN) technology. Apart from LAN installations, Ethernet also became attractive for other application areas such as industrial control, automotive, and avionics. In traditional LAN design, the objective is to minimize the network deployment cost. However, in embedded networks, additional constraints and ambient conditions add to the complexity of the problem. In this paper, the authors propose a Simulated Annealing (SA) algorithm to optimize the physical topology of an embedded Ethernet network. The various constraints and ambient conditions are modeled by a cost map. For networks with small number of nodes and/or switches, the authors were able to find the optimal solutions using adapted algorithms. These solutions will serve as a lower bound for the solutions obtained via the SA algorithm. However, the adapted algorithms are time consuming and application specific. The paper shows that the SA algorithm can be applied in all cases and finds (near-) optimal solutions.

Today, SDL and SystemC are two very popular languages for embedded systems modeling. SDL has specific advanced features that make it good for reflection of the multi-object systems and interactions between modules. It is also good for system model validation. The SystemC models are better suitable for tracing internal functions of the modeled modules. The hypothetical possibility of combined use of these two languages promises a number of benefits for researchers. This article specifically addresses and discusses the integration of SDL and SystemC modeling environments, exchange the data and control information between the SDL and SystemC sub-modules and the real-time co-modeling aspects of the integrated SDL/SystemC system. As a result, the mechanisms of SDL/SystemC co-modeling is presented and illustrated for an embedded network protocols co-modeling case study. The article gives an overview and description of a co-modeling solution for embedded networks protocols simulation based on experience and previous publications and research.

Resource constrained Wireless Sensor Networks (WSNs) require an automated firmware updating protocol for adding new features or error fixes. Reprogramming nodes manually is often impractical or even impossible. Current update protocols require a large external memory or external WSN transport protocol. This paper presents the design, implementation, and experiments of a Program Image Dissemination Protocol (PIDP) for autonomous WSNs. It is reliable, lightweight and it supports multi-hopping. PIDP does not require external memory, is independent of the WSN implementation, transfers firmware, and reprograms the whole program image. It was implemented on a node platform with an 8-bit microcontroller and a 2.4 GHz radio. Implementation requires 22 bytes of data memory and less than 7 kilobytes of program memory. PIDP updates 178 nodes within 5 hours. One update consumes under 1‰ of the energy of two AA batteries.

Geographic Forwarding (GF) algorithms typically employ a neighbor discovery method to maintain a neighborhood table that works well only if all wireless links are symmetric. Recent experimental research has revealed that the link conditions in realistic wireless networks vary significantly from the ideal disk model and a substantial percentage of links are asymmetric. Existing GF algorithms fail to consider asymmetric links in neighbor discovery and thus discount a significant number of potentially stable routes with good one-way reliability. This paper introduces Asymmetric Geographic Forwarding (A-GF), which discovers asymmetric links in the network, evaluates them for stability (e.g., based on mobility), and uses them to obtain more efficient and shorter routes. A-GF also successfully identifies transient asymmetric links and ignores them to further improve the routing efficiency. Comparisons of A-GF to the original GF algorithm and another related symmetric routing algorithm indicate a decrease in hop count (and therefore latency) and an increase in successful route establishments, with only a small increase in overhead.

Section 2
Model-Based Testing of Embedded and Real-Time Communication Systems

Requirements traceability enables the linkage between all development artifacts during the development process. Within model-based testing, requirements traceability links the original requirements with test model elements and generated test cases. Current approaches are either not practical or lack the necessary formal foundation for generating requirements-based test cases using model-checking techniques involving the requirements trace. This paper describes a practical and formal approach to ensure requirements traceability. The descriptions of the requirements are defined on path fragments of timed automata or timed state charts. The graphical representation of these paths is called a computation sequence chart (CSC). CSCs are automatically transformed into temporal logic formulae. A model-checking algorithm considers these formulae when generating test cases.

This article reports the results of an industrial case study demonstrating the efficacy of a model-based testing process in assuring the quality of highly configurable systems from the automation domain. Escalating demand for flexibility has made modern embedded software systems highly configurable. This configurability is often realized through parameters and a highly configurable system possesses a handful of those. Small changes in parameter values can account for significant changes in the system's behavior, whereas in other cases, changed parameters may not result in any perceivable reaction. This case study addresses the challenge of applying model-based testing to configurable embedded software systems to reduce development effort. As a result of the case study, a model-based testing process was developed and tailored toward the needs of the automation domain. This process integrates existing model-based testing methods and tools, such as combinatorial design and constraint processing. The testing process was applied as part of the case study and analyzed in terms of its actual saving potentials, which reduced the testing effort by more than a third.

Introduction of formal model-based practices into the development process of a product in a company implicates changes in the verification and validation activities. A testing process that focuses only on code is not comprehensive in a framework where the building blocks of development are models, and industry is currently heading toward more effective strategies to cope with this new reality. This paper reports the experience of a railway signalling manufacturer in changing its unit level verification process from code-based testing to a two-step approach comprising model-based testing and abstract interpretation. Empirical results on different projects, on which the overall development process was progressively tuned, show that the change paid back in terms of verification cost reduction (about 70%), bug detection, and correction capability.

Monitoring services are essential for advanced, reliable NoC systems. They should support traffic management, system reconfiguration and fault detection to enable optimal performance and reliability of the system. The paper presents a thorough description of NoC monitoring structures and studies earlier works. A distributed monitoring structure is proposed and compared against the structures presented in previous works. The proposed distributed network monitoring system does not require centralized control, is fully scalable and does not cause significant traffic overhead to the network. The distributed structure is in line with the scalability and flexibility of the NoC paradigm. The paper studies the monitoring structure features and analyzes traffic overhead, monitoring data diffusion, cost and performance. The advantages of distributed monitoring are found evident and the limitations of the structure are discussed.

Efficient runtime resource management in multi-processor systems-on-chip (MPSoCs) for achieving high performance and low energy consumption is one of the key challenges for system designers. OSIP, an operating system application-specific instruction-set processor, together with its well-defined programming model, provides a promising solution. It delivers high computational performance to deal with dynamic task scheduling and mapping. Being programmable, it can easily be adapted to different systems. However, the distributed computation among the different processing elements introduces complexity to the communication architecture, which tends to become the bottleneck of such systems. In this work,

The ability to restore a virtual platform from a previously saved simulation state can considerably shorten the typical edit-compile-debug cycle for software developers and therefore enhance productivity. For SystemC based virtual platforms (VP), dedicated checkpoint/restore (C/R) solutions are required, taking into account the specific characteristics of such platforms. Apart from restoring the simulation process from a checkpoint image, the proposed checkpoint solution also takes care of re-attaching debuggers and interactive GUIs to the restored virtual platform. The checkpointing is handled automatically for most of the SystemC modules, only the usage of host OS resources requires user provision. A process checkpointing based C/R has been selected in order to minimize the adaption required for existing VPs at the expense of large checkpoint sizes. This drawback is overcome by introducing an online compression to the checkpoint process. A case study based on the SHAPES Virtual Platform is conducted to investigate the applicability of the proposed framework as well as the impact of checkpoint compression in a realistic system environment.

Preface

The number of embedded systems around us and affecting our daily lives increases tremendously every year. On the other hand, we see ever more capable smartphones entering the market at a steady rate; devices that allow us to fulfill the traditional cell phone functions of placing calls and sending text messages, but which in addition serve as secondary personal computers for purposes like sending and receiving electronic mail, accessing social networks and browsing the World Wide Web. Even more, modern smartphones are capable of providing navigation services for drivers and pedestrians alike, and they can serve as still photo cameras, high definition video cameras, music players, TV remote controllers, and even health and wellness monitors for athletes. And, on the other hand, we see also the increase in the number of ubiquitous and embedded computing systems around us: smart systems that perform computing tasks without the users considering the functionality a result of a computing task. Examples of such systems are modern household appliances, intelligent ventilation, heating and lighting systems for apartments and houses, and medical systems like pacemakers that perform long-term heart rate data acquisition while executing their principal function. In practice, all the previously mentioned computing systems either require or at least would benefit from a communication link to other devices to fully take advantage of the system's capabilities and intended functionality.

The need for communication capabilities for such systems results in the systems becoming embedded and real-time communication systems instead of traditional stand-alone systems. Also the user perspective drives the development to this direction: more and more users nowadays wish that the system they use has the ability to connect to the Internet and even see their system completely unusable or crippled if the Internet connection is not available – despite the fact that the system, for example a smartphone, is in most cases still usable in its principal function (in the case of a smartphone, placing calls) even if there is no Internet connection. The convergence of communication technologies is paving a road for devices that support multiple communication standards for different applications and use scenarios: Near-Field Communication (NFC) for devices within touch range of each other for example for transferring files or collaborating in a game; Bluetooth for Personal Area Networking (PAN) to for example transfer files within a 10 meter range, connecting hands-free systems or wireless microphones and loudspeakers; Wireless Local Area Networks (WLAN) for connecting devices within a single space (for example an apartment) to each other and optionally (and typically) to an Internet access point; and 3G and LTE cellular connections when moving around in urban areas. From the user's perspective, the embedded communication system is expected to support a large variety of different communication technologies and standards, and to be capable of switching easily to the best available communication standard for the current application. From the manufacturer's point of view, the problem is not trivial: first of all, all relevant standards need to be supported by the system, which causes design challenges for device size, power consumption, and manufacturing price. Intelligent hardware design is not enough, but the software running on the device must also be intelligent in its decisions to switch from one communication standard to another, and in switching on and off the different communication subsystems to conserve battery power and increase stand-by time. Constantly developing improved and more efficient specification, modeling, design and verification flows and methodologies is thus in the focal point of manufacturers trying to meet consumer requirements within a sustainable development and manufacturing time and cost.

The difficult design constraints for embedded and real-time communication systems are extremely visible in designing and implementing wireless sensor networks. The sensors need to use their communication features to form a network with each other and the master node that collects the data from the deployed sensors. At the same time, the sensors need to be small and they need to stay operational for lengthy times with their own power source. Depending on the type of sensor network in question, it may be possible to recharge the power source while deployed, for example using solar energy or kinetic energy, but this is only possible if the target application of the sensor network is such that the operating environment and the design constraints allow for such functionality. Planning the network implementation and topology between the sensor nodes is also a problem far from trivial. Many different network topologies and communication standards can be and have been applied to wireless sensor networks, but still more research is needed. A promising approach could be based on implementing dynamically ad-hoc style network connections between neighboring nodes and then to build intelligently and collaboratively a (small) forwarding table between the sensor nodes so that each node learns how to contact the master node. The forwarding table must naturally be balanced so that no single node is overloaded as a relay station between a sensor node and the master node. In such a network, the collaborative intelligence must be able to recover from broken connections between nodes for example in cases where a node fails or runs out of power. The approach is beneficial in terms of overall functionality of the network: nodes that are furthest from the master node do not need to transmit information at full transmission power, but the power consumption is distributed across the network. With this approach, the network is likely to remain fully functional longer than in a situation where the far away nodes consume their power source entirely much before the nodes that are closer to the master node.

Often the term network is considered to mean a wired or wireless network link between two or more devices, but from the point of view of embedded and real-time communication systems, specific attention must today be paid to networks that reside inside devices. Embedded networks are typically networks that connect different modules of a system to each other, making it possible for the modules to communicate with each other. For example, a modern automobile has embedded networks built in the car to make it possible for different master computation modules to gather data from sensors and dispatch operative commands to different subsystems of the automobile depending on the use scenario. For example, Powertrain Control Modules (PCM) control how the engine and transmission work based on the received sensor readings and resulting computations, and the Anti-lock Brake System (ABS) uses sensory input from each wheel accessed through an embedded network to control how the brakes work in situations where one or more wheels of the automobile are about to stop rotating during heavy braking, thus allowing the driver to control the direction of the automobile while making an emergency braking. Besides automobiles, also smaller devices like smartphones, tablets, televisions, and Blu-Ray disc players contain many different types of internal embedded networks for connecting the different subsystems. In all these applications of embedded networks, it is evident that an embedded network must meet heavy design constraints set by the target application for proper functionality. For example in the case of an automobile, the powertrain and ABS control modules need to receive data within real-time constraints to ensure safe functionality of the entire system, and while watching high definition video on a modern television, an ensured data throughput speed between different submodules is needed to provide a pleasant and jitter-free visual experience to the viewer. Clearly, embedded networks are a very important contemporary research area in embedded and real-time communication deserving scholarly attention.

Going one step deeper into a device, that is, inside one of the modules, we find another type of internal network: the Network-on-Chip (NoC). More and more often today a microprocessor controlling the intelligence of the entire system actually consists of multiple processor cores or processing elements built into one component. Such chips are designed to meet a particular usage scenario: the chip may be used as a central processing unit in a computer, in which case the chip would contain several programmable general purpose processor cores, or it may serve as a processing element optimized from some particular application domain in a system. In the latter case, the chip would include one, or perhaps a few, programmable general purpose processor cores accompanied with more specialized cores like a digital signal processor core for DSP functionality or a protocol or network processor core for optimized protocol processing functionality. In any case, the different cores and processing elements must be interconnected so that they are able to communicate with each other to achieve the intended functionality of the chip. The communication between cores and processing elements is organized into an on-chip communication network with an interconnection topology, signal routing ability, and a communication protocol. Microchips structured in this way and connected with such a network are called Networks-on-Chip. The design of an on-chip network is a challenging task and each different NoC or MPSoC (Multi-Processor System-on-Chip) has its own requirements for the communication network infrastructure to be used on the chip. Huge challenges are brought to the network design with the ever-decreasing line widths of chip manufacturing technologies: it is not enough to specify connection links and protocols, but careful consideration must be given to signal quality deterioration, fault tolerance, synchronization and power consumption. Network-on-Chip is a very important research area in embedded and real-time communication systems currently and will be so for the foreseeable future.

Too often information security related aspects are a bit overlooked in designing computing systems. This holds also for embedded and real-time communication systems; returning to the automobile example discussed previously, it is a well-known fact that the embedded networks of modern automobiles are poorly protected. For example, in some cases, an intruder with expert knowledge of a particular make and model of a car could be able to access the embedded network by removing a side mirror and taking advantage of the wiring revealed this way. Depending on the make and model of the car, the intruder could perform a variety of things after gaining access to the network ranging from unlocking the power door locks to disabling some critical functionality in the engine or brake system. Security is many times considered a patch that is built on the final product every time some security related problem is detected in-house or reported from the field, or as an area requiring a "plug-in" from outside of the original design.

This type of approach is seen for example in designs that implement security using a co-processor, probably designed and implemented by a third party. Such an approach also limits the future expandability of the designed system: the co-processor is likely to not scale up to future communication requirements. Also overall system robustness may be lower with this kind of approach. An improved approach would allow for some programmability in the security implementation, for example a programmable co-processor. With this approach, some scalability might be gained. The proper way of taking security into consideration would be to include system security as an integral part of the system from the beginning of the planning, specification and design processes. In this kind of approach, the security features of the system would be put to test and verified in the same design phases where other system components are tested and verified. A deeper look into the essence of information security reveals that it is not just one piece of functionality, but actually a complex concept that has many different aspects the designer must deal with.

At a very high abstraction level, a proper design methodology must take security into account in at least three different forms: (1) security of external communication, (2) security of internal communication, and (3) security of the information stored and processed in the device. The first form is the one that is perhaps most often regarded as information security: securing the communication channel between separate devices. This type of security could be met for example in encrypted cellular telephone calls, protected communication of sensor networks or encrypted Wireless Local Area Network (WLAN) connections. As this form of security is the most commonly considered one, it is no surprise that it is quite often encountered as a piece of functionality in communication systems. However, for example the security failure of the Wired Equivalent Privacy (WEP) WLAN encryption method shows that security was not thoroughly enough considered throughout the specification and design process of such networks. The second mentioned form of security to be considered in designing embedded and real-time communication systems is found in the internal communication structures of devices. Even if the communication takes place within a device, it still needs protection against potential misuse. Electromagnetic signals leaking from the internal communication may be captured and analyzed for example to catch data or, in the case of encrypted communication, to find the encryption key. Internal communication protection could also mean an authentication scheme between the modules of the system to detect if one module has been changed in the system and used for a masquerade style attack for capturing manufacturer-specific information flowing in the system. In designing embedded and real-time communication systems, it is imperative to also consider the possibility of someone trying to gain access to the data transfers occurring inside the system.

Finally, in the third type of security to be considered, the question is about protecting the digital content stored in the device in case the device were for example to fall into wrong hands. Modern digital equipment is nowadays often full of private information that is not intended to be disclosed. For example, a smartphone can be used as a secondary computer for viewing and editing a variety of office file formats. Such files can be received as electronic mail attachments, and the attachments are usually downloaded to the device and stored locally, Similarly, all the contact information found in smartphone address books and electronic mail accounts accessible in the device are information that most users wish to keep private and protected. Unfortunately, no single industry-standard method for protecting the data stored in the device exists. The biggest challenge here lies in usability: many users are willing to trade data security for convenient and easy use of the system. Creating a cross-platform data security framework that is unbreakable but convenient and easy to use from one system to another is a challenge both for scientists and engineers of embedded and real-time communication systems, requiring multi-disciplinary collaboration.

The need for a methodological approach of integrating security into the system design process has been identified already about ten years ago, but a comprehensive embedded system design methodology with incorporated security design support remains to be seen. This is clearly an area in embedded and real-time communication system design where research emphasis is needed in the forthcoming years. One step towards an integrated methodology would be to study embedded communication system platforms that would have security functionality built in. For example, an on-going research effort at the University of Turku, Finland is focusing on developing an integrated security-enabled embedded communication system platform. The work has it foundations in research done on protocol processors supporting Software

Defined Radio (SDR). The Software Defined Radio paradigm is aimed at extensively solving the need for supporting multiple radio interfaces and communication standards. However, the extremely severe concern of information security in modern communication is not adequately addressed by the SDR concept. SDR capable devices should also incorporate built-in security technologies for communication standards supported by the SDR chip. Specification, design, and research effort is needed to find a novel solution in which SDR functionality is accompanied with programmable and hardware-level optimized protocol processing and security algorithm support and functionality. At University of Turku, such a platform is called the Software Defined Secure Communication (SDSC) platform. SDSC should be seen as the next milestone in the evolution of SDR, and more generally, communication systems. Designing a platform capable of hardware acceleration of heterogeneous functionality like protocol processing, SDR baseband processing, cryptography, and security algorithms in a constrained environment is a challenging task, calling forth optimization techniques like cross-domain functionality integration for tighter reuse of shared resources, and utilization of the target applications' inherent parallelism. Meeting all these goals produces vast research challenges to scholars and engineers working in the embedded and real-time communication system field for years to come.

This book is a summation volume of articles published in the *International Journal of Embedded and Real-Time Communication Systems* (IJERTCS) in its second volume year, 2011. The journal has an interdisciplinary scope, binding together research from different disciplines with focus on how the disciplines converge to embedded and real-time systems for the communication application domain. The subject coverage of the journal is broad, which enables a clear presentation of how the research results presented in the journal benefit the convergence of embedded systems, real-time systems and communication systems. The journal is aimed to benefit scientists, researchers, industry professionals, educators, and junior researchers like PhD students in the embedded systems and communication systems sector. An important aim is to provide the target audience with a forum to disseminate and obtain information on the most recent advances in embedded and real-time communication systems research: to give the readers the opportunity to take advantage of the research presented in the journal in their scientific, industrial or educational purposes.

As a journal in the focal point of disciplines such as computer science, computer engineering, telecommunication and communication engineering, the *International Journal of Embedded and Real-Time Communication Systems* is positioned well to provide its readership with interesting and well-focused articles based on recent high-quality research. The journal's coverage in topics from embedded systems, real-time systems, and communications system engineering, and especially how these disciplines interact in the field of embedded and real-time systems for communication, offers its readership both theoretical and practical research results facilitating the convergence of embedded systems, real-time computing, and communication system technologies and paradigms.

This book is organized into three thematic sections. The section on *Designing and Engineering Embedded and Sensor Networks* focuses on novel technological advances for embedded and special-purpose networks. Many different aspects of technological advances are covered, for example embedded network topology optimization, high abstraction level modeling of embedded networks, protocol design for updating sensor network firmware, and improving the reliability and performance of ad-hoc routing in wireless sensor networks. The selection of topics clearly highlights the multidisciplinary nature of this research area: embedded and real-time communication systems are positioned in the focal points of research of several different information technology disciplines.

The second thematic section of this book is *Model-Based Testing of Embedded and Real-Time Communication Systems*. As discussed in the previous pages, developing new design and verification methodologies and flows is essential to efficiently engineer more and more complex embedded and real-time communication systems. The classical way of first developing a system and then testing it at the end of a development cycle does not meet the requirements put on system quality and reliability in a short development time. Lately an interest in academia and industry has been attracted to different model based approaches. Modeling techniques help in improving the correctness and quality of complex system designs, and also improve productivity by for example providing graphical support to modeling languages. Model based testing (MBT) uses abstract specifications of the systems under test and their operating environments. These are used for checking the implementation of the system against its original specifications. The specifications are used for automatically designing, deriving, validating and optimizing tests.

The third thematic section is *Technologies and Design Methods for Network-on-Chip Communication*, where focus is on the research problems regarding reliable communication inside a chip, for example between the different on-chip cores that together form a Multi-Processor System-on-Chip (MPSoC). An on-chip communication implementation itself can be seen as an embedded and real-time communication system, but very often today for example MPSoCs are an essential building part of some larger embedded communication system. The efficient and reliable implementation of on-chip communication is essential for ensuring excellence in device performance and variety in its operating capabilities, and is thus in the heart of the research areas needing scholarly focus for future communication systems and environments.

The first thematic section of this book, Designing and Engineering Embedded and Sensor Networks, consists of four chapters. Chapter 1, titled *Cost-Based Topology Optimization of Embedded Ethernet Networks* is written by Jörg Sommer, Elias A. Doumith, and Andreas Reifert from University of Stuttgart (Germany). They propose a simulated annealing algorithm for optimizing the physical topology of an embedded Ethernet network. Ethernet networks are typically found in offices for implementing local area network functionality. However, recently Ethernet has become an attractive alternative also for embedded applications such as industrial control, automotive applications and avionics. In these types of embedded Ethernet networks, special constraints and ambient conditions add to the complexity of the problem and need to be addressed in designing the physical network topology. The authors propose the use of a cost map to model the different constraints and conditions and are able to find optimal link cost solutions using adapted algorithms. These algorithms are still deemed time-consuming and application-specific, for which reason the author's show that the simulated annealing algorithm can be applied generally to find near-optimal solutions efficiently to minimize the total link cost. An interesting continuation to the research, as pointed out by the authors, is to take into account also upper layer constraints and thus reach even more cost-efficient topologies.

Chapter 2 is titled *Co-Modeling of Embedded Networks Using SystemC and SDL* and is written by Sergey Balandin and Michel Gillet from Nokia Research Center (Finland), and Irina Lavrovskaya, Valentin Olenev, Alexey Rabin and Alexander Stepanov from St. Petersburg State University of Aerospace Instrumentation (Russia). This chapter focuses on the problem of modeling embedded systems using SDL and SystemC. Both modeling environments have their own benefits and specialties that may provide important advantages in one project whereas may play a miniscule role in some other project. Hence, the research in this article aims to combine SDL and SystemC into a co-modeling environment, where the benefits of both would be available to designers. Specific attention is given to integration of SDL and SystemC modeling environments, exchanging the data and control information between the SDL and SystemC sub-modules, and the real-time co-modeling aspects of the integrated SDL/SystemC system. The co-modeling environment is put to test by finding a solution for embedded network protocols simulation. The authors reach the conclusion that of the different ways of co-modeling using SDL and SystemC, a solution where SDL modules are included in SystemC is the most viable modeling option. The authors state that SDL and SystemC co-modeling is a very perspective and interesting area of research and can result in improved quality in facilitation of the specification, modeling and verification work.

Chapter 3 is titled *Design and Implementation of a Firmware Update Protocol for Resource Constrained Wireless Sensor Networks* and is written by Teemu Laukkarinen, Lasse Määttä, Jukka Suhonen, Timo D. Hämäläinen, and Marko Hännikäinen from Tampere University of Technology (Finland). As wireless sensor networks are a very resource-constrained low-cost device environment, they pose high demands for energy efficiency throughout the distributed system. Due to the nature of the systems, it is also a big challenge to update the software in the network nodes. Current update protocols rely on a large external memory or on external transport protocols. The authors present the design, implementation and experiments of a lightweight and reliable Program Image Dissemination Protocol (PIDP) for autonomous multihop wireless sensor networks. The proposed PIDP does not require external memory, is independent of the wireless sensor network implementation, transfers the program image, and reprograms the whole program image. As energy is an important issue in wireless sensor networks, one update only consumes under 1‰ of energy of two AA batteries in a typical network node, and updates a deployment of 178 nodes in 5 hours.

Chapter 4 ends the first thematic section of this book. The chapter is titled *Asymmetric Geographic Forwarding: Exploiting Link Asymmetry in Location Aware Routing* and it is written by Pramita Mitra and Christian Poellabauer from University of Notre Dame (USA). This chapter introduces a location aware routing approach called A-GF (asymmetric geographic forwarding). The approach takes advantage of asymmetric links in increasing the reliability and performance of ad-hoc routing. Link asymmetry in wireless transmission occurs when the transmission ranges of communicating parties are asymmetric: the station with a wider transmission range is able to successfully reach its peer station, but the peer station's transmission range is not wide enough to reach the other station. In this situation, only one-way transmissions are possible. Existing geographic forwarding algorithms assume that their environment consists of symmetric links. By discovering and exploiting also the asymmetric links the routing hop count is reduced, latencies decrease and routing reliability is improved. In making routing decisions, A-GF uses stability and minimum latency as metrics. The results of the work encourage the authors to further investigate the possibilities of combining stability and energy efficiency in asymmetric wireless transmission as their future work.

The second thematic section of this book, Model-Based Testing of Embedded and Real-Time Communication Systems, consists of three chapters. Chapter 5 is titled *Requirements Traceability within Model-Based Testing: Applying Path Fragments and Temporal Logic* and it is written by Vanessa Grosch from University of Ulm (Germany). The author presents a formal approach for tracing requirements in model-based testing in which requirements-based test cases are generated using model-checking techniques. In this approach, requirements are specified using path fragments of timed automata or timed state charts. The graphical representation of these paths is transformed automatically into temporal logic formulae, which in turn are used by a model-checking algorithm for deriving test cases. The suggested approach was applied in practice and evaluated in the automotive domain.

Chapter 6 is titled *Model-Based Testing of Highly Configurable Embedded Systems in the Automation Domain* and it is written by Detlef Streitferdt (Ilmenau University of Technology, Germany), Florian Kantz, Philipp Nenninger, Thomas Ruschival, Holger Kaul (ABB Corporate Research, Germany),

Thomas Bauer, Tanvir Hussain, and Robert Esbach (Fraunhofer IESE, Germany). The authors present a practical instantiation of the model-based testing paradigm based on an industrial case study in the automotive domain. The article addresses the challenge of applying model-based testing to configurable embedded software systems in order to reduce the development efforts, while measuring the benefits (development quality, test case quality, and test effort) against traditional testing. The model-based testing support described in the article integrates existing model-based testing methods and tools such as combinatorial design and constraint processing. The main conclusion of the study was that by using model-based testing techniques the testing effort could be reduced by more than a third compared to the traditional testing process.

Chapter 7 is titled *Adoption of Model-Based Testing and Abstract Interpretation by a Railway Signalling Manufacturer* and it is written by Alessio Ferrari, Alessandro Fantechi (University of Florence, Italy), Gianluca Magnani, Daniele Grasso, and Matteo Tempestini (General Electric Transportation Systems, Italy). The authors present the adoption of model-based testing in conjunction with abstract interpretation techniques for the verification of railway signalling systems. The focus of the article is on evaluating, against two industrial case studies, the benefits of moving the testing process from a code-centric to a model-centric one. The conclusions of the study show that the model-based testing process became more efficient after several iterations. In addition, it resulted not only in a significant reduction in verification costs, but also in more efficient bug detection and correction capabilities.

The third thematic section of this book, Technologies and Design Methods for Network-on-Chip Communication, consists of five chapters. Chapter 8 is titled *Analysis of Monitoring Structures for Network-on-Chip: A Distributed Approach* and it is written by Ville Rantala, Teijo Lehtonen, Pasi Liljeberg, and Juha Plosila from University of Turku (Finland). This chapter focuses on Network-on-Chip monitoring structures. Such on-chip monitoring services are needed for traffic management, system reconfiguration and fault detection to enable optimal performance and reliability of the entire NoC. The authors propose a distributed monitoring system where no centralized control is needed. This approach is meets the Network-on-Chip paradigm's scalability and flexibility concepts without causing significant traffic overhead in the on-chip network. Also system wide fault tolerance is improved as the monitoring system is distributed and thus a faulty component in the monitoring system is likely to not affect system functionality substantially. The authors note based on their cost analysis that the proposed structure has a lower cost of probes but an increased cost in the monitoring unit and registers and conclude that in large and complex systems it might be beneficial to have both centralized and distributed structures.

Chapter 9 is titled *Optimized Communication Architecture of MPSoCs with a Hardware Scheduler: A System-Level Analysis* and it is written by Diandian Zhang, Han Zhang, Jeronimo Castrillon, Torsten Kempf, Gerd Ascheid, Reiner Leupers (RWTH Aachen University, Germany), and Bart Vanthournout (Synopsys Inc., Belgium). Efficient runtime resource management in MPSoCs for achieving high performance and low energy consumption is one of the key challenges for system designers. The authors of this chapter base their proposed solution to this problem on OSIP, an operating system application-specific instruction-set processor, together with its well-defined programming model. They highlight the vital importance of the communication architecture for OSIP-based systems and optimize the communication architecture. Furthermore, they investigate the effects of OSIP and the communication architecture jointly from the system point of view, based on a broad case study for a real life application (H.264) and a synthetic benchmark application.

Chapter 10 is titled *Implementation and Evaluation of Skip-Links: A Dynamically Reconfiguring Topology for Energy-Efficient NoCs* and it is written by Simon J. Hollis and Chris Jackson from University of Bristol (UK). The authors target their work to decreasing the switching activity in many-core systems. The proposed Skip-link architecture reconfigures Network-on-Chip (NoC) topologies dynamically, allowing the creation of long-range skip-links at runtime to reduce the logical distance between frequently communicating nodes. The architecture monitors traffic behavior and optimizes the mesh topology without prior analysis of communications behavior, and is thus applicable to all applications. Their technique does not utilize a master node, and each router acts independently. The architecture is thus scalable to future manycore networks. The bottom line is that the technique provides potential for up to 10% energy reduction, based on experiments on a 16×16 node network using synthetic traffic patterns.

Chapter 11 is titled *Self-Calibrating Source Synchronous Communication for Delay Variation Tolerant GALS Network-on-Chip Design* and it is written by Alessandro Strano, Davide Bertozzi (University of Ferrara, Italy), Carles Hernández, and Federico Silla (Universidad Politécnica de Valencia, Spain). Multiprocessor SoCs are largely relying on networked interconnects. The authors are addressing the vulnerability of source synchronous links in Networks-on-Chip (NoC). In particular, they are discussing the challenge of designing a process variation and layout mismatch tolerant link for synchronizer based Globally Asynchronous, Locally Synchronous (GALS) NoCs by implementing a self-calibration mechanism. They present a variation detector which guarantees the reliability of NoC source synchronous interfaces under high variability. The variability detector is placed in front of the regular synchronizers in the communication architecture. They consider different cases of mismatch: misalignment between data and clock, misalignment between wires of the data link, and even random process variability of the detector's own logic cells. Analysis of the comprehensive metric of skew tolerance for a mesochronous synchronizer has revealed that the insertion of the variation detector has advantages in most cases and for the main synchronization architectures of practical interest (loosely vs. tightly coupled with the NoC). Even for non-variability-dominated links, the architecture with a variability detector can better cope with the layout constraints of the link.

Chapter 12 is titled *Checkpointing SystemC-Based Virtual Platforms* and it is written by Stefan Kraemer, Rainer Leupers (RWTH Aachen University, Germany), Dietmar Petras, Thomas Philipp, and Andreas Hoffmann (Synopsys Inc., Germany). The authors introduce checkpointing to virtual platform simulations, enabling to restore a virtual platform from a previously saved simulation state. Virtual platforms are executable software models of hardware systems, particularly useful for starting the software development in an early design phase when the computing hardware is not yet available. Checkpointing can considerably shorten the edit-compile-debug cycle for software developers in such cases. For a practically useful checkpoint/restore technique, a number of issues have to be taken into account, including reliability, transparency, performance, support for external applications such as graphical user interfaces and debuggers, and support for operating system (OS) resources. The authors present a novel checkpoint framework for virtual platforms, based on user level process checkpointing. The framework for handling external communication channels and OS resources during the checkpoint and restore procedure allows the integration of user modules, debuggers and GUIs. The user-defined modules are notified about checkpoint and restore events by inheriting from a special observer object. All the OS resources and external connections have to be released in order to store the status of the platform in a safe and restorable state. The performance was evaluated using an ARM-based platform. In order to mitigate the large checkpoint size, the checkpoints can be compressed before being stored. On average a compression of 10x is reached.

Seppo Virtanen
University of Turku, Finland

Acknowledgment

As Editor-in-Chief of the *International Journal of Embedded and Real-Time Communication Systems* (*IJERTCS*), I wish to extend my sincerest thanks to all individuals who have contributed to the scientific content of this book by contributing to the second volume of *IJERTCS*, including all contributing authors, members of *IJERTCS* editorial review board and all guest reviewers. More specifically, I would like to extend my warmest thanks and appreciation to Professor Jari Nurmi (Tampere University of Technology, Finland and *IJERTCS* Associate Editor), Dr. Ina Schieferdecker (Technical University Berlin, Germany), Dr. Colin Willcock (Nokia Siemens Networks, Germany) and Dr. Dragos Truscan (Åbo Akademi University, Finland and *IJERTCS* Associate Editor) for their hard work and effort as Guest Editors or Guest co-Editors of journal issues during the second volume year of *IJERTCS*. This volume year of *IJERTCS* could not have been a success without their contributions.

Section 1
Designing and Engineering Embedded and Sensor Networks

Chapter 1
Cost-Based Topology Optimization of Embedded Ethernet Networks

Jörg Sommer
University of Stuttgart, Germany

Elias A. Doumith
University of Stuttgart, Germany

Andreas Reifert
University of Stuttgart, Germany

ABSTRACT

During past decades, Ethernet progressively became the most widely used Local Area Network (LAN) technology. Apart from LAN installations, Ethernet also became attractive for other application areas such as industrial control, automotive, and avionics. In traditional LAN design, the objective is to minimize the network deployment cost. However, in embedded networks, additional constraints and ambient conditions add to the complexity of the problem. In this paper, the authors propose a Simulated Annealing (SA) algorithm to optimize the physical topology of an embedded Ethernet network. The various constraints and ambient conditions are modeled by a cost map. For networks with small number of nodes and/or switches, the authors were able to find the optimal solutions using adapted algorithms. These solutions will serve as a lower bound for the solutions obtained via the SA algorithm. However, the adapted algorithms are time consuming and application specific. The paper shows that the SA algorithm can be applied in all cases and finds (near-) optimal solutions.

DOI: 10.4018/978-1-4666-2776-5.ch001

INTRODUCTION

Today, Ethernet is the predominant Local Area Network (LAN) technology and is considered as a *de facto* standard for network infrastructure. The flexibility and the plug-and-play feature of Ethernet are its key to success. Thanks to the wide availability of its components, its large bandwidth, its reliability, and its backward-compatibility, Ethernet has become an attractive option in many application areas (Sommer et al., 2010).

A prominent example of Ethernet's application area is the industrial domain where Ethernet is gaining ground over traditional fieldbuses (Felser, 2005; Decotignie, 2005). Another area is avionics, where today's Ethernet proved its ability to fulfill real-time requirements needed in such an environment (ARINC 664, 2003). Last but not least, the automotive industry is investigating Ethernet as a suitable in-car network technology (Rahmani, Hillebrand, Hintermaier, Bogenberger, & Steinbach, 2007; Rahmani, Steffen, Tappayuthpijarn, Steinbach, & Giordano, 2008; Rahmani, Tappayuthpijarn, Krebs, Steinbach, & Bogenberger, 2009).

A motivation to use Ethernet in embedded networks is the use of commercial off-the-shelf components (Sommer et al., 2010). This allows integrators to cut down the development cost as well as the time needed to build new components. Furthermore, the large number of Ethernet vendors and the wide range of products promote the competition to provide the best equipment at the lowest price.

In contrast to traditional LANs, an embedded network has to meet additional requirements. For example, in an aircraft or a car, the installation space and the maximum weight tolerated are limited. Furthermore, the ambient conditions impose constraints on the network deployment. For instance, at some places, we need a better cable shielding due to electromagnetic interference or an extra heat-resisting cable due to the environment temperature. At other places, it might be extremely expensive or even impossible to deploy a cable. Moreover, at some places we need additional cables for the power supply of the switches while at other places such as at the positions of the nodes (devices), power supplies are already available. These constraints add to the design complexity of an embedded network.

Ethernet evolved from a bus topology to a so-called micro-segmented network with full duplex links between nodes and switches. On the one hand, deploying a single switch and connecting all the nodes directly to it has the lowest switch cost. However, such a star topology leads to wire bundles, which increase the wiring cost. On the other hand, deploying multiple switches, which increases the switching cost, alleviates the wiring layout, and consequently reduces the wiring cost proportional to the total wire length. Hence, the resulting problem is a multi-objective optimization problem where the optimal solution is a trade-off between the cost penalty due to the additional number of switches and the cost benefit due to the reduction of the total wiring length.

In this paper, we propose a Simulated Annealing (SA) algorithm to optimize the physical topology by minimizing the total link cost of an embedded Ethernet network. We take into account the aforementioned constraints and ambient conditions by modeling them with a cost map. Assuming a constant cost map (fixed cost over all the environment space), we are able to find the optimal solution by means of a Mixed-Integer Linear Program (MILP). For special cases, a Minimum Spanning Tree (MST) algorithm and an exact algorithm for the Steiner Tree (ST) problem can be used. For these cases, we show that the proposed SA algorithm provides optimal solutions in short running time. For general cost map models, the complexity of the problem increases drastically and the ST algorithm is unable to give a solution within reasonable time on today's computers. If we restrict the switches to be placed at the same positions as the nodes, the MILP and the MST algorithms are still able to find optimal solutions, while the proposed SA algorithm is the only general algorithm that keeps its pace with the increasing complexity of the problem.

The rest of the paper is organized as follows. First, we describe the problem and introduce the cost map model. Then, we present in Section *The Simulated Annealing Algorithm* the proposed SA algorithm, and discuss in the following section other related algorithms that find optimal solutions, and state the cases where they can be applied. In Section *Performance Evaluation*, we evaluate the performance of the SA algorithm and compare two embedded network designs. Finally, we conclude the paper and give an outlook for future work.

Related Work

Topology optimization is a frequently encountered problem in network design and planning. Since the early nineties, the problem of topology optimization has been investigated in many fields of application. For instance, networks-on-chip require the reduction of the wire lengths in order to optimize for speed and power consumption. This can be achieved by keeping the most critical interconnections as short as possible through system partitioning and block placement in the layout (Ahonen, Sigüenza-Tortosa, Bin, & Nurmi, 2004). Power distribution system is another field where topology optimization plays an important role. For instance, the authors of (Wakileh & Pahwa, 1996) investigated the problem of restoring power to a distribution system that has experienced a long-duration outage. This was formulated as an optimization problem where the objective is to find the size of the transformers as well as the number of switches that minimize an annual cost function. However, the most-known application area of topology optimization is the design of transport networks such as mobile networks (GSM, UMTS, WiMax, etc.), fixed and wireless access networks, and optical core networks. In Harmatos, Jüttner, and Szentesi (1999), the authors propose a topology optimization problem for UMTS transport networks. The proposed method optimizes the number and locations of base station controllers together with the transmission network topology. The method is based on an SA algorithm combined with a greedy algorithm. In Pióro and Mehdi (2004), the authors list many other works for a wide variety of topology design problems.

Topology optimization problems can be classified in capacity problems, commonly known as network dimensioning problems, and location problems. The network dimensioning problems consist of determining the minimum capacity requirements that allow the network to satisfy a given traffic matrix. This involves determining a route for each traffic demand and computing the maximum number of channels required between the switches. For example, in Mukherjee, Banerjee, Ramamurthy, and Mukherjee (1996), the authors minimize the network-wide average packet delay for a given traffic matrix and maximize the scale factor by which the traffic matrix can be scaled up. Location problems involve determining where to place particular components and how to connect them. The proposed approaches to solve this problem determine the cost of transmission and the cost of switching, and thereby determine the optimum connection matrix and the location of switches and concentrators while guaranteeing a minimal connectivity. In Chardaire and Lutton (1993), the authors present an SA algorithm that optimizes the cost of telecommunication networks. In this work, the problem consists in finding the number of concentrators, their locations, and the connection of the terminals to concentrators while minimizing the total cost of the network. However, this approach assumes a predefined set of possible locations of the concentrators and the algorithm has to choose among these locations. Moreover, all the concentrators are connected to a central point that simplifies even more the problem of interconnecting the concentrators between them. The connection cost is assumed to be the Euclidean distance. Thus, it does not take into account ambient conditions. Finally, only heuristics were considered and the authors never assessed the quality of the obtained results by comparing them with the optimal results.

Problem Description

In this paper, we propose an SA algorithm to optimize the physical topology of an embedded Ethernet network without redundant paths. Our approach takes into account constraints and ambient conditions modeled by a cost map. The resulting topology guarantees low network cost, but is unable to recover from a link cut or a switch failure. In such topologies, each node is directly connected to a single switch via a point-to-point link. The switches are interconnected in a tree topology. Deploying additional switches gives rise to further optimization issues such as the optimal number of switches, their optimal positions, and the connectivity between them.

In this work, we consider two network designs, based on a tree topology, that differ in the acceptable positions of the switches.

- In the first design referred to as *Integrated Switches*, the switches can be placed only at the same positions as the nodes (cf., Figure 1 a.). This design is motivated by the increasing number of devices with a built-in switch available on the market. These integrated switches use the same power supply as the nodes and require minimal additional installation space.
- In the second design referred to as *Self-contained Switches*, the switches can be placed anywhere in the environment space (cf., Figure 1 b.). Hence, the position of the switches has a higher degree of freedom and thus can achieve a reduced link deployment cost.

It is up to the manufacturers to decide which design is suitable for their embedded networks.

Figure 1. Embedded Ethernet network designs: a.) Tree with integrated switches b) Tree with self-contained switches

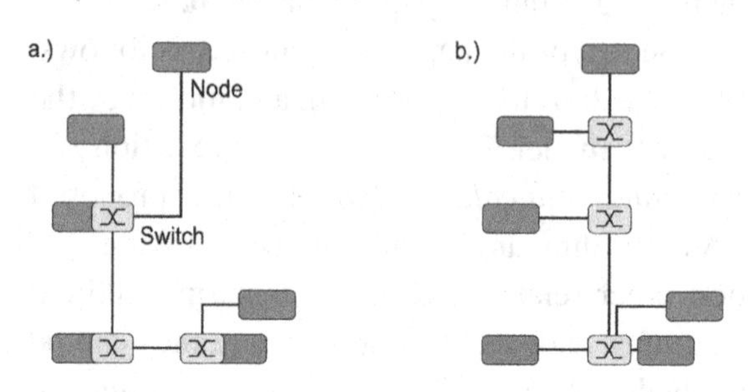

It is to be noted that this problem is harder to solve than traditional network dimensioning problems (Pióro & Mehdi, 2004) where the position of the switches is given and the connectivity between them has to be optimized. It is also harder to solve than already investigated location problems where the position of the switches/concentrators is limited to a small set of possible locations and the connectivity between them is achieved using a central point. In our case, only the nodes' positions are given and we have to optimize the switches' positions as well as the connectivity between them and towards the nodes using a tree topology. When the number of switches is not limited, this problem is reduced to the ST problem that is NP-complete (Garey, Graham, & Johnson, 1977).

We formulate the total network deployment cost C_{Net} of an Ethernet network without redundant paths

Cost Modeling

$$C_{Net} = \underbrace{2\cdot(n+m-1)\cdot C_C + m\cdot C_s +}_{\text{Position independent cost}} \underbrace{C}_{\text{Position independent cost}} \tag{1}$$

where

- n is the number of nodes and thus the number of full duplex links used to connect the nodes to the switches.

- m is the number of switches used. Consequently, in a tree topology, $m-1$ is the number of full duplex links used to connect the switches between them.
- C_C is the cost of a connector (endpoint) composed of a physical layer device and a Medium Access Control (MAC) unit. Due to the micro-segmentation of the network, there exist only point-to-point (full duplex) links. Therefore, we have to deploy a connector at both ends of each link. In our topology, we have a total number of $n+m-1$ full duplex links that require $2 \cdot (n+m-1)$ connectors.
- C_S is the cost of a switching fabric (e.g., a relay unit) without the physical layer devices and the MAC units.
- C is the cost of deploying full duplex links that connect the nodes to the switches and the switches between them. This cost is proportional to the length of the links used to interconnect the nodes and switches, and depends on the positions where these are deployed. In the next sections, we introduce the principle of a cost map and detail how to compute the cost C.

Through all this paper, we assume that we deploy only full duplex links. For sake of simplicity, from now we will use the term *link* to denote a *full duplex link*.

Cost Map

As introduced previously, due to the ambient conditions, the cost of deploying a link depends on its position. In order to take into account this cost variation, we introduce the principle of a cost map. For this purpose, we assume that the environment is reduced to a two-dimensional plane of size $u \cdot v$ rasterized into small areas or pixels. The value assigned to a given pixel represents the cost of deploying a link segment at that particular position. From this cost map, we can compute the cost of any link connecting any two nodes/switches in the network as the sum of the cost of the underlying contiguous pixels where this link is deployed.

Typically, we can derive the cost map from any environment skeleton where already existing link ducts are represented by pixels with the lowest cost. However, this information is application specific and is, in most cases, proprietary for the owner of the embedded environment. For this reason, we chose to use randomly generated maps to test our SA algorithm. We want to point out, though, that the performance of our algorithm is independent of the model used to create the cost map. For the sake of completeness, we will present in the following our approach for generating the cost map.

In embedded environments, low cost areas mostly have low cost neighboring areas. The same usually holds for high cost areas. Thus, pure random maps are not sufficient, they must exhibit a certain autocorrelation structure given by an autocorrelation matrix $\underline{\mathcal{G}}$. Such random maps with a given autocorrelation structure are used in mobile radio network simulators to model the *shadowing* or *shadow fading* effect (Cai & Giannakis, 2003; Patzold & Nguyen, 2004; Forkel, Schinnenburg, & Ang, 2004).

The basic idea of the cost/fading map finds its origin in image processing, where we apply (convolute) a predefined blur filter $\underline{\mathcal{F}}$ to a given image $\underline{P}$. The filtered image $\underline{\Gamma}$ is given by:

$$\underline{\Gamma} = \underline{\mathcal{F}} * \underline{P} \tag{2}$$

We must determine the filter $\underline{\mathcal{F}}$ in such a way that the filtered image $\underline{\Gamma}$ has a predefined autocorrelation matrix $\underline{\mathcal{G}}$.

Figure 2 illustrates the complete cost map generation process. The process starts with a given uncorrelated random cost map $\underline{P}$, where all values are uniformly and independently chosen

Figure 2. Generating the cost map $\underline{\Gamma}$

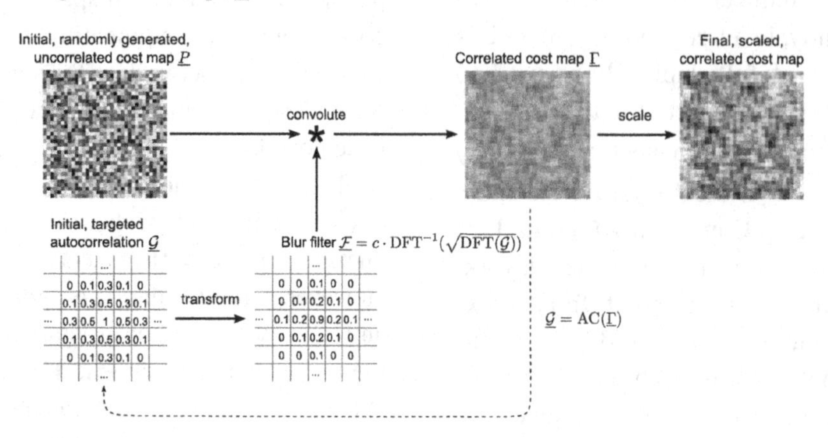

from the interval $[-1,1]$. Dark fields represent small values close to 0. We also provide the targeted autocorrelation matrix $\underline{\mathcal{G}} = (g_{i,j})$, with entries

$$g_{i,j} = \mathrm{e}^{-(|i|+|j|)} \tag{3}$$

From $\underline{\mathcal{G}}$ we construct the filter $\underline{\mathcal{F}} = (f_{i,j})$ using the 2-dimensional Discrete Fourier Transformation (DFT).

$$\underline{\mathcal{F}} = \sqrt{\tfrac{3}{uv}} \cdot \mathrm{DFT}^{-1}\left(\sqrt{\mathrm{DFT}(\underline{\mathcal{G}})}\right) \tag{4}$$

We show in the Appendix that $\underline{\mathcal{F}}$ exhibits the required property. The reader finds there the complete derivation of this formula and the definitions of the necessary functions. We want to point out that for reasons of convenience, Figure 2 shows $\underline{\mathcal{F}}$ and $\underline{\mathcal{G}}$ with index (0, 0) in the center. The derivation in the Appendix assumes the index (0, 0) is in the upper left corner due to simpler notation. Both approaches are equivalent, though.

Finally, we apply $\underline{\mathcal{F}}$ to $\underline{P}$ and obtain the correlated cost map $\underline{\Gamma}$. Its entries are within the range $\left[-\sum_{i,j} f_{i,j}, \sum_{i,j} f_{i,j}\right]$. As we require positive cost values as an input for the Floyd-Warshall algorithm (Cormen, Leiserson, Rivest, & Stein, 2001), we linearly scale the values of the resulting map $\underline{\Gamma}$ into the interval $[0,1]$. Again, darker fields in the matrix representations correspond to small values close to 0.

In the sequel, we will refer to the cost matrix $\underline{\Gamma}$ generated using this approach as *Arbitrary Cost Map* in contrast to *Constant Cost Map* where the cost matrix has the same value for all entries.

Problem Formulation

For a given set of n nodes $N_i(x_i, y_i)$ $(i = 1,\ldots,n)$ and a given number m of switches, the network deployment cost C_{Net} (*cf.* Equation (1)) is a function of the link cost C. As stated before, in an embedded environment the link cost depends on the position where the links are deployed.

Our objective is thus to minimize C by positioning m switches $S_j(\alpha_j, \beta_j)$ $(j = 1,\ldots,m)$, connecting the nodes to the switches, and connecting the switches between them. For an instance of this problem, we define the following matrices:

- $\underline{\Gamma} = (\gamma_{i,j})$ is a $u \times v$ matrix representing the discrete cost map where $\gamma_{i,j}$ is a non-negative real value in the interval $[0,1]$ specifying the cost of deploying a link segment at position (i, j).

- $\underline{\Omega} = (\omega_{I,J})$ is a $uv \times uv$ matrix representing the minimum link cost required to connect a device at position $I(x_I, y_I)$ to another device at position $J(x_J, y_J)$. The elements $\omega_{I,J}$ are non-negative real values derived from the matrix $\underline{\Gamma}$ by applying the Floyd-Warshall algorithm (Cormen et al., 2001). It should be noted that the links can only be deployed along either the *x-axis* or the *y-axis*. Figure 3 shows an example of a link connecting the top left corner with the bottom right corner. The values in the different fields represent the cost $\gamma_{i,j}$ of deploying a full link segment at position (i, j). At the endpoints of the link we assume that only half of a link segment is required, which leads to half of the cost there. Two endpoints in the same field require no link. The cost of the considered link is the sum of the incurred cost of the contiguous fields where it is deployed. Consequently, the link plotted in Figure 3 has a cost of $(1/2 \cdot 0.4) + 0.2 + 0.3 + 0.1 + (1/2 \cdot 0.2) = 0.9$. When considering a constant cost map $\underline{\Gamma} = (c > 0)_{i,j}$ the minimum cost to connect a device at position $I(x_I, y_I)$ to another device at position $J(x_J, y_J)$ reduces to:

$$\omega_{I,J} = c \cdot d_{I,J} = c \cdot \left(| x_I - x_J | + | y_I - y_J | \right), \tag{5}$$

where $d_{I,J}$ is the Manhattan distance between I and J (Black, 2006). In this case, the link plotted in Figure 3 would have a cost of $4c$. Consequently, deploying a link between two endpoints assuming an arbitrary cost map may have higher or lower cost values than the physical length of this same link computed using the Manhattan distance and assuming a constant cost map.

- $\underline{\Psi} = (\psi_{i,j})$ is an $n \times m$ matrix representing the connectivity between the nodes and the switches where $\psi_{i,j}$ is a binary variable specifying the presence or the absence of a link between node N_i and switch S_j.

$$\psi_{i,j} = \begin{cases} 1 & \text{if } N_i \text{ is connected to } S_j, \\ 0 & \text{otherwise.} \end{cases} \tag{6}$$

- $\underline{\Phi} = (\phi_{i,j})$ is an $m \times m$ matrix representing the connectivity between the switches where $\phi_{i,j}$ is a binary variable specifying the presence or the absence of a link between switch S_i and switch S_j. Note that in case of full duplex links, the matrix $\underline{\Phi}$ is symmetric $(\phi_{i,j} = \phi_{j,i})$.

$$\phi_{i,j} = \begin{cases} 1 & \text{if } S_i \text{ is connected to } S_j, \\ 0 & \text{otherwise.} \end{cases} \tag{7}$$

Depending on the positions of the switches, we calculate the cost of the individual links. For this purpose, we also define:

- $\underline{\Delta} = (\delta_{i,j})$ is an $n \times m$ matrix derived from the matrix $\underline{\Omega}$. It contains the costs of the links needed to connect the nodes and the switches where $\delta_{i,j}$ is a non-negative real value equal to the minimal link cost between node N_i and switch S_j.
- $\underline{\Lambda} = (\lambda_{i,j})$ is an $m \times m$ matrix also derived from the matrix $\underline{\Omega}$. It contains the costs of the links needed to interconnect the switches where $\lambda_{i,j}$ is a non-negative real value equal to the minimal link cost between the switches S_i and S_j.

The link cost C results in:

$$C = \sum_{\substack{i=1,\ldots,n \\ j=1,\ldots,m}} \psi_{i,j} \cdot \delta_{i,j} + \sum_{\substack{i=1,\ldots,m-1 \\ j=i+1,\ldots,m}} \phi_{i,j} \cdot \lambda_{i,j} \quad (8)$$

The Simulated Annealing Algorithm

In this section, we present our Simulated Annealing (SA) algorithm that solves the complex problem formulated in the previous section. Kirkpatrick et al. (1983) introduced SA as a randomized local search algorithm to solve combinatorial optimization problems (Kirkpatrick, Gelatt, Jr., & Vecchi, 1983). Today, it is commonly known that SA is able to find solutions with high quality in a short time (Glover & Laguna, 1997; Aarts & Lenstra, 2003; Gonzales, 2007). The basic idea of SA is to improve a given initial solution iteratively by applying defined rearrangement operations. We describe the complete SA algorithm in Algorithm 1.

As an initial solution S_0 of the SA algorithm, we place the switches randomly within a predefined solution space, connect each node to its nearest switch, i.e., the switch that requires the lowest link cost, and interconnect the switches using an MST algorithm (Kruskal, 1956; Prim, 1957). For given switch positions, such a link scheme guarantees the minimal link cost. Let C_0 be the cost of this initial/current solution.

By applying defined rearrangement operations to the current solution, also called *perturbations*, a new solution S_x with cost C_x is obtained (Line 15 of Algorithm 1). In the case of integrated switches, the solution space is restricted to the node positions, and the perturbation moves a randomly selected switch from its current node position to the position of another selected node randomly. In contrast to this, in the case of self-contained switches, the perturbation moves a randomly selected switch to any other position on the cost map. In both cases, the nodes are connected to their nearest switch, while the switches are interconnected using the MST algorithm.

Figure 3. Path calculation based on a discrete cost map

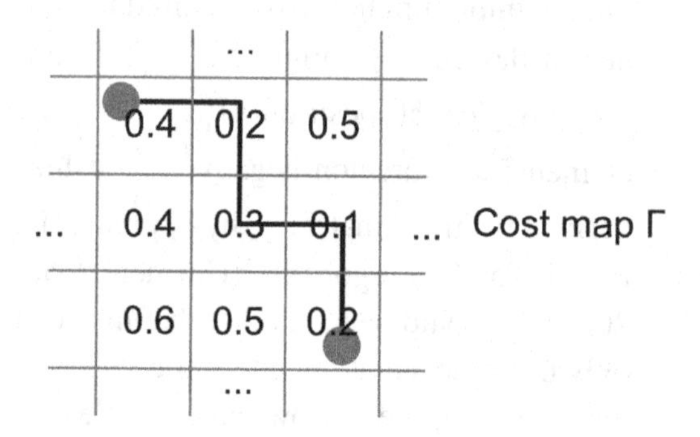

New solutions with lower cost than the current solution are accepted automatically (Line 19 of Algorithm 1). In order to avoid local minima, solutions with higher cost than the current solution are accepted with a probability determined by a system control temperature T. However, the probability that these more expensive solutions are chosen decreases as the algorithm progresses in time to simulate the *cooling* process associated with annealing. This probability is based on a negative exponential function and is inversely proportional to the difference between the cost of the current solution and the cost of the new solution (Line 29 of Algorithm 1). If the cost of the new solution is equal to the cost of the current solution, one can randomly choose to accept the new solution as the current solution or rejected it (Line 26 of Algorithm 1).

By iteratively applying the rearrangement operations and appropriately updating the current solution, the SA algorithm searches the solution space for a solution with minimal cost. The best solution obtained by the SA algorithm during its run is stored in S_{best} with corresponding cost C_{best}.

It is to be noted that we delete unnecessary switches in a post-processing step. A switch is unnecessary, if its non-existence does not affect the link cost. These are switches to which only two full duplex links are connected. Normally, it

Algorithm 1. SA algorithm

1:	**define**
2:	mIter ▷ maximum number of consecutive iterations without any cost improvement
3:	mImpr ▷ maximum number of cost improvements
4:	mAtmp ▷ maximum number of attempts
5:	T ▷ initial temperature
6:	p ▷ cooling factor
7:	**end define**
8:	**procedure** SA ($N_i(x_i, y_i)$, m, $\underline{\Omega}$)
9:	Construct an initial solution S_0 ; $C_0 \leftarrow$ Cost(S_0)
10:	$(S_{\text{best}}, C_{\text{best}}) \leftarrow (S_0, C_0)$
11:	$\mathrm{iter} \leftarrow 0$
12:	**while** $\mathrm{iter} < \mathrm{mIter}$ **do**
13:	$\mathrm{impr} \leftarrow 0$; $\mathrm{atmp} \leftarrow 0$
14:	**while** $(\mathrm{impr} < \mathrm{mImpr}) \wedge (\mathrm{atmp} < \mathrm{mAtmp})$ **do**
15:	$S_x \leftarrow$ Perturbation(S_0) ; $C_x \leftarrow$ Cost(S_x)
16:	atmp++
17:	Generate a random number $r \in [0,1)$
18:	**if** $C_x < C_0$ **then** ▷ improved solution
19:	$(S_0, C_0) \leftarrow (S_x, C_x)$
20:	impr++
21:	**if** $C_x < C_{\text{best}}$ **then**
22:	$(S_{\text{best}}, C_{\text{best}}) \leftarrow (S_x, C_x)$

continued on following column

Algorithm 1. Continued

23:	$\mathrm{iter} \leftarrow 0$
24:	**end if**
25:	**else if** $C_x = C_0$ **then** ▷ equivalent solution
26:	**if** $r < 0.5$ **then**
27:	$(S_0, C_0) \leftarrow (S_x, C_x)$
28:	**end if**
29:	**else if** $r < e^{-\frac{(C_x - C_0)}{C_0 T}}$ **then** ▷ worse solution
30:	$(S_0, C_0) \leftarrow (S_x, C_x)$
31:	**end if**
32:	**end while**
33:	**if** atmp = mAtmp **then** ▷ maximum number of attempts reached
34:	iter++
35:	**end if**
36:	$T \leftarrow pT$
37:	**end while**
38:	**return** (S_{best}, C_{best})
39:	**end procedure**

is possible to delete one or more switches if the number of switches is close to the number of nodes $(m \simeq n)$.

Related Reference Algorithms

In the previous section, we presented the SA algorithm that is a powerful stochastic search method (Vidal, 1993). In this section, we introduce exact algorithms that return the optimal solution for particular networks. However, due to the complexity of the problem, these algorithms are only able to solve *small* instances of the problem with limited number of nodes and switches. These optimal solutions, when computable, will serve as a lower bound for the SA's solutions and will help us to evaluate their quality. In the general case of self-contained switches and arbitrary cost map, to the best of our knowledge, there is not one algorithm that finds the optimal solution in an acceptable running time.

Integrated Switches

Limited Number of Switches. In this case, we are able to find the optimal positions of a given number of switches that guarantees the minimal link deployment cost when these switches can only be placed at the same positions as the nodes. This algorithm is based on a MILP formulation that can be solved by means of linear solvers such as Scip (Zuse Institute Berlin (ZIB), 2009). The MILP problem is formulated as:

Given

- An $n \times n$ matrix $\Theta = (\theta_{i,j})$ containing the cost of the links needed to interconnect the nodes where $\theta_{i,j}$ is a non-negative real value equal to the minimal link cost between the node $N_i(x_i, y_i)$ and the node $N_j(x_j, y_j)$ $(i, j = 1, \ldots, n)$. This matrix is derived from Ω and can be obtained for arbitrary cost map and constant cost map as described in Section *Cost Map*.
- The number m of the switches to be deployed.

Variables

The binary variables $\psi_{i,j}$ (*cf.* Equation (6)) representing the connectivity between the nodes and the switches are slightly modified and are now defined as: t

$$\psi_{i,j} = \begin{cases} 1 & \text{if } N_i \text{ is connected to a switch at } N_j, \\ 0 & \text{otherwise.} \end{cases} \tag{9}$$

The binary variables o_i $(i = 1, \ldots, n)$

$$o_i = \begin{cases} 1 & \text{if a switch is built-in } N_i(x_i, y_i), \\ 0 & \text{otherwise.} \end{cases} \tag{10}$$

The binary variables $\phi_{i,j}$ (*cf.* Equation (7)) representing the connectivity between the switches are slightly modified and are now defined as: $i, j = 1, \ldots, n$

$$\phi_{i,j} = \begin{cases} 1 & \text{if a switch at } N_i \text{ is connected to a switch at } N_j, \\ 0 & \text{otherwise.} \end{cases} \tag{11}$$

The binary variables $\xi_{k,l}^{i,j}$ $(i = 1, \ldots, n-1, j = i+1, \ldots, n, k, l = 1, \ldots, n)$. $\xi_{k,l}^{i,j} = 1$, if the link between the node N_k and the node N_l is used when sending data from a switch placed at node N_i to another switch placed at node N_j. $\xi_{k,l}^{i,j} = 0$, otherwise.

Constraints

We have to choose at most m nodes where to place the switches.

$$\sum_{i=1}^{n} o_i \le m \tag{12}$$

The interconnection between the switches is based on the principle of *Flow Conservation* (Pióro & Mehdi, 2004; Atallah & Blanton, 2009) commonly used in the field of *Network Dimensioning*.

The basic idea is to assume that every switch has to send some data to all the other switches. For a given data flow between a source switch and a destination switch, the flow conservation constraint states that the data flow is only leaving the source switch, and is only entering the destination switch. At all the other switches, the amount of data entering the switch is equal to the amount of data leaving the same switch. However, in our formulation, we do not know the exact position of the switches but we know their possible locations (at the nodes). Hence, the flow conservation constraint assumes that some data has to be sent between two locations if both locations contain a switch.

$\forall i = 1,\ldots,n-1$

$\forall j = i+1,\ldots,n$

$\forall k = 1,\ldots,n$

$$\sum_{\substack{l=1,\ldots,n \\ l\neq k}} \xi_{k,l}^{i,j} - \sum_{\substack{l=1,\ldots,n \\ l\neq k}} \xi_{l,k}^{i,j} = \begin{cases} = 0 & \text{if } k \neq i \text{ and } k \neq j, \\ \geq o_i + o_j - 1 & \text{if } k = i, \\ \leq -o_i - o_j + 1 & \text{if } k = j. \end{cases} \tag{13}$$

The connectivity between the switches must contain all the links needed to connect any two switches.

$\forall i = 1,\ldots,n-1,\ \forall j = i+1,\ldots,n$

$\forall k = 1,\ldots,n,\ \forall l = 1,\ldots,n$

$$\phi_{k,l} \geq \xi_{k,l}^{i,j} \tag{14}$$

The connectivity between the switches is symmetric. $\forall k = 1,\ldots,n-1,\ \forall l = k+1,\ldots,n$

$$\phi_{k,l} = \phi_{l,k} \tag{15}$$

Every node has to be connected to exactly one position of the solution space. $\forall i = 1,\ldots,n$

$$\sum_{j=1}^{n} \psi_{i,j} = 1 \tag{16}$$

However, this position is only valid, if it contains a switch. $\forall i = 1,\ldots,n$ and $\forall j = 1,\ldots,n$

$$\psi_{i,j} \leq o_j \tag{17}$$

Objective

The objective defined in Equation (8) can be expressed as

$$C = \sum_{\substack{i=1,\ldots,n \\ j=l,\ldots,n}} \psi_{i,j} \cdot \theta_{i,j} + \sum_{\substack{i=1,\ldots,n-1 \\ j=i+1,\ldots,n}} \phi_{i,j} \cdot \theta_{i,j} \tag{18}$$

Unlimited Number of Switches. It is obvious that, for small number of switches, increasing the number of switches will reduce the total link cost. The lower bound on the total link cost can be obtained by the MST algorithm. The MST connects all the nodes of the embedded environment with a tree-like topology having the lowest possible cost. Once the MST is computed, every node that is connected to at least two other nodes is assumed to have a built-in switch. As the number of switches is only determined afterwards, this algorithm is not suited for networks with a limited number of switches. In the worst case, the maximum number of required switches is equal to the number of nodes minus two.

There are two commonly used algorithms to compute the MST: *Prim's* algorithm (Prim, 1957) and *Kruskal's* algorithm (Kruskal, 1956). Both are greedy algorithms that run in polynomial time.

Self-Contained Switches

Limited Number of Switches. When the switches can be placed anywhere in the embedded environment space, the solution space for the position of the switches becomes large. Consequently, if we use the same MILP formulation as in the case of integrated switches, the number of variables and constraints becomes extremely large, which makes it impossible to get any solution in acceptable running time even for small problem instances. However, in the special case of a constant cost map, the minimum link cost to connect two nodes/switches at different positions is proportional to the Manhattan distance between these two positions. The Manhattan distance is not a linear function but can be linearized by means of additional variables and constraints. The linearized model of the problem is formulated as (Sommer & Doumith, 2008).

Given

- The position of n nodes $N_i(x_i, y_i), i = 1,\ldots,n$.
- The number m of the switches to be deployed.
- The parameter $\aleph$ defined as follows:

$$\aleph \gg 1 \tag{19}$$

- The constant cost c of the fields in the cost map.

Variables

- The binary variables $\psi_{i,j}$ (*cf.* Equation (6)) representing the connectivity between the nodes and the switches.
- The non-negative real variables $\delta_{i,j}^X$ and $\delta_{i,j}^Y$ $\left(i = 1,\ldots,n,\ j = 1,\ldots,m\right)$ containing the distances between the nodes and the switches along the *x-axis* and the *y-axis*, respectively.
- The non-negative real variables $\delta_{i,j}$ $\left(i = 1,\ldots,n,\ j = 1,\ldots,m\right)$ containing the Manhattan distances between the nodes and the switches.
- The binary variables $\phi_{i,j}$ (*cf.* Equation (7)) representing the connectivity between the switches.
- The non-negative real variables $\mu_{i,j}$ $\left(i = 1,\ldots,n,\ j = 1,\ldots,m\right)$ equal to the product of the two variables $\delta_{i,j} \cdot \psi_{i,j}$.
- The abscissae α_j $\left(\alpha_j \in \mathbb{R}^+\right)$ and the ordinates β_j $\left(\beta_j \in \mathbb{R}^+\right)$ of the switches $\left(j = 1,\ldots,m\right)$.
- The binary variables $\xi_{j,k}^i$ $\left(i = 2,\ldots,m,\ j,k = 1,\ldots,m\right)$. $\xi_{j,k}^i = 1$, if the link between the switch S_j and the switch S_k is used when sending data from the switch S_1 to the switch S_i. $\xi_{j,k}^i = 0$, otherwise.
- The non-negative real variables $\lambda_{i,j}^X$ and $\lambda_{i,j}^Y$ $\left(i = 1,\ldots,m-1,\ j = i+1,\ldots,m\right)$ containing the distances between the switches along the *x-axis* and the *y-axis*, respectively.
- The non-negative real variables $\lambda_{i,j}$ $\left(i = 1,\ldots,m-1,\ j = i+1,\ldots,m\right)$ containing the Manhattan distances between the switches.
- The non-negative real variables $\nu_{i,j}$ $\left(i = 1,\ldots,m-1,\ j = i+1,\ldots,m\right)$ equal to the product of the two variables $\lambda_{i,j} \cdot \phi_{i,j}$.

Constraints

The interconnection between the switches is also based on the principle of *Flow Conservation*. In contrast to the previous section, the switches can be placed anywhere in the environment space. In this case, it is obvious that if all switches are able to exchange data with switch S_1, then any two switches can exchange data between them. Consequently and without loss of generality, we can reduce the number of constraints to only insure that node S_1 is able to send data to all the other nodes.

$\forall i = 2,\ldots,m$

$\forall k = 1,\ldots,m$

$$\sum_{\substack{j=1,\ldots,m \\ j\neq k}} \xi_{k,j}^{i} - \sum_{\substack{j=1,\ldots,m \\ j\neq k}} \xi_{j,k}^{i} = \begin{cases} = 0 & \text{if } k \neq 1 \text{ and } k \neq i, \\ +1 & \text{if } k = 1, \\ -1 & \text{if } k = i. \end{cases} \tag{20}$$

The connectivity between the switches must contain all the links needed to connect any two switches.

$\forall i = 2,\ldots,m$

$\forall j = 1,\ldots,m, \forall k = 1,\ldots,m$

$$\phi_{j,k} \geq \xi_{j,k}^{i} \tag{21}$$

The connectivity between the switches is symmetric. $\forall j = 1,\ldots,m-1, \forall k = j+1,\ldots,m$

$$\phi_{j,k} = \phi_{k,j} \tag{22}$$

- The distance between the switches along the *x-axis* is given by: $\forall j = 1,\ldots,m-1, \forall k = j+1,\ldots,m$

$$\begin{aligned} \lambda_{j,k}^{X} &= \mid \alpha_j - \alpha_k \mid \cdot c \\ &= \max\left\{(\alpha_j - \alpha_k)\cdot c,\ (\alpha_k - \alpha_j)\cdot c\right\} \end{aligned} \tag{23}$$

In linear form, this equation is equivalent to: $\forall j = 1,\ldots,m-1, \forall k = j+1,\ldots,m$

$$\begin{aligned} \lambda_{j,k}^{X} &\geq (\alpha_j - \alpha_k)\cdot c \\ \lambda_{j,k}^{X} &\geq (\alpha_k - \alpha_j)\cdot c \end{aligned} \tag{24}$$

Similarly, the distance between the switches along the *y-axis* is given by: $\forall j = 1,\ldots,m-1, \forall k = j+1,\ldots,m$

$$\begin{aligned} \lambda_{j,k}^{Y} &\geq (\beta_j - \beta_k)\cdot c \\ \lambda_{j,k}^{Y} &\geq (\beta_k - \beta_j)\cdot c \end{aligned} \tag{25}$$

Consequently, the Manhattan distances of the links between the switches is equal to: $\forall j = 1,\ldots,m-1, \forall k = j+1,\ldots,m$

$$\lambda_{j,k} = \lambda_{j,k}^{X} + \lambda_{j,k}^{Y} \tag{26}$$

The product $\mu_{j,k}$ is then given by: $\forall j = 1,\ldots,m-1, \forall k = j+1,\ldots,m$

$$\begin{aligned} \mu_{j,k} &= \phi_{j,k}\cdot\lambda_{j,k} \\ &= \min\left\{\lambda_{j,k}^{X} + \lambda_{j,k}^{Y},\ \aleph\cdot\phi_{j,k}\right\} \end{aligned} \tag{27}$$

In linear form, this equation is equivalent to: $\forall j = 1,\ldots,m-1, \forall k = j+1,\ldots,m$

$$\begin{aligned} \mu_{j,k} &\leq \lambda_{j,k}^{X} + \lambda_{j,k}^{Y} \\ \mu_{j,k} &\leq \aleph\cdot\phi_{j,k} \\ \mu_{j,k} &\geq \lambda_{j,k}^{X} + \lambda_{j,k}^{Y} + \aleph\cdot(\phi_{j,k} - 1) \end{aligned} \tag{28}$$

Every node has to be connected to exactly one switch. $\forall i = 1,\ldots,n$

$$\sum_{j=1}^{m} \psi_{i,j} = 1 \tag{29}$$

The linear formulation of the distance along the *x-axis* between the nodes and the switches is given by: $\forall i = 1,\ldots,n, \forall j = 1,\ldots,m$

$$\begin{aligned} \delta_{i,j}^{X} &\geq (x_i - \alpha_j)\cdot c \\ \delta_{i,j}^{X} &\geq (\alpha_j - x_i)\cdot c \end{aligned} \tag{30}$$

Similarly, the linear formulation of the distance along the *y-axis* between the nodes and the switches is given by: $\forall i = 1,\ldots,n, \forall j = 1,\ldots,m$

$$\begin{aligned} \delta_{i,j}^{Y} &\geq (y_i - \beta_j)\cdot c \\ \delta_{i,j}^{Y} &\geq (\beta_j - y_i)\cdot c \end{aligned} \tag{31}$$

Consequently, the Manhattan distances of the links between the nodes and the switches is equal to: $\forall i = 1,\ldots,n, \forall j = 1,\ldots,m$

$$\delta_{i,j} = \delta_{i,j}^{X} + \delta_{i,j}^{Y} \tag{32}$$

The product $\nu_{i,j}$ is then given by: $\forall i = 1,\ldots,n, \forall j = 1,\ldots,m$

$$\begin{aligned} \nu_{i,j} &= \psi_{i,j}\cdot \delta_{i,j} \\ &= \min\left\{\delta_{i,j}^{X} + \delta_{i,j}^{Y}, \aleph\cdot \psi_{i,j}\right\} \end{aligned} \tag{33}$$

In linear form, this equation is equivalent to: $\forall i = 1,\ldots,n, \forall j = 1,\ldots,m$

$$\begin{aligned} \nu_{i,j} &\leq \delta_{i,j}^{X} + \delta_{i,j}^{Y} \\ \nu_{i,j} &\leq \aleph \cdot \psi_{i,j} \\ \nu_{i,j} &\geq \delta_{i,j}^{X} + \delta_{i,j}^{Y} + \aleph\cdot(\psi_{i,j} - 1) \end{aligned} \tag{34}$$

Objective

The objective as defined in Equation (8) can now be expressed as

$$C = \sum_{\substack{i=1,\ldots,n \\ j=1,\ldots,m}} \nu_{i,j} + \sum_{\substack{i=1,\ldots,m-1 \\ j=i+1,\ldots,m}} \mu_{i,j} \tag{35}$$

As stated before, the MILP formulation for a given number of switches and a constant cost map can be solved by means of linear solvers such as Scip (Zuse Institute Berlin (ZIB), 2009).

Unlimited Number of Switches. For small number of switches, increasing the number of switches will reduce the total link cost. However, this cost cannot be reduced indefinitely, and the lower bound on the total link cost can be obtained in this case by solving the ST problem. The classical ST problem consists in connecting all the nodes of the embedded environment with a tree-like topology that has the lowest possible cost while adding additional points, called *Steiner points* (Du & Hu, 2008). These Steiner points as well as the nodes with two or more connections represent in our case the positions of the switches. As the number of switches is only determined afterwards, this algorithm is not suited for networks with a limited number of switches. In the worst case, the maximum number of required switches is equal to the number of nodes minus two.

As we mentioned before, the ST problem is NP-complete. In practice, it is solved by means of heuristics. In the special case of a constant cost map, the GeoSteiner package (Skiena, 2009) gives the optimal solution for rectilinear ST problems.

Performance Evaluation

We presented an SA algorithm to minimize the total link cost of an embedded Ethernet network. We also introduced algorithms that find the globally optimal solutions for *small* instances of the problem. In Figure 4, we classify these algorithms according to their use cases.

In this section, we evaluate the performance of the SA algorithm in terms of the quality of the obtained final solution as well as the running time. In order to find the optimal solution by means of the exact algorithms, which we discussed in the previous section, we have to consider problem instances with small number of nodes and/or switches. For this reason, we consider two different sets of 15 and 20 nodes, respectively. For each set of nodes, we consider two cost maps: a constant cost map and an arbitrary cost map. In both cases, the minimum cost to connect any two positions of the embedded environment is pre-computed beforehand and stored in the matrix $\underline{\Omega}$. We point out that all the discussed algorithms (SA, MILP for integrated switches, ST, and MST) make use of this matrix, and thus the pre-computation step does not affect the comparisons' solutions. In the case of the MILP with self-contained switches, and due to the large solution space, some elements of the matrix $\underline{\Omega}$ are computed online during the simulation as in Equations (24), (25), (30), and (31). The parameters of the SA algorithm are $\mathrm{mIter} = 500$, $\mathrm{mAtmp} = 500$, $\mathrm{mImpr} = 100$, $T = 1$ (initial temperature), and $p = 0.9$ (cooling factor).

Figure 4. Classification and accordingly application area(s) of the algorithms

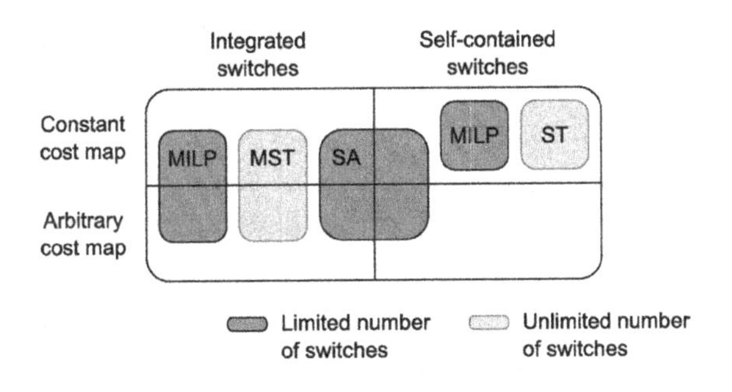

In the case of a constant cost map and the integrated switches design, the SA algorithm finds in all cases the same solution as the MILP (the optimal solution). As shown in Table 1, the running time of the SA is always less than one minute while the running time of the MILP increases with the increasing number of nodes and switches and exceeds two weeks in some cases. For large number of switches (number of nodes minus two), the SA algorithm maintains its excellent performance and finds the same solution as the MST in a comparable running time.

By replacing the constant cost map with an arbitrary cost map with the same average cost of 0.5, all the algorithms exhibit an increased running time, except for the SA algorithm. As shown in

Table 1. Integrated Switches: SA vs. optimal solutions using a constant cost map (size 50×50)

# switches		15 nodes		20 nodes	
		C	t [min]	C	t [min]
5	MILP	112.0	$\simeq 52$	115.5	$\simeq 412$
	SA	112.0	< 1	115.5	< 1
10	MILP	93.5	$\simeq 70$	99.0	> 30*d*
	SA	93.5	< 1	99.0	< 1
15	MILP	93.5	$\simeq 5$	98.0	> 30*d*
	SA	93.5[1]	< 1	98.0[2]	< 1
unlimited	MST	93.5	< 1	98.0	< 1
$\max = n - 2$	SA	93.[1]	< 1	98.0[2]	< 1

Table 2, the performance of the SA algorithm has not changed. In fact, it still finds the globally optimal solutions within less than one minute.

By considering the self-contained switches design, the solution space for the positions of the switches becomes large. In this case, the optimal link deployment cost can only be obtained for the constant cost map. For small number of switches, as shown in Table 3, the SA algorithm finds the same solutions as the MILP for self-contained switches in considerably reduced running time. Even for large number of switches (number of nodes minus two), the SA algorithm keeps its pace and finds optimal solutions, too.

Network Design Evaluation

In this section, we compare two network designs mentioned in Section *Problem Description*. For this purpose, we consider several cost maps. For each of them, we consider a set of 50 nodes and we distribute them uniformly on the map. As

Table 2. Integrated Switches: SA vs. optimal solutions using an arbitrary cost map (size 50×50*)*

# switches		15 nodes		20 nodes	
		C	t [min]	C	t [min]
5	MILP	99.0	$\simeq 68$	104.4	$\simeq 471$
	SA	99.0	< 1	104.4	< 1
10	MILP	83.5	$\simeq 40$	90.2	> 30*d*
	SA	83.5	< 1	90.2	< 1
15	MILP	83.5	$\simeq 5$	88.9	> 30*d*
	SA	83.5[1]	< 1	88.9[3]	< 1
unlimited	MST	83.5	< 1	88.9	< 1
$\max = n - 2$	SA	83.[1]	< 1	88.9[3]	< 1

Table 3. Self-contained Switches: SA vs. optimal solutions using a constant cost map (size 50×50*)*

# switches		15 nodes		20 nodes	
		C	t [min]	C	t [min]
2	MILP	139.5	< 1	164.5	< 1
	SA	139.5	< 1	164.5	< 1
3	MILP	122.0	$\simeq 4$	132.0	$\simeq 17$
	SA	122.0	< 1	132.0	< 1
5	MILP	107.0	> 7*d*	111.5	> 7*d*
	SA	107.0	< 1	111.5	< 1
unlimited	ST	83.0	< 1	83.0	< 1
$\max = n - 2$	SA	83.	< 1	83.0[4]	< 1

depicted in Figure 4, only the SA algorithm is able to find the (near-) optimal solutions in an acceptable amount of running time for such large networks. In order to ensure the best quality of the obtained solution, we make 25 independent runs for each case and record the lowest cost. Due to the complexity of the problem instances, we do not know the globally optimal solutions. However, the proposed SA algorithm has proven its excellent performance for small network scenarios and it is expected to keep such performance for larger problem instances. Therefore, we use the term *(near-)optimal* solutions to qualify the solution of the SA algorithm as there is no way to guarantee that the obtained result is optimal.

As expected, the link cost in the case of self-contained switches design is smaller than the link cost in the case of the integrated switches design. This is because the solution space of the latter design is included in the former one. However, in most cases the difference between the two designs is less than 5%. Only in the cases with unlimited number of switches $\left(m = n - 2\right)$. the difference can exceed 10%. Moreover, by deploying additional switches, we can reduce the total link cost. However, as shown in Table 4 and 5, the higher the number of switches, the lower the total link cost reduction.

Finally, we evaluate the impact of the ambient conditions on the embedded network design problem. For this purpose, we consider an arbitrary cost map with an average value of approximately 0.5 and a constant cost map where all the fields are equal to $c = 0.5$. As shown in Table 4 and Table 5 (constant vs. arbitrary), the SA algorithm benefits from the map fields with low cost to

Table 4. Comparison of networks with $n = 50$ nodes, a large number of integrated switches, and different cost maps

Map size	m = 10		m = 25		$m = n - 2$[5]	
	constant	arbitrary	constant	arbitrary	constant	arbitrary
50×50	221.0	203.6	167.0	157.9	158.0	150.3
75×75	328.5	298.5	243.5	227.5	230.0	215.9
100×100	401.5	345.7	310.5	274.9	294.0	262.4

Table 5. Comparison of networks with $n = 50$ nodes, a large number of self-contained switches, and different cost maps

Map size	m = 10		m = 25		$m = n - 2$[6]	
	constant	arbitrary	constant	arbitrary	constant	arbitrary
50×50	216.5	195.5	163.5	151.5	141.5	133.0
75×75	316.5	286.7	234.0	219.1	207.0	197.3
100×100	392.0	339.4	297.0	263.1	258.5	237.1

[1] Post-Processing switch deletion: 10 switches needed. [2] Post-Processing switch deletion: 12 switches needed.
[3] Post-Processing switch deletion: 14 switches needed. [4] Post-Processing switch deletion: 17 switches needed.
[5] Post-Processing switch deletion: 38 switches needed. [6] Post-Processing switch deletion: 47 switches needed.

interconnect the nodes and the switches. These solutions result mostly in lower link cost for the arbitrary cost map.

CONCLUSION AND OUTLOOK

During the last decade, apart from LAN installations, Ethernet became attractive for other application areas such as industrial control and avionics. In these areas, we have to consider additional constraints and ambient conditions, while reducing the deployment cost. In this paper, we proposed an SA algorithm that minimizes the total link cost of an embedded Ethernet network. We modeled the ambient conditions by means of a cost map and used it for deriving the link cost.

The proposed SA algorithm is applicable for two network designs that are relevant in embedded networks: integrated switches and self-contained switches. We evaluated the performance of the SA algorithm by comparing its solutions with the optimal solutions for small networks or simplified network designs. We have shown that, in our comparisons, the SA algorithm achieves optimal solutions in short running time.

Part of our ongoing work is to extend the SA algorithm. We are working on further perturbations. For example, we decrease the size of the space where a switch can move when it is subject to a perturbation. The aim is to approximate a local optimum carefully. Besides, we are extending the SA algorithm in order to consider ducts. These ducts are used to protect the cables that pass through and can carry one or more links. In this case, we also have to consider junction points where a link joins/leaves a duct. The duct cost depends on the number of links. With this enhancement, we will further reduce the link cost.

Until now, we focused on the physical topology. We did not take into account higher layer constraints. In the future, we will extend the SA algorithm to take into account a given traffic demand matrix, while optimizing the topology. The resulting topology should fulfill the traffic demands at a low cost

ACKNOWLEDGMENT

The authors would like to thank Matthias Kaschub for many fruitful discussions and Quentin Duval for his contribution to the implementation of the SA algorithm and the cost map generator.

REFERENCES

Aarts, E. L., & Lenstra, J. K. (2003). *Local search in combinatorial optimization*. Princeton, NJ: Princeton University Press.

Ahonen, T., Sigüenza-Tortosa, D. A., Bin, H., & Nurmi, J. (2004). *Topology optimization for application-specific networks-on-chip*. In *Proceedings of the 2004 international workshop on System Level Interconnect Prediction (SLIP '04)* (pp. 53-60). New York: ACM.

ARINC 664. (2003). *Aircraft Data Network, Part 7: Deterministic Networks* (Computer software manual).

Atallah, M. J., & Blanton, M. (2009). *Algorithms and Theory of Computation Handbook* (*Vol. 2*). Boca Raton, FL: CRC Press LLC.

Black, P. E. (2006, May). *Manhattan distance*. Retrieved from http://www.itl.nist.gov/div897/sqg/dads/

Cai, X., & Giannakis, G. (2003, November). A two-dimensional channel simulation model for shadowing processes. *IEEE Transactions on Vehicular Technology*, *52*(6), 1558–1567. doi:10.1109/TVT.2003.819627

Chardaire, P., & Lutton, J. L. (1993). In Vidal, R. V. (Ed.), *Applied Simulated Annealing* (pp. 173–199). New York: Springer Verlag.

Cormen, T. H., Leiserson, C. E., Rivest, R. L., & Stein, C. (2001). *Introduction to Algorithms* (2nd ed.). Cambridge, MA: MIT Press.

Decotignie, J.-D. (2005, June). Ethernet-based real-time and industrial communications. In *Proceedings of the IEEE*, Neuchatel, Switzerland (Vol. 93, p. 1102-1117). Washington, DC: IEEE.

Du, D., & Hu, X. (2008). *Steiner Tree Problems in Computer Communication Networks*. New York: World Scientific. doi:10.1142/9789812791450

Felser, M. (2005, June). Real-Time Ethernet – Industry Prospective. *Proceedings of the IEEE*, *93*(6), 1118–1129. doi:10.1109/JPROC.2005.849720

Forkel, I., Schinnenburg, M., & Ang, M. (2004, September). Generation of Two-Dimensional Correlated Shadowing for Mobile Radio Network Simulation. In *Proceedings of the 7th International Symposium on Wireless Personal Multimedia Communications (WPMC)*, Abano Terme (Padova), Italy (p. 5).

Garey, M. R., Graham, R. L., & Johnson, D. (1977, June). The complexity of computing Steiner minimal trees. *SIAM Journal on Applied Mathematics*, *31*(4), 835–859. doi:10.1137/0132072

Glover, F., & Laguna, M. (1997). *Tabu Search*. Dordrecht, The Netherlands: Kluwer Academic Publisher.

Gonzales, T. F. (Ed.). (2007). *Handbook of Approximation Algorithms and Metaheuristics*. Boca Raton, FL: Chapman & Hall CRC.

Harmatos, J., Jüttner, A., & Szentesi, A. (1999, September). Cost-based UMTS Transport Network Topology Optimisation. In *Proceedings of the International Conference on Computer Communication (ICCC'99)*, Tokyo, Japan.

Kirkpatrick, S., Gelatt, C. D. Jr, & Vecchi, M. P. (1983). Optimization by simulated annealing. *Science*, *220*, 671–680. doi:10.1126/science.220.4598.671

Kruskal, J. B. (1956, February). On the Shortest Spanning Subtree of a Graph and the Traveling Salesman Problem. *Proceedings of the American Mathematical Society*, *7*(1), 48–50. doi:10.1090/S0002-9939-1956-0078686-7

Mukherjee, B., Banerjee, D., Ramamurthy, S., & Mukherjee, A. (1996). Some principles for designing a wide-area WDM optical network. [TON]. *IEEE/ACM Transactions on Networking*, *4*(5), 684–696. doi:10.1109/90.541317

Patzold, M., & Nguyen, V. (2004, September). A spatial simulation model for shadow fading processes in mobile radio channels. In *Proceedings of the 15th IEEE International Symposium on Personal, Indoor and Mobile Radio Communications (PIMRC 2004)* (Vol. 3, pp. 1832-1838).

Pióro, M., & Mehdi, D. (2004). *Routing, Flow, and Capacity Design in Communication and Computer Networks*. San Francisco: Morgan Kaufmann Publishers.

Prim, R. (1957). Shortest connection networks and some generalizations. *The Bell System Technical Journal*, *36*, 1389–1401.

Rahmani, M., Hillebrand, J., Hintermaier, W., Bogenberger, R., & Steinbach, E. (2007, May). A Novel Network Architecture for In-Vehicle Audio and Video Communication. In *Proceedings of the 2nd IEEE/IFIP International Workshop on Broadband Convergence Networks (BcN '07)* (pp. 1-12).

Rahmani, M., Steffen, R., Tappayuthpijarn, K., Steinbach, E., & Giordano, G. (2008, February). Performance Analysis of Different Network Topologies for In-vehicle Audio and Video Communication. In *Proceedings of the 4th International Telecommunication Networking Workshop on QoS in Multiservice IP Networks* (pp. 179-184).

Rahmani, M., Tappayuthpijarn, K., Krebs, B., Steinbach, E., & Bogenberger, R. (2009). Traffic

Shaping for Resource-Efficient In-Vehicle Communication. *IEEE Transactions on Industrial Informatics.*

Skiena, S. (2009). *GeoSteiner: Software for Computing Steiner Trees.*

Sommer, J., & Doumith, E. A. (2008, July). Topology Optimization of In-vehicle Multimedia Communication Systems. In *Proceedings of the First Annual International Symposium on Vehicular Computing Systems (ISVCS),* Dublin, Ireland.

Sommer, J., Gunreben, S., Feller, F., Köhn, M., Mifdaoui, A., Saß, D., et al. (2010). Ethernet – a survey on its fields of application. *IEEE Communications Surveys & Tutorials, 12*(2).

Vidal, R. V. (Ed.). (1993). *Applied Simulated Annealing* (*Vol. 1*). New York: Springer Verlag.

Wakileh, J., & Pahwa, A. (1996, November). Distribution system design optimization for cold load pickup. *IEEE Transactions on Power Systems, 11*(4), 1879–1884. doi:10.1109/59.544658

Zuse Institute Berlin (ZIB). (2009). *SCIP: Solving Constraint Integer Programs.*

APPENDIX: PRINCIPLE OF AUTOCORRELATION MASK GENERATION

Before going into the detail of the construction of the Filter $\underline{\mathcal{F}}$, let us introduce some notations and definitions. Unless stated otherwise, all the matrices are two dimensional $u \times v$ matrices with indices in the range $I = [0, u-1] \times [0, v-1]$ (index (0, 0) corresponds to the upper left entry of the matrix). All following operations are cyclic in contrast to using 0-padded matrices. The infinitely wrapped-around matrix $\underline{A}^E = (a^E_{x,y})$ with infinite index ranges $E = [-\infty, \infty] \times [-\infty, \infty]$ is defined as:

$$a^E_{x,y} = a_{x \bmod u, y \bmod v} \qquad \forall (x, y) \in E \tag{36}$$

The point inverted matrix $\underline{A}^- = (a^-_{x,y})$ of a matrix $\underline{A} = (a_{x,y})$ is given by

$$a^-_{x,y} = a_{-x \bmod u, -y \bmod v} \tag{37}$$

The cyclic convolution $\underline{C} = (c_{x,y}) = \underline{A} * \underline{B}$ of two matrices $\underline{A} = (a_{x,y})$ and $\underline{B} = (b_{x,y})$ of the same size is given by:

$$c_{x,y} = \sum_{(k,l) \in I} a_{k,l} \cdot b^E_{x-k, y-l} \qquad \forall (x, y) \in I \tag{38}$$

$$b_{x,y} = \sum_{(k,l) \in I} \overline{a}_{k,l} \cdot a^E_{k+x, l+y} \qquad \forall (x, y) \in I \tag{39}$$

The 2-dimensional Discrete Fourier Transform $\underline{B} = (b_{x,y})$ of a matrix $\underline{A} = (a_{x,y})$, noted $\underline{B} = \mathrm{DFT}(\underline{A})$, and its inverse, noted $\underline{A} = DFT^{-1}(\underline{B})$. are matrices of the same size:

$$b_{x,y} = \sum_{(k,l) \in I} a_{k,l} \cdot \mathrm{e}^{-2\pi i \left(\frac{x \cdot k}{u} + \frac{y \cdot l}{v}\right)} \qquad \forall (x, y) \in I \tag{40}$$

$$a_{k,l} = \frac{1}{uv} \sum_{(x,y) \in I} b_{x,y} \cdot \mathrm{e}^{2\pi i \left(\frac{x \cdot k}{u} + \frac{y \cdot l}{v}\right)} \qquad \forall (k, l) \in I \tag{41}$$

$$\mathrm{DFT}(\underline{A} * \underline{B}) = \mathrm{DFT}(\underline{A}) \odot \mathrm{DFT}(\underline{B}) \tag{42}$$

In our case, the values of the original map $\underline{P}$ are uniformly and independently chosen in the interval $[-1,1]$. Hence, the map $\underline{P}$ has an average value $\mu = 0$, thus a second moment equal to the variance $\sigma^2 = 1/3$. The autocorrelation matrix $\underline{\mathcal{R}} = \mathrm{AC}(\underline{P})$ has a maximum value equal to $uv\sigma^2$ at index (0, 0). The remaining values are all 0 as the values of $\underline{P}$ are independent of each other. Consequently, $\mathrm{DFT}(\underline{\mathcal{R}})$ is a matrix with all its values equal to $uv\sigma^2$.

We want to apply a filter $\underline{\mathcal{F}}$ to the original map $\underline{P}$ such that the resulting map $\underline{\Gamma} = \underline{\mathcal{F}} * \underline{P}$ has a predefined autocorrelation matrix $\underline{\mathcal{G}}$. By applying the DFT transformation, we get:

$$\begin{aligned}
\mathrm{DFT}(\underline{\mathcal{G}}) &= \mathrm{DFT}\left(\overline{(\underline{\mathcal{F}} * \underline{P})^{-}} * (\underline{\mathcal{F}} * \underline{P})\right) \\
&= \mathrm{DFT}\left(\overline{\underline{\mathcal{F}}^{-}}\right) \odot \mathrm{DFT}\left(\overline{\underline{P}^{-}}\right) \odot \mathrm{DFT}\left(\underline{\mathcal{F}}\right) \odot \mathrm{DFT}\left(\underline{P}\right) \\
&= \mathrm{DFT}\left(\overline{\underline{P}^{-}} * \underline{P}\right) \odot \mathrm{DFT}\left(\overline{\underline{\mathcal{F}}^{-}}\right) \odot \mathrm{DFT}\left(\underline{\mathcal{F}}\right) \\
&= uv\sigma^2 \cdot \mathrm{DFT}\left(\overline{\underline{\mathcal{F}}^{-}}\right) \odot \mathrm{DFT}\left(\underline{\mathcal{F}}\right)
\end{aligned} \tag{43}$$

Among the many possible filters $\underline{\mathcal{F}}$ that satisfy this equation, we can choose the one that is point invertible to its conjugate $\underline{\mathcal{F}} = \overline{\underline{\mathcal{F}}^{-}}$. We can directly calculate it via

$$\underline{\mathcal{F}} = \frac{1}{\sigma\sqrt{uv}} \cdot \mathrm{DFT}^{-1}\left(\sqrt{\mathrm{DFT}(\underline{\mathcal{G}})}\right) \tag{44}$$

This work was previously published in the International Journal of Embedded and Real-Time Communication Systems, Volume 2, Issue 1, edited by Seppo Virtanen, pp. 1-23, copyright 2011 by IGI Publishing (an imprint of IGI Global).

Chapter 2
Co-Modeling of Embedded Networks Using SystemC and SDL

Sergey Balandin
Nokia Research Center, Finland

Michel Gillet
Nokia Research Center, Finland

Irina Lavrovskaya
St. Petersburg State University of Aerospace Instrumentation, Russia

Valentin Olenev
St. Petersburg State University of Aerospace Instrumentation, Russia

Alexey Rabin
St. Petersburg State University of Aerospace Instrumentation, Russia

Alexander Stepanov
St. Petersburg State University of Aerospace Instrumentation, Russia

ABSTRACT

Today, SDL and SystemC are two very popular languages for embedded systems modeling. SDL has specific advanced features that make it good for reflection of the multi-object systems and interactions between modules. It is also good for system model validation. The SystemC models are better suitable for tracing internal functions of the modeled modules. The hypothetical possibility of combined use of these two languages promises a number of benefits for researchers. This article specifically addresses and discusses the integration of SDL and SystemC modeling environments, exchange the data and control information between the SDL and SystemC sub-modules and the real-time co-modeling aspects of the integrated SDL/SystemC system. As a result, the mechanisms of SDL/SystemC co-modeling is presented and illustrated for an embedded network protocols co-modeling case study. The article gives an overview and description of a co-modeling solution for embedded networks protocols simulation based on experience and previous publications and research.

DOI: 10.4018/978-1-4666-2776-5.ch002

INTRODUCTION

The embedded systems design is a labor-consuming and complicated process. Modeling is used to simplify it and speed it up. This article gives an overview of a modeling application in the general embedded systems design flow and shows scenarios, in which modeling is especially efficient to be adopted into the development process. We selected two widely used modeling languages SDL and SystemC and started with an overview and comparison of their key features (Gipper, 2007; Fonseca, 2008). The comparison shows that both languages have a number of disadvantages that could be compensated by the joint use of SystemC/SDL. Thus, we propose three approaches for the SystemC and SDL co-modeling and compare them. Analysis of all factors shows that one of the approaches is better for the co-modeling application and we propose two ways for implementation of the approach. In the end, we give an example practical application for the introduced way of co-modeling.

EMBEDDED SYSTEMS DESIGN FLOW

An *embedded system* is a combination of computer hardware and software (could have fixed capability or be programmable) that is specifically designed for a particular function. This makes it a system dedicated for an application(s), its specific parts or parts of a larger system. An embedded system is built to control a function or range of functions and usually have strict real-time computational restrictions (Heath, 2003; Barr & Massa, 2006; Kamal, 2008).

Embedded systems span all aspects of modern life and there are many examples of their use. They are cash dispenses, payment terminals, handheld computers, mobile phones, telecommunication equipment and so on.

Many different activities are required to carry a complex electronic system from initial idea to the physical implementation. Performance modeling helps to understand and establish the major functional and nonfunctional characteristics of the product. Functional modeling results in a specification of the functional behavior of the product. Design and synthesis refine the product specification into a sequence of design descriptions that contain progressively more design decisions and implementation details. Validation and verification hopefully ensure that the final implementation behaves as specified. All these activities operate on models and not on the real physical object. One obvious reason for using a model is that the real product is not available before the development task is completed (Jantsch, 2004).

But the embedded systems design encounters a number of difficulties caused by increasing complexity of projects, increasing requirements to products reliability, power consumption and demand to speed-up the project design phase. The modern approach to the system design is illustrated by Figure 1 and includes the following stages:

- **Conceptual System Design:** The primary goal of this stage is the formulation of the main system design choice, analysis of the system design and development of the specification draft;
- **Specification:** This stage is targeted to get the final version of the system specification and benefits from having the model done in a high-level language, usually in C/C ++, SystemC, SDL;
- **Logical Design:** Converting the executable project specification to the register transfer level, i.e., the result is a specification implementation in Verilog/VHDL and further at the gate level. *In this context can be also called – architectural design*;

Figure 1. General design flow

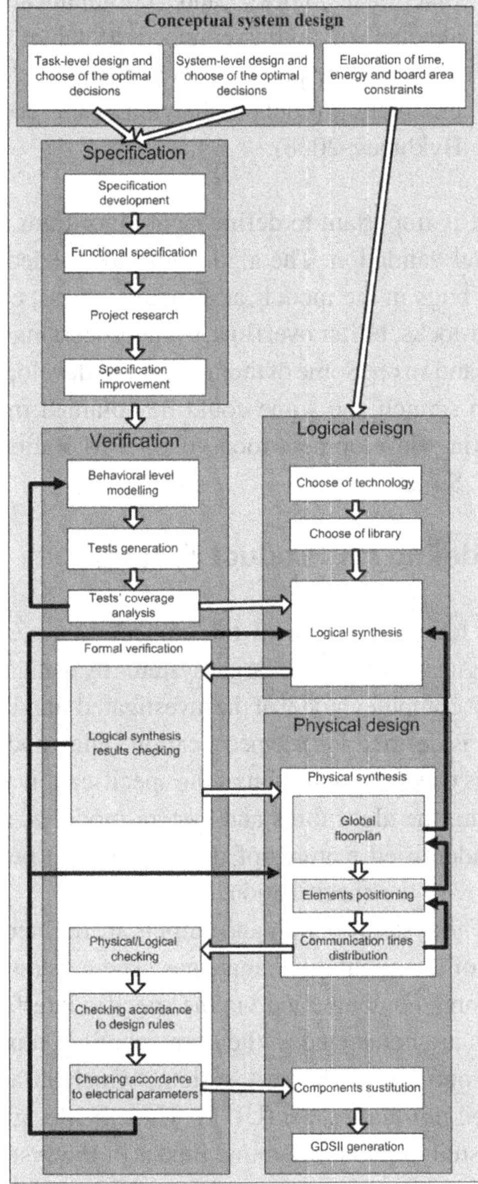

- **Project Verification:** Verification of the design decisions in conformity to the specification and other requirements, in the course of design and detailed elaboration process;
- **Physical Design:** This stage begins from the selection of technological and library basis, and it is completed when there is the final specification of the project in GDSII format. GDSII is a binary file format representing planar geometric shapes, text labels, and other information about the layout in hierarchical form.

The main purpose of Figure 1 is to show the modeling role in the general design flow. Also this picture proves that modeling could be used on every design stage.

The conceptual system design is the most critical step for estimation of the general characteristics of the resulted embedded system. At this phase, the first drafts of the system specification are created, which allows one investigating and estimating different design decisions in order to select the optimal one for system implementation (Olenev, Sheynin et al., 2009).

MODELING FOR THE SPECIFICATION PROCESS

The main objectives of the project specification process are definition and specification of the basic functions of the system and the development of executable system model. This system model is used to verify correctness of the system operation from the functional point of view, estimate required hardware resources and define the system architecture. It is used also to check various mechanisms and approaches defined in the specification for the internal correctness and compatibility.

At this stage the following problems have to be solved:

- Creation of the system functional model, i.e., description of the target system in terms of algorithms and functions, which should be implemented, but without binding to any specific implementation techniques;

- Modeling of the system in its operational environment using real data and signals;
- Specification of the system architecture in terms of required resources and definition of the hardware and software implementation of the functional model.

It is also important to note that to speed up the design process some further stages of design, verification and topological implementation of the system could be done in parallel to the executable system specification, the behavioral models and general architecture (Olenev, Sheynin et al., 2009; Bykhteev, 2008).

Modeling for the Validation

With continues growth of complexity of the systems, functional verification becomes more and more important in the design process. Before, the design was seen as project development at register transfer level with verification done by the logical modeling tools. Nowadays, the verification process starts at the behavioral stage of the project specification development.

The main requirements for the structure of functional design and verification tools are:

- Analysis of the system architecture, its performance and other parameters;
- Hardware and software design of the systems, when possibility joint development and verification of the hardware and embedded software;
- System design with processor blocks, use of processor models in hardware and embedded software development;
- Uniform design of the environment starting from the system-level to the register transfer level and then to the gate level, by using C, C++, SystemC languages and hardware description languages such as VHDL and Verilog;
- Libraries and high-level constructs for functional blocks and communication channels, including connectivity tables;
- Tools for the project data control and documentation (Olenev, Sheynin et al., 2009; Bykhteev, 2008).

It is important to define right algorithms for model validation. The algorithms are needed to find bugs in the models and specifications, e.g., dead-locks, buffer overflow, event processing errors and so on. Some of them should be developed from scratch and some could be obtained from existing development tools (e.g., IBM Rational SDL Suite).

Modeling for Product Testing

The important stage of the embedded systems design is testing of the target systems by using the exact computer model of the investigated standard as it is defined in the specification. This modeling is used for validation of the specification and testing the algorithms and system modules. All the identified features of the model must be in line with the specification.

The system's inputs and outputs are defined in one of the simulation languages. The generators run on a computer and via the specific interface units are connected to the corresponding inputs and outputs of the real system, which is also called *unit under test* (UUT). Thus the designer can study various operating modes of the system via the software and test responses of the attached device and its behavior in simple point-to-point connections and network structures. By using the model of the interface and network standard, the environment for testing, validation and certification can be feasibly built. Though there is no standard for building such testing environments, the approaches are well known. An example of the *Tester* architecture is shown in Figure 2.

The four-layer protocol is taken as an example. The Service Action Points (SAPs) are defined between every layer for the proper handshaking. For each pair of its layers, an additional inter-layer part of the tester entity is specified. A given layer can control everything that goes between upper layer and lower layer, can independently send data directly to layers and takes care of the event log. It can generate data and dispatch it directly to any of the adjacent layers; erroneous data also could be generated for testing.

The highest level is occupied by the test generation software. By means of this software one can generate tests for the underlying model using some customary programming language. Data passed through the model goes from Tester into the Interconnection entity and through it to a UUT – the device prototype for testing, device for verification, etc. We can also connect via interconnection entities a couple of testers (e.g., for standards with multipoint nodes) and use such an installation for the validation of the specification itself and of its model. The tester could also contain mechanisms for detection of industrial defects.

In a nutshell the Tester allows to:

- Validate the standard specification, e.g., during the standard development and evolution;
- Validate the model for correspondence to the specification;
- Test prototypes or boards in production;
- Certify products, verify products for conformance to the standard.

Proper modeling of the Tester is a challenging task that requires the use of flexible programming languages or a better combination of the languages and should be based on the well developed models of Testing entity, Interconnection entity and Test Generator (Olenev, Sheynin et al., 2009; Gillet, 2008; Suvorova, 2007).

Figure 2. Generalized architecture of a Tester

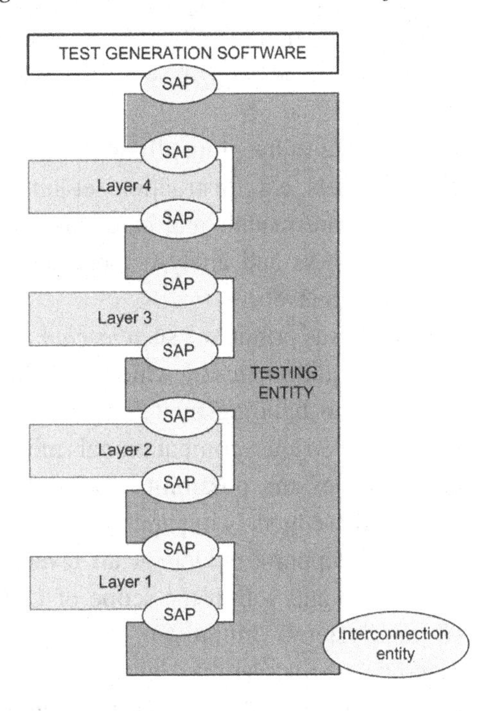

OVERVIEW OF THE SYSTEMC MODELING LANGUAGE

The *SystemC* (Open SystemC Initiative, 2005) modeling becomes one of the most efficient and widely used methods for studying, analysis and constructing multi-component systems, such as stacks of protocols, embedded networks of a large number of nodes, systems-on-chip, networks-on-chip, etc (Gipper, 2007).

SystemC is a set of C++ classes and macros that provide an event-driven simulation engine. It is specifically designed for modeling parallel systems. This library allows describing multi-component systems and program components, and modeling their operation. By using the internal mechanism of events it allows to model operations distributed in time of the modeled system. The SystemC kernel defines basic data types. It also provides such kernel units as *processes*, *events* and *channels*. Elementary channels are defined to support the interaction between parallel objects (Olenev et al., 2008).

SystemC qualifies as a language for specification, modeling, design and verification of systems. It has the following advantages:

- It is feasible to integrate all kinds of C++ based models, e.g., instruction set simulators, software models and hardware models, to process and analyze modeling results at high level;
- SystemC uses primitives such as *channels*, *interfaces* and *methods*, which give high flexibility in building models that can be based on various computational models and provides the possibility to integrate and use these models in parallel;
- SystemC supports models at all levels of abstraction and within the scope of OSCI (Open SystemC Initiative), standard semantics at transactions level and API, which guarantees the possibility of simultaneous work with different abstraction levels of the models;
- C/C++ languages are the most widely known by both software and hardware experts, which significantly facilitates cooperation in HW and SW co-design;
- SystemC supports modern methodologies for verification by means of SystemC multilevel verification library;
- SystemC has been created to facilitate development of IP-blocks models, reuse of the design and verification components, based on C++ and OSCI components interaction standards at the transactions level;
- SystemC supports hardware modeling, when detailing the project to RTL level;
- Modeling of the on-chip communications is naturally supported in SystemC (Black & Donovan, 2004; Nemydrov & Martin, 2004; Swan, 2003).

But the SystemC language has a number of disadvantages also:

- SystemC is not the optimal decision for the gate-level design (Nemydrov & Martin, 2004);
- SystemC does not provide a higher level of abstraction for modeling as used from VHDL or Verilog when targeting automated synthesis. All high-level language constructs like non-primitive channels that are offered by SystemC must be refined manually, down to a cycle accurate hardware model that uses signals for communication. This is not just a specific problem of the language itself but also a result of existing synthesis (Grimpel et al., 2002);
- SystemC also has disadvantages of low simulation speed and duplication of effort.

OVERVIEW OF SDL MODELING LANGUAGE

The SDL (Specification and Description Language) (ITU-T, 2000) is a language for unambiguous specification and description of the telecommunication systems behavior. The *"specification"* term means the system requirements defined before its creation; *"description"* of a system means the implementation description (ITU, 2002).

The SDL model covers the following five main aspects: structure, communication, behavior, data, and inheritance. The behavior of components is captured by partitioning the system into a series of hierarchies. Communication between the components takes place through gates connected by channels. The channels have delayed channel type, so communication is usually asynchronous, but when the delay is set to zero the communication is synchronous.

SDL language is intended for description of structure and operation of the distributed real-time systems. The description of a SDL system defines through the dependence of the output signals sequence from the input signals sequence. Thus, first of all, SDL is intended for a description of the systems containing components interpretable in a form of finite state machines. Therefore, the interaction between these components as state machines is very important.

The SDL state machine is represented by SDL *processes*. It is one of the main structural components of the SDL language that describes the behavior of the SDL system and operates as a state machine.

Sets of processes are grouped into *blocks*. These blocks, in turn, are connected by channels with each other and with the environment. SDL processes communicate with each other by means of SDL signals. The architectural view of SDL system (IBM, 2009) is presented in Figure 3.

SDL structuring features are its key advantages as they allow simplifying the description of large and compound systems. They allow splitting the system into separate units, which could be studied independently. Derived units could be split into subunits and so on. In the end, it leads to creation of a multilevel hierarchical structure. The description of system operation could be located at bottom layers.

SDL can be used at various stages of system creation: from its initial design to the development and maintenance (Stepanov, 2009; IBM, 2009; Karabegov & Ter-Mikaelyan, 1993).

Figure 3. General structure of SDL system

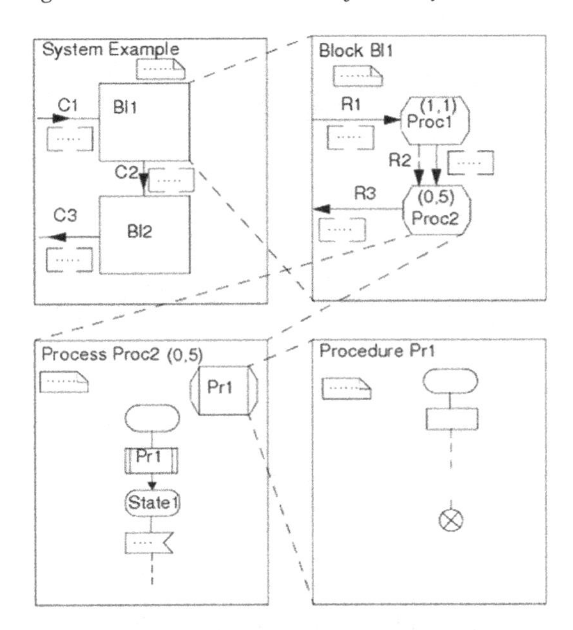

COMPARISON OF SDL AND SYSTEMC FOR THE PROTOCOLS SIMULATION

The selection of the right modeling language is the key factor of success and must be carefully thought before the system modeling starts. This choice depends on a large number of factors, which is primary driven by the vision of what kind of problems need be solved by the developed model.

If the model is targeted to check correctness of the developed protocol then it is convenient to use SystemC. Such model done in SystemC could be used for testing devices operation logic, finding errors and ambiguous situations. However, if the required model is intended not so much for the specification testing, but for its visualization then it is better to use graphic modeling languages and SDL would be one of the best choices. SDL has a graphical representation feature so it gives an ability to visually estimate the created system hierarchy, modules, which it consists of and the communication mechanisms between them.

Figure 4 shows general route modeling differences for SystemC and SDL.

Use Cases for SystemC and SDL in Protocols Simulation

The applicability difference of SDL and SystemC languages for creation of protocol models is also an important point. Both languages could be potentially used for solving any modeling problem, but the efficient use requires a number of features to be taken into account in each concrete situation.

There are a number of approaches to modeling the data transmission protocol. The first suggests per layer modeling of the protocol stack as shown in Figure 5. This requires splitting the protocol into layers, define rules of interaction between them and description of the internal operation

Figure 4. General modeling differences for SystemC and SDL

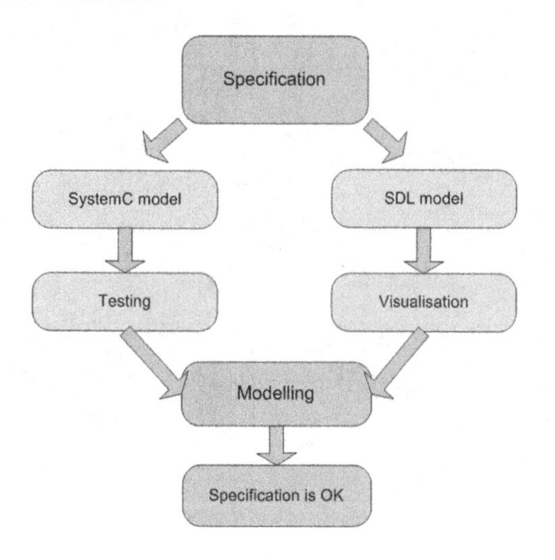

logic for each layer also (Olenev, 2009). SDL is the most reasonable solution in the case, when description of system modules and interactions between them are most important. In this case the system is divided into blocks, which corresponds to certain layers of the protocol stack. Each of these blocks contains one or several sub-blocks or processes, which defines specific functions of the corresponding layer. The channels and lists of signals transmitted via these channels define all possible interactions between parts of the model. In this case the description of the protocol functionality could also be described in SDL language. This modeling approach is very useful for illustration of specifications (Karabegov & Ter-Mikaelyan, 1993).

If the main purpose of per layers protocol modeling is to the model internal logic of the protocol, while the visual representation of the model structure is less significant, then SystemC model is more convenient. In this case the protocol structure is determined by the set of modules and connections between them. So the modules become the basic design blocks. Interaction between the modules is defined through interfaces and channels. Definition of the modules function can be performed as a structural one. In this case one module is split to a set of modules at the lower level of hierarchy. Thus the model written in SystemC is not inferior to the SDL model due to its structuring features. Plus this approach simplifies creation of a functional part of the protocol.

The second approach to protocol modeling assumes modeling of a system of devices, like shown in Figure 6. In this case the devices interact with each other via the network protocol. Let us take the network of devices that exchange data packets. In this case the interaction of the components inside the device is not important, e.g. between layers of the protocol stack. But more important here is the mechanism of communication between the devices, e.g., packet exchange, routing, etc. (Olenev, 2009). Using SDL in this case is not efficient. Of course SDL allows defining data types, so to create a network of similar devices it is enough to define the device type once and have each unique device as an instance of this type. However, the interaction between the devices, i.e. between instances of the structuring object, should be specified completely manually. Consequently if the task is to model large network, modeling in SDL will require large amount of time and efforts to describe interaction between the devices.

In SystemC, the designer can describe all nodes as instances of one class, which is similar amount of efforts as for SDL. But to create interconnections between the devices it is needed only to bind all ports and there will be no need to describe each connection individually.

Figure 5. Example of per layer modeling of the protocol

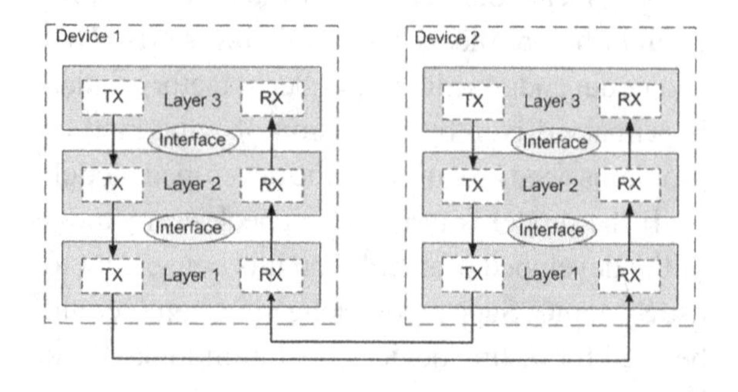

Dynamic Object Creation

Another factor to be considered is a possibility to dynamically determine parameters of the system in the second modeling approach. Assume there is a need to create a model of a switch to simulate a network of devices. SDL language unfortunately does not allow giving number of switch ports as a parameter. Designer has to define different switches with various numbers of ports. While in SystemC the programmer can define the switch class and set the number of ports as a parameter. Thus, in this case SystemC is better fulfilling the system of devices description.

Simulation Time

The next criterion to take into account is different options for the simulation of time use. SDL has two data types that can be used to work with time, i.e., *Time* and *Duration*. The first one is used to define the "point of time" and the second for "time interval". Generally both types are used in conjunction with timers. There is only one operator "now" that returns value of type Time. This value could be used by setting the timer if a delay with type Duration is added. Note that SDL specification does not define the system time unit, so it depends on the implementation of specific SDL environment. Thus both of these data types are types with floating point (IBM, 2009).

In SystemC modeling the absolute integer time is used. The modeling time is encoded by an unsigned integer of at least 64 bits. In contrast to the continuous astronomical time the modeling time is discrete. Modeling time starts at 0 and increments at the course of simulation. There are two parameters related to the modeling time, they could be configured by the developer. These parameters are time resolution and the default time unit. Time resolution defines the time interval that corresponds to one reading of modeling time. All events could be planned only for the ends of these intervals. Thus the potential

Figure 6. Modeling of a system of devices

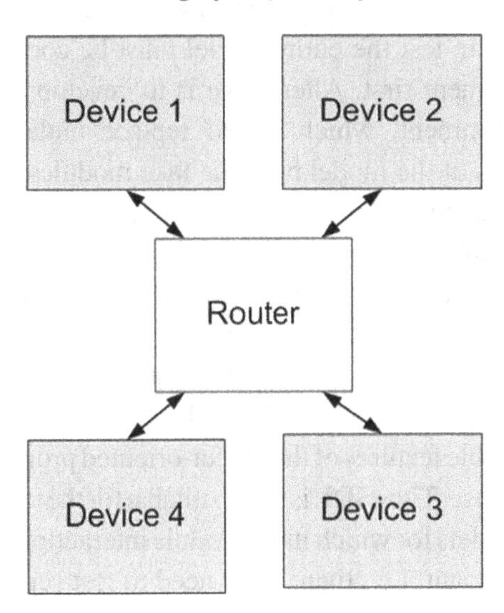

accuracy of the modeling depends on the size of the interval. But on the other hand the maximal model time in the given modeling environment is smaller. The default value for time resolution in SystemC is one picosecond. The default time unit corresponds to one nanosecond.

The modeling time in SystemC models provide broader opportunities for application. In addition to the definition of timers similarly to SDL, the SystemC give a possibility of modeling per cycles, setting of clock signal for all blocks of modeling system. It is possible to set the clock signal for each block separately and the sensitivity of the block can be put to its positive or negative front. Duration of the clock cycle can also be specified for each device differently. In addition it is possible to start system modeling for the time specified in time parameter by setting of *sc_start (time, ...)* function parameters. It significantly simplifies testing of the whole model and its elements.

Testing of the Models

The last point of our comparative analysis is possibilities for testing of the models. SDL has a requirement that every planned interaction should be done between all parts of the model under test.

Regardless of what part of model the designer is going to test the entire model must be correctly implement first. Alternative is to develop a test environment, which would replace undefined blocks of the model by some fake modules.

The SystemC model has to be competently done by a programmer for testing. The developer will be able to test different parts of the model independently if the rational partition of a model to blocks, declaration of the required constants, classes and their methods will be properly done by available features of the object-oriented programming use. Thus SDL is more suitable for the testing of models for which inter-module interactions are significant, i.e., there is no need to test separate parts of the model. Whereas SystemC models are better suitable for testing in all other types of cases. However, the role of proper competent code writing is also stronger increasing in this case.

Summarizing the above we can say that the choice of modeling language in each concrete case primary depends on the modeling purposes. In addition it is usually useful to specify main stages of work and ways of dealing with them before actual development starts. For instance, a number of modeled objects, timing, and further testing's implementation way are among the most important factors. Each of these factors can have a significant impact on the decision to favor one or another language (Stepanov, 2009).

Comparison of SDL and SystemC – Conclusion

In conclusion we would like to point out the main goals of the SDL/SystemC comparison:

- SystemC model is targeted to check correctness of the developed protocol;
- SystemC is better applicable for internal logic of the protocol modeling;
- SystemC gives an ability to use IP blocks;
- Dynamic object creation makes SystemC better for the large systems modeling;
- SystemC models are better suitable for testing in all types of cases.
- SDL model is intended not so much for the specification testing, but for its visualization;
- SDL is better applicable in cases, when description of system modules and interactions between them are more important;
- SDL is better suitable for the testing of models for which inter-module interactions are significant.

So the SystemC and SDL joint use could give the benefits of both languages to the finite model. Let us consider how SDL and SystemC could be combined and what results could it give for the model user.

OVERVIEW OF RELATED STUDIES

Unfortunately there are not many publications in the field of SystemC and SDL co-modeling, however a few good papers can be found. For example, there is a good publication by M. Haroud and L. Blazevic which gives superficial comparison of SDL and SystemC languages and proposes a method for SDL translation into SystemC code (Haroud & Blazevic, 2006). This method is used to translate the SDL implementation of protocol specification into the SystemC model. It allows using rigorous protocol modeling and verification provided by SDL and FPGA netlist synthesis enabled by SystemC. This article gives a comparison and defines general use of SystemC and SDL for the protocols modeling. Another interesting paper was published by T. Josawa at.el. from Nokia Research Center (Jozawa et al., 2006). This work is more closely correlated to this study and the proposed methods for the co-modeling have a lot in common with our "SystemC and SDL parallel use" co-modeling approach. These methods are illustrated by Figure 7.

Figure 7. Two approaches for SDL and SystemC co-modeling proposed in Jozawa et al. (2006)

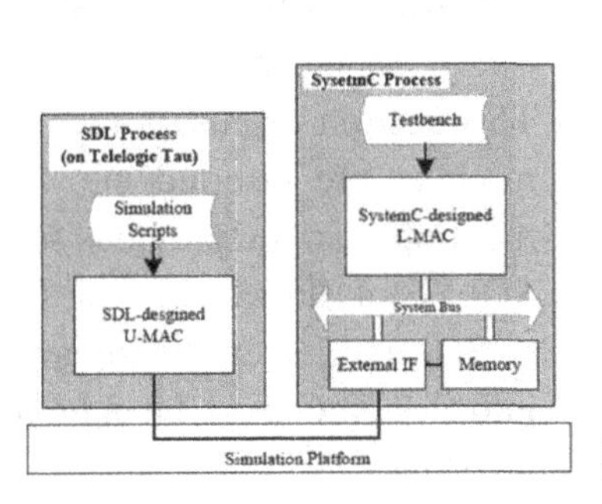

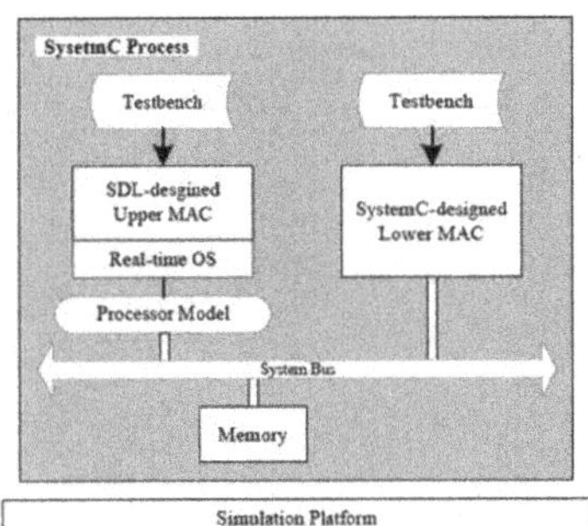

THREE MAIN METHODS OF CO-MODELING IN SYSTEMC AND SDL

As we discussed before, SDL and SystemC are very popular tools for modeling embedded networks and both have clear strong sides, but are there any chance to use these two languages together and combine the benefits? The next question, if joint use even is possible will it give an ability to use both modules at the same time by setting rules of direct interaction and internal operations of the simulation?

We see three possible co-modeling solutions for SDL and SystemC. The first approach is to insert SystemC modules into the SDL model by including the corresponding C header files into SDL model. The second approach is to insert SDL module into SystemC by "teaching" the SDL model how to process the requests and commands of the SystemC model. The third approach assumes running both SDL and SystemC independently in parallel in the operating environments and make a special tool that will interface SDL and SystemC. In this chapter we compare these approaches specifically stressing their advantages and disadvantages.

Inserting SystemC modules into SDL (SystemC→SDL)

This section discusses the approach when SystemC modules are inserted into SDL. Some SDL frameworks, for example Telelogic TAU understands C code by means of IBM Rational SDL Tool. So SDL and SystemC models integration can be divided into the following stages:

- Prepare SDL model of the target part of the system;
- Prepare SystemC model of the target part of the system;
- Prepare SystemC channel;
- Prepare SDL interlayer;
- Prepare a patch for converting SDL data types to C types;
- Prepare C code implementation of interface between interlayer and channel;
- Include the prepared C code (*.*h* files) to SDL interlayer;
- Include the prepared C code (*.*h* files) to SystemC channel.

The overall procedure for the first approach is illustrated by Figure 8.

This approach has some implementation restrictions. First, SDL could execute file written only in pure C and use of classes, inheritance and other features of C++ and SystemC imposes additional complications. But even in this case SDL data types cannot be easily converted to C types. As a consequence, joint use of two languages requires implementation of a kind of "type converter". This converter has to convert arrays, structures and signals into corresponding C structures. These result in significant complication of using additional C and C++ libraries for SDL.

Figure 8. Inserting of SystemC into SDL (SystemC→SDL)

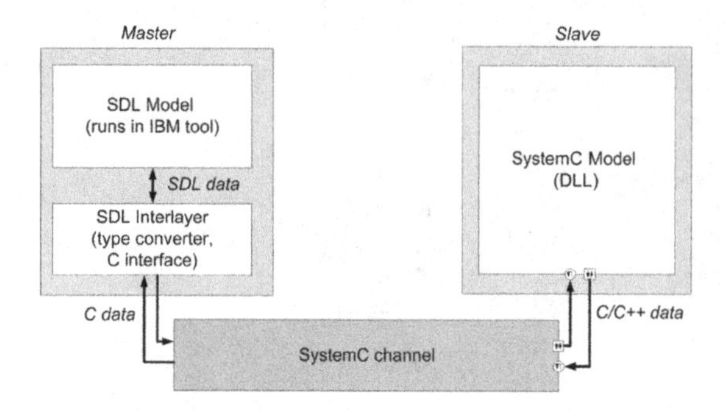

But potentially it is still possible to implement such an interlayer that converts data types and so let to include SystemC modules.

Another disadvantage of this implementation is related to usability. Any change in SDL or SystemC model that result in a change on the interface, in turn will force update in the interlayer.

Another problem is that this scenario of co-modeling allows seeing the results of only SDL operations. All SystemC side results will be hidden from the designer and consequently debugging of SystemC and C parts becomes very difficult. The only way how SystemC results can become available for the designer is by setting in the code special commands to write most relevant data to log-files.

At the same time this method has a number of clear benefits. It gives convenient way to organize SystemC and SDL interaction and it is the easiest way to implement joint use of both languages. Also in this approach SystemC model is used as a library, so all the function calls are performed from SDL model. As a result this approach simplifies SDL to SystemC synchronization issues.

Also this approach allows using SDL Simulator, as SDL model will be the master. So the designer can observe all interactions using SDL model (Olenev, Rabin et al., 2009).

Inserting SDL Modules into SystemC (SDL→SystemC)

This section discusses the approach when SDL modules are inserted into SystemC. As it was mentioned in the previous section IBM Rational SDL Tool allows creating C code on the basis of SDL model. So the integration of SDL model into SystemC model can be divided into following stages:

- Prepare SDL model of the target part of the system;
- Generate C code on the basis of created SDL model;
- Insert this C code to predefined SDL kernel;
- Prepare SystemC model of the target part of the system;
- Prepare SystemC channel;
- Integrate SDL kernel (generated C code) into SystemC model.

The overall procedure for the second approach is illustrated by Figure 9.

SystemC model is a master and SDL model is a slave. SystemC model interacts with C version of SDL model by means of SystemC channels. This approach has several disadvantages. To make

Figure 9. Insert of SDL into SystemC (SDL→SystemC)

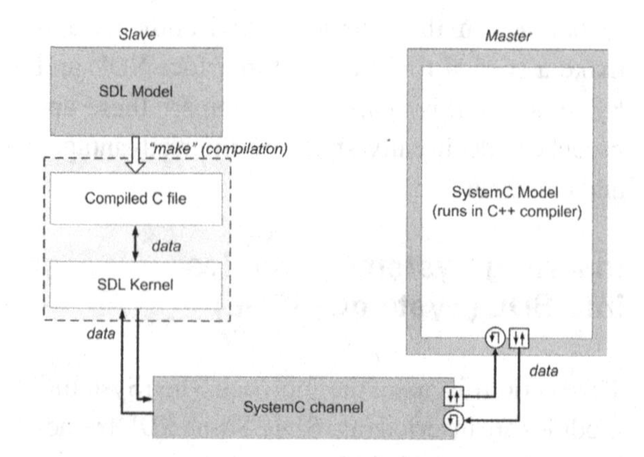

any change in SDL model designer has to make necessary modifications in SDL code and then rebuild it to C. Another disadvantage is a need to create C++ interface to connect C analogue of SDL model and SystemC channel. This way of co-modeling has a number of advantages. Thanks to the fact that SystemC model is a master it is possible to work directly with the code of SystemC model. It means that the designer can easily perform step by step debugging of the model. Moreover this co-modeling solution allows using SystemC clocking. Also the SDL Simulator can be used to see SDL work process. Finally this approach allows working with C code only on "SDL side" (Olenev, Rabin et al., 2009).

Parallel Use of SDL and SystemC (SDL↔SystemC)

The third approach discussed in this article is based on parallel use of SDL and SystemC independently and connecting them through the special manager and passing through a file. The process of writing results to file and reading them is clock-depend. This approach consists of the several steps:

- Prepare SDL model of the target part of the system;
- Prepare SystemC model of the target part of the system;
- Create the file for communication between two models;
- Prepare special tool for managing interface between two models.

The overall procedure for the third approach is illustrated by Figure 10.

This approach requires introducing a special tool for managing SDL and SystemC models. The tool handles point-to-point communication between SDL and SystemC models and provides access to all required information about interactions between them. At this point one should remember that both models have own clocks and the main role of manager is to provide the clock synchronization. This approach has several advantages comparing to the previous two. First of all functionality of both models could be fully observed. It combines benefits of SystemC and SDL languages. The manager modules simplifies monitoring of the data transmitted through the channel. But this approach has a number of restrictions. As SDL and SystemC have different notations of time the synchronization procedure is very complicated. Proper development of the manager module is a very challenging problem in itself.

Generally we can say that this approach is applicable in less number of cases than previous two (Olenev, Rabin et al., 2009).

CONCLUSIONS ON CO-MODELING OF SYSTEMC AND SDL

Earlier we considered three approaches to co-modeling of SystemC and SDL. The summary of their distinct features is provided in Table 1.

The applicability of each approach depends on how well its features match to the developer requirements. For example, SystemC→SDL approach inherits maximum of SystemC advantages, i.e. clocking and threads. In the SDL↔SystemC approach both languages get complete set of tools to exchange data, control and synchronization information, including clock signals. However analysis of all factors shows that inserting of SDL into SystemC gives the best

Figure 10. Parallel use of SDL and SystemC (SDL/SystemC)

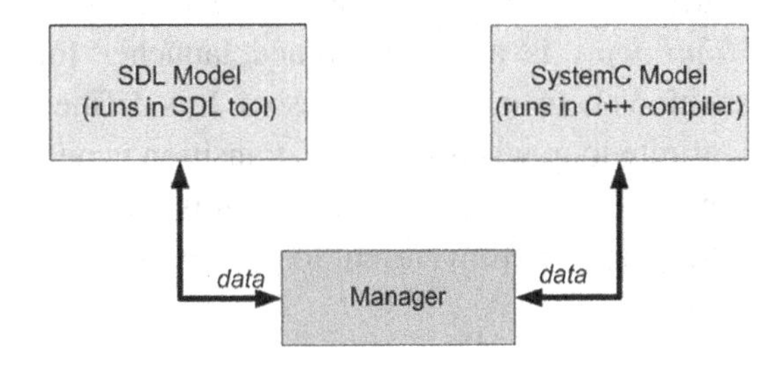

Table 1. Specific features of methods

Feature	SystemC → SDL	SDL → SystemC	SDL ↔ SystemC
Master or Slave	SDL is master, SystemC is slave	SystemC is master, SDL is slave	No master/slave relation, Manager is the master for both
Clocking	SDL clocking	SystemC clocking	Independent clocks
Cost of Resource	Operation with SDL Tool	Operation with C++ compiler	Parallel operation with SDL Tool, C++ compiler and management tool
Visibility	Only SDL interactions are visible	Both models parts are visible, but SDL has limited functionality	Functionality of processes in both models is visible

solution for co-modeling of SystemC and SDL. So let us consider SDL→SystemC approach in details (Olenev, Rabin et al., 2009).

TWO WAYS TO ADDRESS THE SDL→SYSTEMC CO-MODELING APPROACH

Principals of Co-Modeling for the SDL→SystemC Approach

IBM Rational SDL Tool allows generating C code based on SDL model and provides environment for SDL model analysis, simulation and validation to the SDL kernel. Now let us step-by-step discuss the process of the model integration.

First of all SDL part of the model should be prepared. Then using the IBM SDL Tool we generate C code out of SDL model and add it to the predefined *SDL kernel*. After that C code is a part of the joint SDL model. SDL kernel includes a number of functions that facilitates simulation of SDL system. The most important are functions *xMainInit()* and *xMainLoop()*. Function *xMainInit()* makes initialization of SDL kernel and it is called before start of the simulation. Function *xMainLoop()* is a scheduler and launcher for transitions between SDL processes. It is defined as infinite loop where one SDL transition is performed per iteration. SDL transition is one step of SDL process from current to the next state. If any output signal is sent during this step then the receiving process will be informed and the next transition of the receiver is scheduled.

Then the SystemC model has to be implemented. It could contain a number of SystemC threads. Couple of these threads is presented in SystemC part of the model and others in the channel part. One thread should be kept to work with SDL (*sdl_thread*). *Sdl_thread* calls *xMainLoop()* function when SDL part gets incoming data events. The corresponding initialization of SystemC requires call of *xMainInit()* function. The clock signal is initialized in SystemC *(sc_clock)* and SDL uses it only for synchronization. Figure 11 shows a simple example of co-modeling SDL and SystemC. This is an example when two nodes working with each other, but one node is done in SDL, another in SystemC.

Figure 11. SDL/SystemC co-modeling example

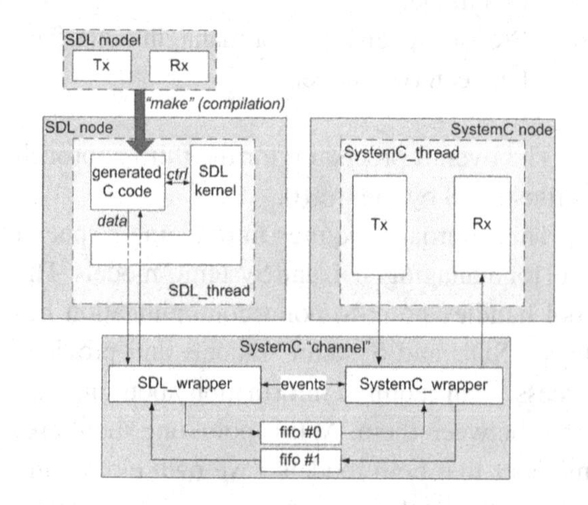

When it is done SDL part of the model is presented by C code and SDL kernel functions. The rest of code is written in SystemC and it also implements "channel" for interaction between the nodes.

Notion of Delta-cycles

Another key aspect for discussing modeling principles of different languages is *delta-cycles*, as this notion is very important for elaboration of co-modeling approaches.

The first solution for organization of delta-cycles is provided in SystemC and requires declaration of a notion for *delta-delay* (Δ), which is used to define cause-effect relation between different events. In terms of modeling time the delta-delay triggers in zero time. If two events occur at the same time, but the first event causes the second one then it is considered as there is "delta-delay" between these events.

Figure 12 shows the cause-effect relation between events in SystemC. The scheduled *event#1* is performed at *T1* modeling time. While the event is in processing the function *notify(ZERO)* for *event#2* is called. In terms of modeling time *event#1* and *event#2* should be processed at *T1* moment. But from the cause-effect relation point of view *event#1* causes *event#2*. So *event#2* should be performed after a delta-delay (at $T1+\Delta$ modeling time).

Implementation of the delta-cycle is defined in the several steps. At the first step a scheduler defines a list of processes, which are to be executed at this moment of modeling time.

The second step is evaluation of the system values. At this step all scheduled processes are executed sequentially. New values are computed and saved but not assigned. All processes use the same values of the objects independently of the processing order.

The third step is update of the values that were computed in the previous step.

Figure 12. Cause-effect relation of events in SystemC

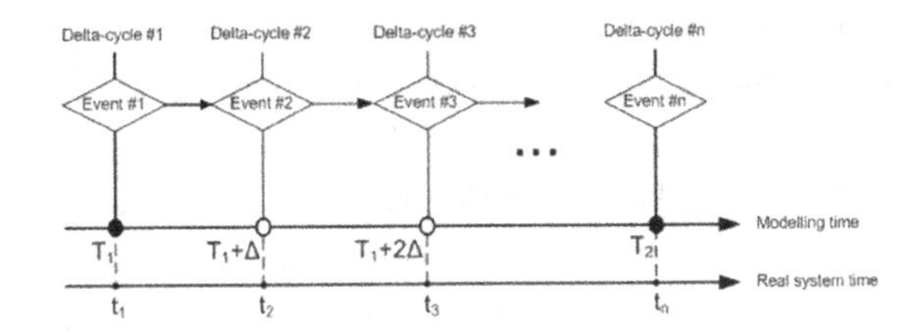

Then the list of new notify events is formed, which were scheduled on the second step to occur at current modeling time, for execution after delta-delay.

If the list of notifies events is not empty after the check for delta-cycles then the sequence of steps is performed again after one delta-delay by setting the next delta-cycle.

The typical flow chart for handling delta-cycles is shown on Figure 13.

Let us take modeling time *T1* at which all scheduled events are performed. Then all events that were scheduled for *T1* with zero delay will be executed at $T1+\Delta$. At the moment $T1+2\Delta$ will be executed all events scheduled at $T1+\Delta$ with

Figure 13. Algorithm of handling delta-cycles

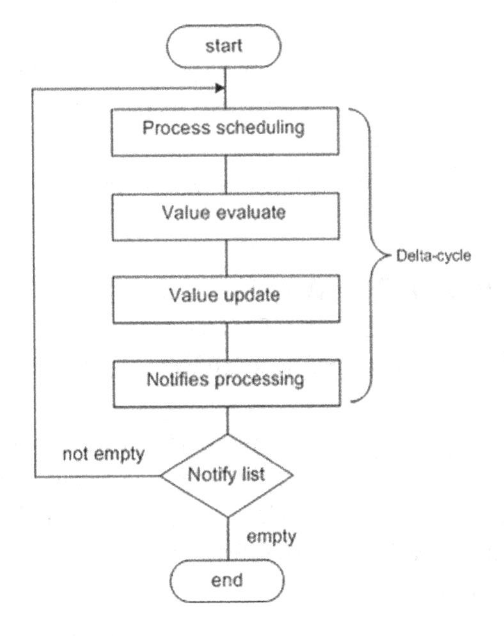

zero delay, and so on. Finally the modeling time is set to *T2* only when all events scheduled for the moment of time *T1* will be completely processed (Suvorova & Sheynin, 2003). SDL specification does not define notions of the delta-cycles and delta-delays. However, we have to introduce these structures to describe the principles of SDL systems modeling using SDL Tool for elaboration of co-modeling approaches.

Let us take certain moment of modeling time, when there is a number of scheduled events, where each SDL event represents transition of the process from one state to another. Each delta-cycle contains one execution of SDL transition and consequently during SDL delta-cycle a set of tasks is performed, one of which is scheduling of the new events. There are two ways for events scheduling: signals and timers. Using of signals means that the event should be performed at the current moment of modeling time. Such event is processed during the next delta-cycles after delta-delay and this delta-delay triggers in a zero time. The timer expiration is scheduled at the different moment of modeling time. So it causes new event, which is processed when all current time events will be performed. We propose two implementations for SDL→SystemC co-modeling approach, which differ by the type of SDL and SystemC interaction. The first one is a solution when one SDL transition performed in one SystemC clock-cycle (*one-per-one*). Another when many SDL transitions performed per one SystemC clock-cycle (*many-per-one*). Let us discuss in details these implementation approaches.

One-Per-One Way for the SDL→SystemC Co-Modeling Implementation

Some changes in SDL modeling principles are necessary for elaboration of the one-per-one way. At each moment of modeling time certain number of delta-cycles has to be performed. This requires modifying the mechanism of output signals processing. In this case using of output signals means that the event should be performed at the next moment of modeling time (after one time unit).

As it was described earlier, SDL part is executed in its own thread. This thread is implemented in such a way that finite number of SDL delta-cycles can be performed during each SystemC modeling cycle. Thus a definite number of SDL transitions are executed during one modeling time clock. All events that are scheduled for the current moment of modeling time are executed one-by-one.

SystemC part of the model uses the standard notion of SystemC delta-cycles. SystemC starts to operate according to the scheduled event of writing data to the channel by SDL part. One should note here that SystemC part should perform all planned actions during one delta-cycle.

Interaction between SDL and SystemC parts is organized in the following way:

- SystemC generates data and sends it to the remote node via channel within one delta-cycle.
- After that it starts to wait for event from the remote node.
- The new delta-cycles start when SystemC get next events.
- SystemC node begins to read the channel and process the input.
- SDL part can spend more than one clock cycle to perform similar actions.
- After performing all these tasks SDL thread tries to perform all possible SDL transitions and only after that SDL thread will wait for event from the remote node.
- Each modeling cycle SDL reads data from the channel, receives and processes data that is sent to the channel by SystemC.

Figure 14 shows an example of interaction between SDL and SystemC using one-per-one co-modeling way.

A description of interaction between SDL and SystemC parts according to the example above is presented in Table 2.

Figure 14. One-per-one co-modeling way

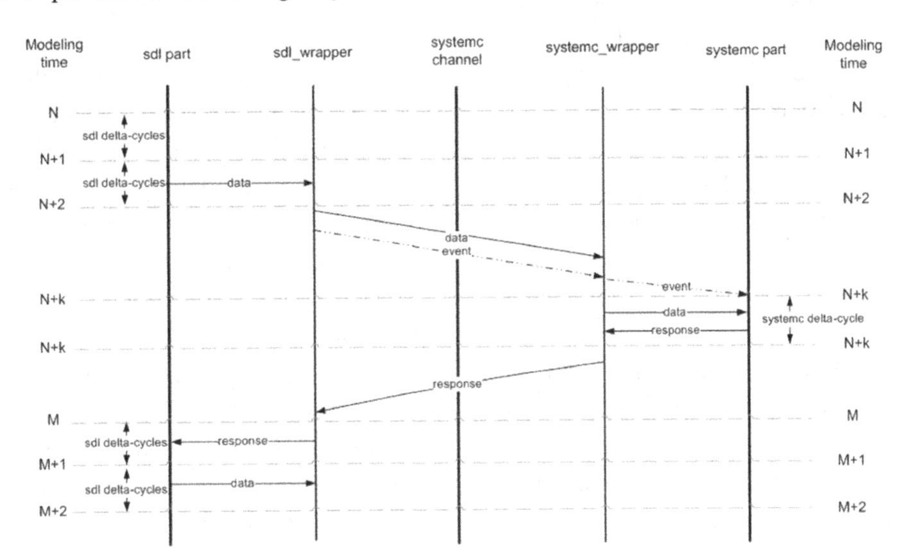

Table 2. Description of SDL and SystemC interaction for the example in Figure 12

Time	SDL part work	SystemC part work
N.. (N+1), (N+1).. (N+2)	SDL thread calls *xMainLoop()* function and so starts to execute SDL transitions. During these clock-cycles SDL system generates data and writes it to *SDL_wrapper*	SystemC part waits for events from the data channel
(N+2).. (N+k)	*SDL_wrapper* sends data to FIFO. The event is sent to *SystemC_wrapper* to inform that data is in FIFO. At each clock-cycle SDL thread calls *xMainLoop()* function and so starts to execute possible SDL transitions and read FIFOs	
(N+k)		*SystemC_wrapper* gets the event, reads data from FIFO and sends event to SystemC part
(N+k)		SystemC part gets event and starts delta-cycle. At these delta-cycles SystemC Rx reads data and Tx generates response. SystemC delta-cycle is finished when the response is sent to *SystemC_wrapper*.
(N+k)..M	At each clock-cycle SDL thread calls *xMainLoop()* function and so starts to execute possible SDL transitions and read FIFOs	SystemC_wrapper sends response to FIFO
M.. (M+1), (M+1).. (M+2)	SDL thread calls *xMainLoop()* function and so starts to execute SDL transitions and read FIFOs. At these clock-cycles SDL system reads response from FIFO, generates data and writes it to *SDL_wrapper*.	

Many-Per-One Way for the SDL→SystemC Co-Modeling Implementation

Many-per-one co-modeling does not require changing the principles of SDL delta-cycles organization. Consequently, all transitions that are scheduled for the current moment of time will be performed during this clock cycle. That means that one clock cycle contains a large number of delta-cycles. The modeling time increments after each event, referred to that moment as executed.

SystemC operates the same way as in case of the first approach. So SystemC execution process goes according to the scheduled events, which are writing data to the channel from SDL part.

Note that SystemC part should perform all planned actions during one delta-cycle.

Interaction between SDL and SystemC parts is organized in the following way:

- Each part generates data and sends it to the remote node via channel at one clock-cycle. Thus SystemC has one delta-cycle per clock-cycle and SDL has a number of delta-cycles.
- After that it starts to wait for an event from the remote node and the delta-cycle is over.
- When it gets the new event, it starts new delta-cycle. Node starts reading the channel and processing input.

Let us consider a description of interaction between SDL and SystemC parts when the communication is based on requests (data) and responses. Figure 15 shows an example of interaction between SDL and SystemC parts using many-per-one co-modeling way.

A description of interaction between SDL and SystemC parts according to the example above is presented in Table 3.

Practical Applicability

As it was mentioned before, modeling is a useful tool for the specification development, protocol validation and device testing. The last point is an area where co-modeling of SystemC and SDL could play the key role. SDL is used to describe specification. The SDL model of specification can be used for verification and validation of the protocol. But as was shown by Figure 2, proper device testing requires not only protocol model, but also testing environment and dynamic modeling software, that cannot be done in SDL, but easily implementable in C/C++. It would be great help for developers to allow co-execution of the code prepared in two different languages.

Moreover, SDL and SystemC models can be originally done for different purposes. For example, SDL model for verification and SystemC model for the studies of further hardware development. In this case it might be useful to allow these models work together to compare results and check on correctness of implementation.

The Protocol Model Tester is a good example of a practical applicability of SystemC/SDL co-

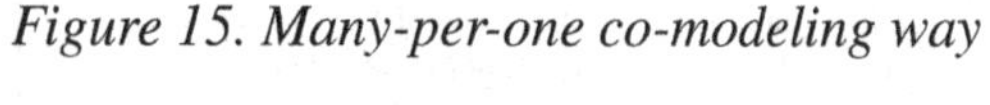

Figure 15. Many-per-one co-modeling way

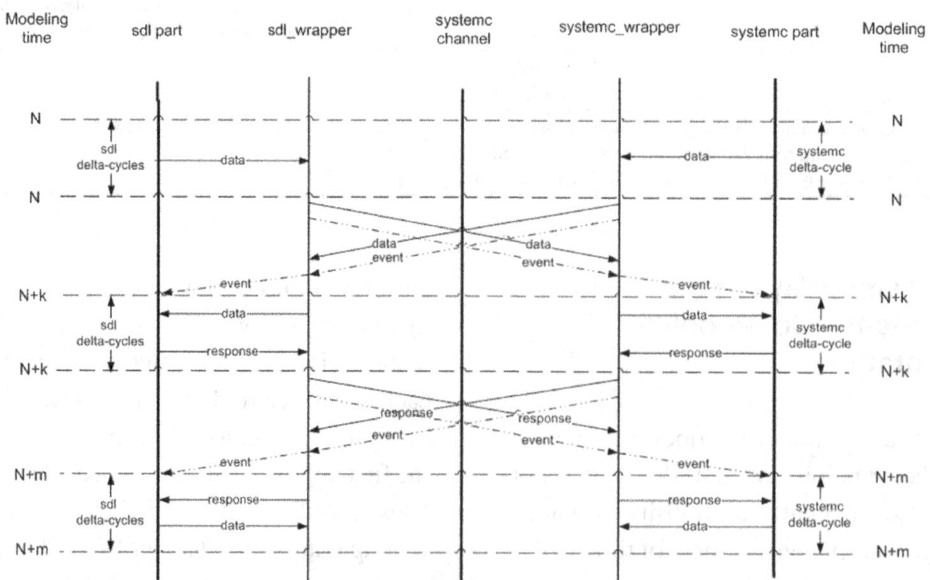

Table 3. Description of SDL and SystemC interaction for the example in Figure 14

Time	SDL part work	SystemC part work
N	SDL thread calls *xMainLoop()* function and so starts the first SDL delta-cycle. During delta-cycles SDL system generates data and writes it to *SDL_wrapper*. When SDL system does not have transitions to execute, SDL delta-cycles are over. After this function *xMainLoop()* is finished	SystemC part starts delta-cycle. At this delta-cycle SystemC Tx generates data. After that data is sent to SystemC_wrapper and SystemC delta-cycle is finished
N.. (N+k)	*SDL_wrapper* sends data to FIFO. The event is sent to *SystemC_wrapper* to inform that data is in FIFO	*SystemC_wrapper* sends data to FIFO. The event is sent to *SDL_wrapper* to inform that data is in FIFO
(N+k)	*SDL_wrapper* gets the event and reads the data from FIFO and sends event to SDL part	*SystemC_wrapper* gets the event and reads the data from FIFO and sends event to SystemC part.
(N+k)	SDL thread gets the event and calls *xMainLoop()* function and so starts the first SDL delta-cycle. During delta-cycles SDL system reads data, generates response and writes it to *SDL_wrapper*. When SDL system does not have any transitions to execute, SDL delta-cycles is over. After this function *xMainLoop()* is finished	SystemC part gets event and starts delta-cycle. During this delta-cycle SystemC Rx reads data and Tx generates response. After that the response is sent to *SystemC_wrapper*. SystemC delta-cycle is finished
(N+k) .. (N+m)	*SDL_wrapper* sends response to FIFO. The event is sent to *SystemC_wrapper* to inform that response is in FIFO	*SystemC_wrapper* sends response to FIFO. The event is sent to *SDL_wrapper* to inform that response is in FIFO
(N+m)	*SDL_wrapper* gets the event and reads the data from FIFO and sends event to SDL part	*SystemC_wrapper* gets the event and reads the response from FIFO and sends event to SystemC part
(N+m)	SDL thread gets the event and calls *xMainLoop()* function and so starts the first SDL delta-cycle. During the delta-cycles SDL system reads response, generates data and writes it to *SDL_wrapper*. When SDL system does not have any transitions to execute, SDL delta-cycles is over. After this function *xMainLoop()* is finished.	SystemC part gets event and starts delta-cycle. During this delta-cycle SystemC Rx reads response and Tx generates data. After that data is sent to SystemC_wrapper. SystemC delta-cycle is finished

modeling. It is used for simulation of data transfer protocols. This example is depicted in Figure 16. The Tester contains two identical protocol stack models and is used for testing the point-to-point connection between two nodes. Every stack represents an abstract four-layered protocol model in SDL language.

The Tester consists of two main parts: User Interface and Test Environment. The User Interface includes Configuration Interface and Monitoring Interface modules. The first one is intended for setting simulation parameters and definition of the test data sequence. Monitoring Interface gives access to the test results. The User Interface part is implemented in SystemC.

The Test Environment is responsible for target model simulation. This model is implemented in SDL compiled to C/C++ code and added to special SDL kernel. Then the resulting compiled C/C++ code together with SDL kernel is used inside the Tester. The simulated SDL model has to satisfy to a number of requirements. Firstly, modeling system should consist of two nodes. Each node represented by the model of corresponding protocol stack. They communicate with each other by use of the underlying layer interface. The nodes also communicate with the corresponding Node Managers. So, second requirement for tested SDL model is that all modules of the system should be able to interconnect with environment via the predefined interfaces by using set of SDL signals.

The Test Environment has to provide following features. Every Node Manager should be able to use the corresponding interface of SDL models upper layer to transmit user test data to the model. Every Medium module should provide the interfaces which are used by bottom layers of nodes for interconnection. Medium is a kind of representation of the channel so it transmits data from one node to another.

Compiled C/C++ code and SDL kernel connect to Test Environment module through SDL/SystemC Wrapper. The Wrapper is used for data conversion between SystemC/C++ and SDL

Figure 16. Protocol Model Tester scheme

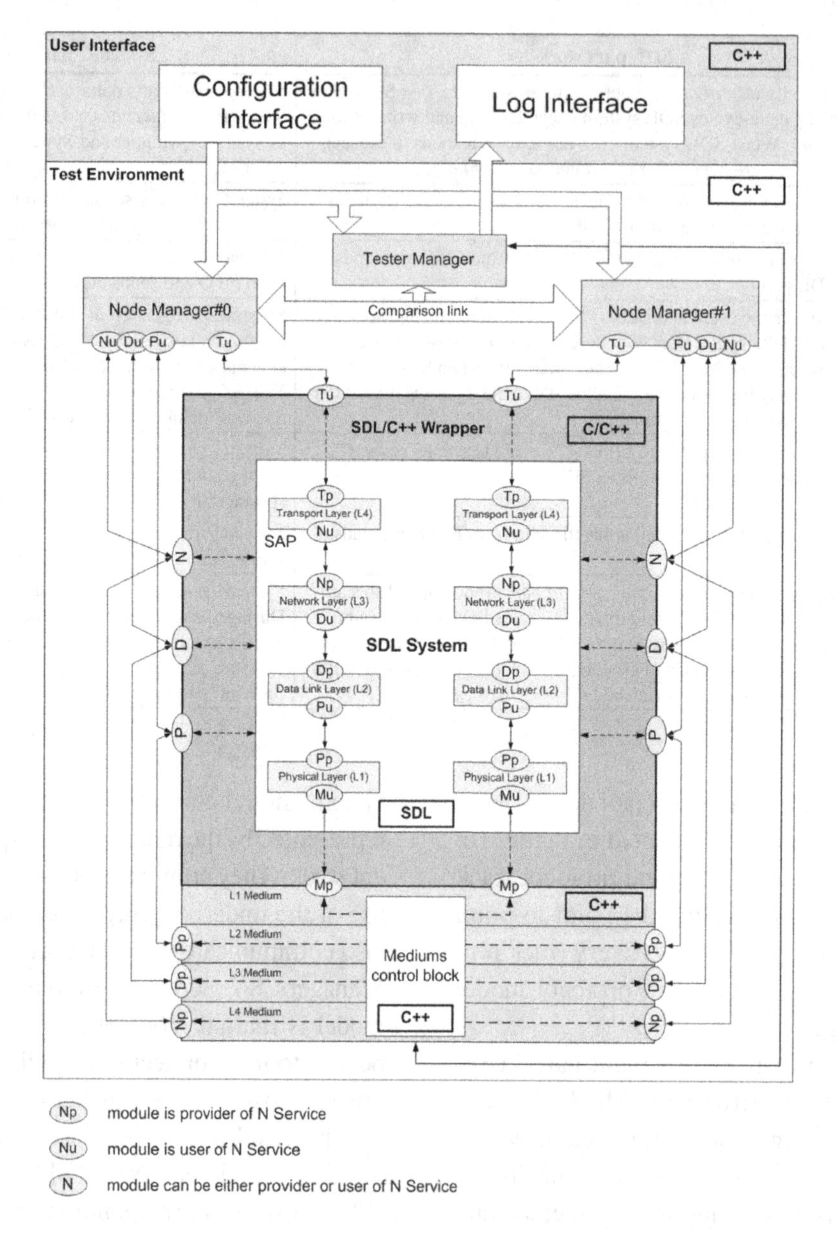

styles, SDL simulation control and configuration of connections between model and Test Environment. The monitoring of data transmitted through the Mediums is implemented by Mediums Manager. This module can perform the following tasks:

- **Log Creation:** Writing to an external stream information about transmitted data.
- **Error Injection:** Changing of transmitted data in accordance with defined parameters.
- **Delay Addition:** Introduce timeout between receiving and transmitting data.

This module is configured by Tester Manager in accordance with user requirements defined through Configuration Interface. The Tester is a mighty tool that could help not only testing of the specification, but also to get the first results of the network communication using the protocol. The analysis of these results could be of a great help for the further protocol design steps.

The proposed co-modeling approaches can be used with wide-variety of tools, as the only strict requirement is that SDL tool should be able to generate C code. For the SystemC part any C++ compiler with SystemC library should be fine.

CONCLUSION

This article gives analysis of the ways in which SDL and SystemC languages can be used together for co-modeling of the embedded systems. We start with introduction of the notions of embedded systems, where the traditional and spiral solutions of embedded systems design are discussed. The article specifically discusses the main problems of the modeling stage of specification development and sets the main requirements for the structure of functional design and verification. Modeling is used for validation of the specification and for algorithms and systems testing. The article provides a general comparison of SDL and SystemC and shows advantages and disadvantages of these languages for modeling. SystemC is a set of C++ classes that is used for modeling parallel systems. The mechanism of internal events allows to efficiently models distributed in time operation of the system under study. The main conclusion of the provided summary is that SystemC qualifies as a proper language for modeling, design and verification of the systems.

The purpose of SDL is to make an unambiguous specification and description of the system behavior. SDL language is intended for the description of the structure and operation of the distributed real-time systems, which has components that are interpretable in a form of finite state machines. The article provides the comparative analysis of both languages with respect to the following characteristics: applicability for modeling protocols; possibility of dynamic object creation; notion of the modeling time; testing of the models. The analysis once more proves that choice of the modeling language depends on the modeling purposes.

The main purpose of this article was to evaluate whether it is possible, and if yes than how to organize the co-modeling of SystemC and SDL in order to take the best from both languages. The following three approaches were proposed and investigated: precompiled SDL model included into SystemC; SystemC modules attached to SDL; and parallel use of SystemC and SDL. The main stages of each approach were identified and discussed in details, including advantages, disadvantages and restrictions of each approach. The study performed by our team in practice discovered that solution with SDL modules included into SystemC is the best option. The selected approach allows two ways of implementation. The first way requires changes in SDL modeling principles. The second does not change the modeling principles, but requires additional SDL model restrictions, i.e., SDL model should be able to split modeling cycle so that consequent transitions allocated to the same moment will be still triggered sequentially in respect of the modeling time.

Again this difference between the ways of implementation shows that each solution is better in different modeling tasks. The second way gives better real timing results, but if it is difficult to implement the system with the pointed out restrictions, than we can use the first way.

We believe that SDL and SystemC co-modeling is very perspective and interesting area of researches that results in a new quality of facilitation of the specification, modeling and verification work. The implementation of the described co-modeling solution is currently under development in our team.

REFERENCES

Barr, M., & Massa, A. (2006). *Programming Embedded Systems: With C and GNU Development Tools, Second edition*. New York: O'Reilly Media, Inc.

Black, D., & Donovan, J. (2004). *SystemC: From the Ground Up*. New York: Springer. doi:10.1007/0-387-30864-4

Bykhteev, A. (2008). Methods and facilities for systems-on-chip design. *ChipInfo microchip manual.* Retrieved from http://www.chipinfo.ru/literature/chipnews/200304/1.html

Fonseca, P. (2008). SDL, a graphical language useful to describe social simulation models. In *Proceedings of the 2nd Workshop on Social Simulation and Artificial Societies Analysis (SSASA'08)*. Retrieved from http://ceur-ws.org/Vol-442/p4_Fonseca.pdf

Gillet, M. (2008). Hardware/software co-simulation for conformance testing of embedded networks. In *Proceedings of the 6th Seminar of Finnish-Russian University Cooperation in Telecommunications (FRUCT) Program*. Retrieved October 31, 2008, from http://fruct.org/index.php?option=com_content&view=article&id=68&Itemid=73

Gipper, J. (2007). SystemC the SoC system-level modeling language. *Embedded computing Design*. Retrieved from www.embedded-computing.com/pdfs/OSP2.May07.pdf

Grimpel, E., Timmermann, B., Fandrey, T., Biniasch, R., & Oppenheimer, F. (2002). SystemC Object-Oriented Extensions and Synthesis Features. In *Proceedings of the European Electronic Chips & Systems design Initiative*. Retrieved from www.ecsiassociation.org/ecsi/projects/odette/files/fdl2002.pdf

Haroud, M., & Blazevic, L. (2006). HW accelerated Ultra Wide Band MAC protocol using SDL and SystemC. In *Proceedings of the Fourth IEEE International Conference on Pervasive Computing and Communications Workshops (PERCOMW'06)*. Retrieved from http://fmv.jku.at/papers/HaroudBlazevicBiere-RAWCON04.pdf

Heath, S. (2003). *Embedded Systems Design* (2nd ed.). New York: Newnes.

IBM. (2009). *SDL Suite and TTCN Suite Help*. IBM Rational SDL and TTCN Suite.

International Telecommunication Union. (2002). *Recommendation Z.100. Specification and Description Language (SDL)*. Geneva, Switzerland: ITU.

ITU-T. (2000). *Specification and Description Language (SDL)*. Retrieved from http://www.itu.int

Jantsch, A. (2004). *Modeling Embedded Systems and SoCs*. San Francisco: Morgan Kaufmann Publishers.

Jozawa, T., Huang, L., Sakai, E., Takeuchi, S., & Kasslin, M. (2006). Heterogeneous Co-simulation with SDL and SystemC for Protocol Modeling. In. *Proceedings of the IEEE Radio and Wireless Symposium, 2006*, 603–606. Retrieved from http://research.nokia.com/node/5789. doi:10.1109/RWS.2006.1615229

Kamal, R. (2008). *Embedded systems: architecture, programming and design* (2nd ed.). New Delhi, India: Tata McGraw-hill Publishing Company Limited.

Karabegov, A., & Ter-Mikaelyan, T. (1993). *Introduction to the SDL language*. Moscow, Russia: Radio and communication.

Nemydrov, V., & Martin, G. (2004). *Systems-on-chip. Design and evaluation problems*. Moscow, Russia: Technosphera.

Olenev, V. (2009). Different approaches for the stacks of protocols SystemC modeling analysis. In *Proceedings of the Saint-Petersburg University of Aerospace Instrumentation scientific conference* (pp. 112-113). Saint-Petersburg, Russia: Saint-Petersburg University of Aerospace Instrumentation (SUAI).

Olenev, V., Onishenko, L., & Eganyan, A. (2008). Connections in SystemC Models of Large Systems. In *Proceedings of the Saint-Petersburg University of Aerospace Instrumentation scientific student's conference* (pp. 98-99). Saint-Petersburg, Russia: Saint-Petersburg University of Aerospace Instrumentation (SUAI).

Olenev, V., Rabin, A., Stepanov, A., & Lavrovskaya, I. (2009). SystemC and SDL Co-Modeling Methods. In *Proceedings of the 6th Seminar of Finnish-Russian University Cooperation in Telecommunications (FRUCT) Program* (pp. 136-140). Saint-Petersburg, Russia: Saint-Petersburg University of Aerospace Instrumentation (SUAI).

Olenev, V., Sheynin, Y., Suvorova, E., Balandin, S., & Gillet, M. (2009). SystemC Modeling of the Embedded Networks. In *Proceedings of 6th Seminar of Finnish-Russian University Cooperation in Telecommunications (FRUCT) Program* (pp. 85-95). Saint-Petersburg, Russia: Saint-Petersburg University of Aerospace Instrumentation (SUAI).

Open SystemC Initiative (OSCI). (2005). *IEEE 1666™-2005 Standard for SystemC*. Retrieved from http://www.systemc.org

Stepanov, A. (2009). Comparison of SDL and SystemC Languages applicability for the protocol stack modeling. In *Proceedings of the Saint-Petersburg University of Aerospace Instrumentation scientific student's conference* (pp. 76-80). Saint-Petersburg, Russia: Saint-Petersburg University of Aerospace Instrumentation (SUAI).

Suvorova, E. (2007). A Methodology and the Tool for Testing SpaceWire Routing Switches. In *Proceedings of the first International SpaceWire Conference*. Retrieved September 19, 2007, from http://spacewire.computing.dundee.ac.uk/proceedings/Papers/Test and Verification 2/suvorova2.pdf

Suvorova, E., & Sheynin, Y. (2003). *Digital systems design on VHDL language*. Saint-Petersburg, Russia: BHV-Petersburg.

Swan, S. (2003). *A Tutorial Introduction to the SystemC TLM Standard*. Retrieved July 7, 2008, from http://www-ti.informatik.uni-tuebingen.de/~systemc/Documents/Presentation-13-OSCI_2_swan.pdf

This work was previously published in the International Journal of Embedded and Real-Time Communication Systems, Volume 2, Issue 1, edited by Seppo Virtanen, pp. 23-48, copyright 2011 by IGI Publishing (an imprint of IGI Global).

Chapter 3
Design and Implementation of a Firmware Update Protocol for Resource Constrained Wireless Sensor Networks

Teemu Laukkarinen
Tampere University of Technology, Finland

Jukka Suhonen
Tampere University of Technology, Finland

Lasse Määttä
Tampere University of Technology, Finland

Timo D. Hämäläinen
Tampere University of Technology, Finland

Marko Hännikäinen
Tampere University of Technology, Finland

ABSTRACT

Resource constrained Wireless Sensor Networks (WSNs) require an automated firmware updating protocol for adding new features or error fixes. Reprogramming nodes manually is often impractical or even impossible. Current update protocols require a large external memory or external WSN transport protocol. This paper presents the design, implementation, and experiments of a Program Image Dissemination Protocol (PIDP) for autonomous WSNs. It is reliable, lightweight and it supports multi-hopping. PIDP does not require external memory, is independent of the WSN implementation, transfers firmware, and reprograms the whole program image. It was implemented on a node platform with an 8-bit microcontroller and a 2.4 GHz radio. Implementation requires 22 bytes of data memory and less than 7 kilobytes of program memory. PIDP updates 178 nodes within 5 hours. One update consumes under 1‰ of the energy of two AA batteries.

DOI: 10.4018/978-1-4666-2776-5.ch003

INTRODUCTION

A Wireless Sensor Network (WSN) consists of autonomous sensor nodes (Akyildiz, Weilian, Sankarasubramaniam, & Cayirci, 2002). The goal of sensor node hardware development is to create tiny battery-powered low-cost disposable nodes. Increasing the performance or memory capacity increases the physical size, energy consumption and manufacturing costs. Thus, nodes are limited in computation, storage, communication and energy resources. These limitations must be addressed when designing and implementing protocols in WSNs.

It is not always possible to physically access the nodes in the field once they are deployed. Yet, adding new features, applications and program error fixes necessitates updating the program image that contains the software and protocols running on a node. The solution is a WSN reprogramming protocol, which is used to inject new software into a WSN.

Five general challenges affecting reprogramming in WSNs can be identified (Wang, Zhu, & Cheng, 2006). First, large program images must be transferred reliably through an error prone medium. Thus, the receiver should be able to detect errors and request the corrupted segments again. Second, processing speed and memory capacity in nodes set limits to the time and space complexity of designed protocols. Third, battery powered WSN nodes inherently require the reprogramming protocols to be energy efficient. Fourth, the reprogramming protocol must be scalable enough to handle WSNs that consist of hundreds or thousands of nodes deployed in varying densities. And fifth, the operating system, which is used in nodes, can set limits on the program image format and the reprogramming protocol.

Several protocols (Wang, Zhu, & Cheng, 2006) have been proposed for reprogramming a WSN. A common approach is to equip each node with external memory storage where the new program image is stored. Once the image has been received and verified, a dedicated image transfer program copies the new program image over the old image. This approach allows uninterrupted operation as the new image is transferred in the background. However, the additional memory increases hardware price and takes place on the circuit board, therefore necessitating expensive or energy consuming platforms that prohibit the vision of long term, disposable nodes. Furthermore, many protocols (Hui & Culler, 2004; Levis, Patel, Culler, & Shenker, 2004; Levis & Culler, 2002) support a particular operating system only.

In this paper we present the design, implementation and experimental results of a Program Image Dissemination Protocol (PIDP) for autonomous adhoc multihop WSNs. PIDP consists of firmware version handshakes between nodes, periodic firmware version advertisements and a reliable program image transfer, as shown in Figure 1. Firmware version advertisements are used between neighboring nodes to advertise and compare firmware versions and check for compatibility. The reliable image transfer is used to transfer program images between nodes and to rewrite the program memory. A small bootloader program locates and executes the loaded program image. PIDP is lightweight, energy efficient, reliable and, unlike other reprogramming protocols, does not require external memory for temporary storage of program images. A PIDP update in one part of the WSN does not disturb the whole network, thus, allowing a continuous operation of the non-affected nodes. Furthermore, PIDP is not restricted to a particular operating system or WSN protocol.

PIDP was evaluated using the TUTWSN prototype (Kuorilehto, Kohvakka, Suhonen, Hämäläinen, Hännikäinen, & Hämäläinen, 2007). TUTWSN is a state of the art adhoc multihop WSN technology for resource-constrained WSNs developed by Department of Computer Systems at Tampere University of Technology. TUTWSN features an energy efficient medium access control (MAC), which uses time-division multiple

Figure 1. The logical structure and the memory layout of PIDP and the WSN stack. PIDP is a separate protocol stack. Firmware version advertisements and handshaking co-operate with the WSN stack to disseminate version information and to begin reliable program image transfer.

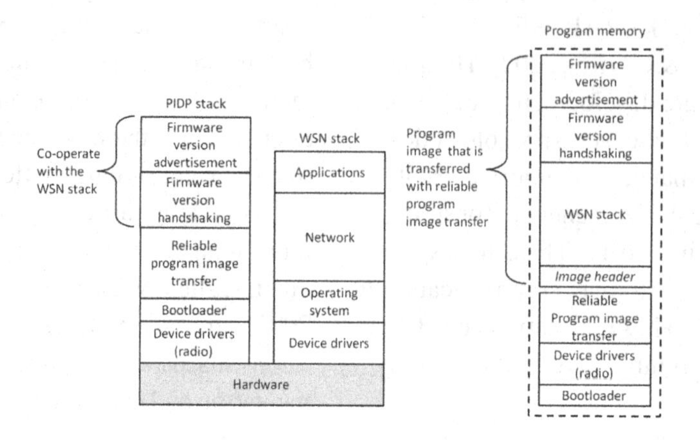

access (TDMA), a cost-aware routing protocol (Suhonen, Kuorilehto, Hännikäinen, & Hämäläinen, 2006) and multiple custom designed hardware platforms. The operating principle of the MAC layer of TUTWSN is similar to the beacon enabled clustered mode in IEEE 802.15.4 (IEEE Standards Association, 2008). Therefore, a similar implementation can be applied to ZigBee (ZigBee Alliance, 2010).

The paper is organized as follows. First, related work is covered. Second, the design of PIDP is presented. Third, implementation is shown. Fourth, evaluation, performance measurements are given. Finally, the paper is concluded.

RELATED WORK

A number of reprogramming protocols for WSN are built on the TinyOS (Hill, Szewczyk, Woo, Hollar, Culler, & Pister, 2000) operating system. TinyOS does not support loadable modules. Thus, a program image must be loaded as a single binary image.

XNP (Crossbow Technologies, 2003) is one of the first reprogramming services for TinyOS and the MICA2 platform. It features a single-hop reprogramming scheme where the program image is sent as unicast to a particular node or broadcasted to a group of nodes. The single-hop nature limits the scalability of XNP and it only serves as an alternative to manual wired reprogramming.

The successor to XNP is Deluge (Hui & Culler, 2004). Deluge is an epidemic multihop protocol that allows nodes to store several different program images in an external EEPROM memory. One of these images can act as the so called Golden image, which is used as a backup image if the main program image is corrupted. The 2.0 version (Hui, 2005) also adds support for resuming incomplete program image downloads and additional program image verification. MOAP (Stathopoulos, Heidemann, Estrin, & SENSING, 2003) is similar to Deluge.

The Maté virtual machine (Levis & Culler, 2002), which is built upon TinyOS, bypasses the lack of loadable modules by presenting a high-level virtual machine instruction set. Maté bytecode programs are smaller than full program images, which lowers the energy cost of disseminating them. The downside is that interpreting the bytecode creates energy overhead. If new software is disseminated only seldom, the energy consumption of the code interpretation is dominant.

Unlike the TinyOS-based approaches, individual applications and services can be loaded individually in the Contiki operating system (Dunkels, Gronvall, & Voigt, 2004; Dunkels,

Finne, Eriksson, & Voigt, 2006). Like Maté, this saves energy as only parts of the whole image need to be disseminated. This dynamic loading only applies to the applications, while the operating system and the protocol stack can only be updated with a separate special image transfer program.

The requirement for external memory storage is common to all these reprogramming protocols, as they use transport layer dissemination protocols to transfer program images. These dissemination protocols are stored within the main program image, which cannot be overwritten as long as it is being executed. This can cause problems e.g. in (Langendoen, Baggio, & Visser, 2006), where unreliable MAC protocol made Deluge useless. As a result, nodes were updated by hand on the deployment site.

An approach to updating based on the differences between the old and the new program image is presented in (Mukhtar, Kim, Kim, & Joo, 2009). The old and new images are analyzed and a model to modify the old one is created. The model and completely new parts are disseminated with any dissemination protocol. Similar approach is presented in (Reijers & Langendoen, 2003), but this is a processor specific solution. These approaches do not efficiently update the image when it is significantly different from the old image. However, these do not require external flash, but they do require a reliable transport protocol layer. Reliable transport protocols for code dissemination have been presented in Stathopoulos, Heidemann, Estrin, and SENSING (2003) and Miller and Poellabauer (2008).

As opposed to Contiki, Maté or difference models, PIDP transfers complete program images. In our experience, the ability to update individual applications is seldom needed as programming error fixes and new features often affect multiple modules of the program image. In addition, loading individual applications requires either support from the operating system or a separate mechanism for handling runtime relocation of modules. PIDP requires no such support and is operating system independent. PIDP does not require an external reliable transport protocol and it can be used to update completely different image to the network.

PIDP DESIGN

PIDP design consists of firmware version handshaking, periodic firmware version advertisements, and reliable image transfer. Figure 2 presents the PIDP design in action. Following sections present the design in detail. PIDP minimizes communication and memory overhead, therefore allowing very resource-constrained implementations. Also, PIDP design includes new firmware injection, security, operating system support, and support for heterogeneous networks.

Firmware Version Handshaking

Program image transfer begins automatically when a node detects that one of its neighbors has a new version of a compatible program, as shown in Figure 2 between the nodes A and B. The node with a lower version number sends a *firmware request* and the other node responds with a *firmware confirmation* at the WSN protocol level. After this, the nodes jump to the reliable image transfer of the PIDP, which is independent of the WSN stack.

PIDP assumes that a node performs a handshake with its neighbors after powering up to find new routes. This is the case in most of the sender decided WSN protocols, such as ZigBee. Version information is exchanged in PIDP when a node exchanges routing information or synchronizes with its neighbors, which adds a small overhead. Either node participating in the handshaking can start the update operation. Both nodes reboot after the update and perform handshaking with their neighbors. This guarantees that program images will propagate epidemically in the WSN.

Figure 2. Node B has a newer version. Node A and Node B execute firmware handshaking on WSN association. Then they start program image update and move to the PIDP reliable image transfer. Meanwhile Node C and D continue normal WSN operation. Eventually, Node A is updated and nodes reboot. Node B associates with Node C, starts the update, and continues disseminating image further.

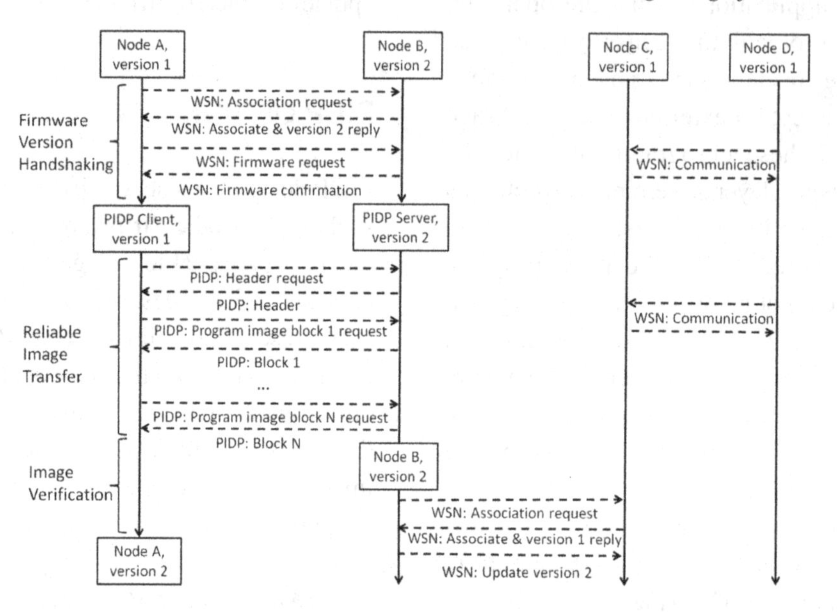

It is important to note that version information is exchanged on a hop-by-hop basis between neighbors without flooding the version information further into the network. If a network contains multiple nodes with incompatible program images, it may limit the propagation of program images.

Periodic Firmware Version Advertisements

Nodes periodically advertise their program image version on an advertisement channel, as shown in Figure 3. After each advertisement the source listens for a reply. The parameter T_a adjusts the interval between these periodic advertisements. The main purpose of the periodic advertisements is to act as a failsafe. If a node encounters a problem while reprogramming or the image transfer is disrupted, the node may listen for the periodic advertisements to find a new source for image transfer as shown in Figure 3. In addition, periodic advertisements allow nodes to perform image acquisition even if their protocol stacks might otherwise be incompatible. As periodic advertisements are only transmitted seldom and nodes do not listen for them during normal operation, they use very little energy. WSNs that use synchronized MAC protocols, such as IEEE 802.15.4, can embed the version information in synchronization beacons, which nodes transmit periodically.

Reliable Image Transfer

Information about the program image is stored in a header, which consist of hardware identifier, firmware version, message authentication code, and valid and dissemination bits. Program images are identified by a combination of the hardware platform identification number and the firmware version number. Platform identification numbers are used to limit the transfer of program images between incompatible sensor nodes. Furthermore, the header contains a valid bit that indicates

Figure 3. Example of periodic advertisements and their functioning as a fail-safe. Node A transmits advertisements with interval T_a. Node B is updating its firmware with Node C at the data channel of Node C. Node B encounters a problem at t_{error}, scans the advertisement channel and begins a new transmission with Node A using the data channel of Node A.

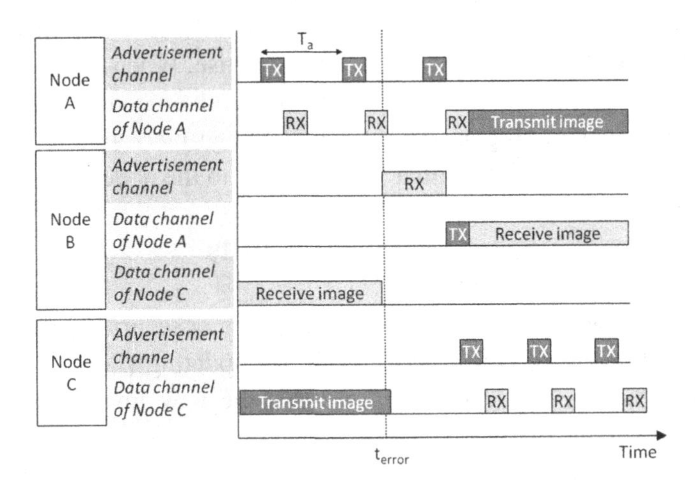

whether or not the program image has been successfully validated and can be safely executed. The dissemination bit decides if the node will disseminate the image to the network.

Unlike other reprogramming proposals, the image transfer in PIDP operates independently of the main WSN stack. This allows the image transfer protocol to achieve a better energy-efficiency, as minimizing the number of protocols layers used in the transfer also minimizes the amount of overhead in the transmission of program images. Furthermore, simple independent stack can be tested thoroughly and possible problems of unreliable transfer protocols cannot prevent program image update. As the WSN protocol stack is part of the updated image, the possible problems on WSN protocols can be fixed with PIDP.

The image transfer protocol follows the general client-server architecture as seen in Figure 2. The receiver of the program image acts as a PIDP client, while the sender acts as a PIDP server. Communication between a PIDP client and a PIDP server is performed at a channel selected by the PIDP server, which is transmitted within the firmware advertisements. PIDP servers may choose to use a single network-wide dedicated channel for the image transfers or they may use an appropriate channel selection algorithm to choose channels that are not being used. Choosing different channels is preferred, as this lowers the chance of collisions with nearby image transfers.

The PIDP client begins by requesting the header of the program image as presented in Figure 2. Once the header is received the PIDP client marks the current header invalid and requests the contents of the program image in blocks. After each block the PIDP client immediately writes the data to the program memory, thus invalidating the previous program image. After the whole image is received the PIDP client validates the program image by calculating the message authentication code and comparing it to the one in the header. If the validation calculation is correct, the header is marked valid. Otherwise, the header remains marked invalid.

After the transfer the PIDP client and the PIDP server reboot and return to the normal WSN operation. The PIDP client uses the PIDP bootloader program to check that the header is valid and begins executing code from the beginning of the program image. If the header is not valid then the PIDP client begins to scan the advertisement channel for firmware advertisements and re-executes the image transfer.

Program Image Injection

Three alternative ways exist for new program image injection with PIDP. First, a new node with a new image may be brought to the coverage area of the WSN. The new image is then disseminated to the network automatically by PIDP. Second, a *PIDP cloner device* may be used to transfer program image to one node in the network, which then starts advertising the new version. Third, the new image can be uploaded to a server, which delivers it to the gateways. The gateways advertise the new image to the network and PIDP will first update the nearest nodes using the gateways as relays.

The PIDP cloner device is a specially programmed node that does not act as a part of the WSN. It only advertises the new image on a special cloning channel. The nodes do not normally listen for this channel. When the node is rebooted while a button is pressed, the node will listen for the cloning channel for a period. If there is a PIDP cloner device nearby and the hardware platform identifier match, the node will start the image transfer. If the dissemination bit is set, the newly programmed node will then start disseminating the image further.

The server injection is presented in Figure 4. The image is first compiled and then modified with a script to a XML file, and the XML file is finally uploaded to the database. The server indicates to the gateways that there is a new image to advertise. When a node notices the new image from the advertisements, the gateway requests the new image piece by piece from the server and relays it to the node.

Security

Three major security questions concern program image updating (Deng, Han, & Mishra, 2006). First, the new image must be from a reliable source. Second, the new image must be valid. Third, the image must be transferred securely to preserve intellectual property. PIDP accepts only images with correct message authentication code. This ensures that unknown source cannot inject a new image to the network and hijack the network. The message authentication code is calculated with a one-way function that uses a secret key, the program image, and a magic number as parameters. As the program image is used in calculation of the message authentication code, the image validity is secured at the same time. The secret key can be used to encrypt and decrypt the program image packets with AES algorithm after the handshaking to prevent stealing the program image with sniffing.

Figure 4. Program image injection starts with compilation of the image from a source code to a hex file, then formatting it to an xml file, and uploading it to the database. The server will retrieve image information and relay it to the gateways. The gateways will advertise the image to the WSN and relay the image, when a node requests the new image.

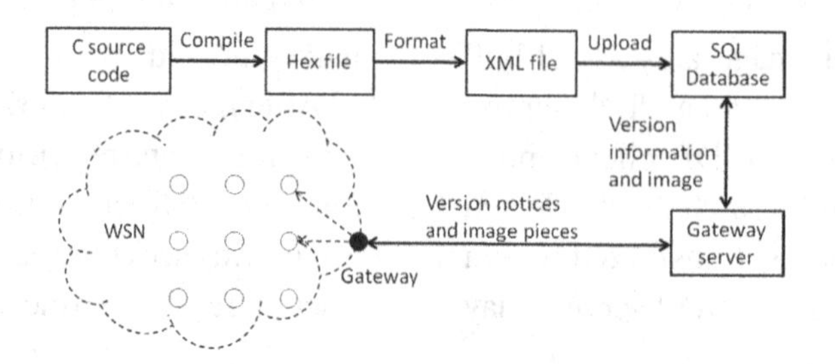

Operating System Support

WSN operating systems have two approaches for updating. The whole image including the application and the operating system are disseminated (TinyOS and Deluge), or only the applications are disseminated to the network (Contiki). PIDP supports both ways as presented in Figure 5. The operating system can be a part of the whole program image as in Figure 5a. This is similar to Deluge and TinyOS. The operating system can be left out of the program image, as in Figure 5b, but the applications are treated as one image. Injecting one new application requires re-injecting all the existing applications.

PIDP allows as many image version headers as there are room in the version advertisement packet. Thus, program image can be split in several parts as presented in Figure 5c. These parts can be separately updated. The selected program image part is indicated in the handshaking between the PIDP client and server. Then, the PIDP will update only the selected part. This can be used to inject new applications to the network without injecting the remaining ones again and the operating system can be updated separately. However, each image requires a new header. The header overhead would increase and the amount of applications would be limited. Furthermore, the applications should always fit inside a certain space and some applications would waste the program memory. Every program image requires known entry functions, which will add some complexity in the development. Multiple program image support is not currently incorporated to the PIDP design nor implemented. It will be implemented in future work. Novel solutions are needed for solving the problems.

PIDP does not restrict the operating system from using its own protocols to update applications. For example, PIDP can update Contiki and Contiki can use its own protocols to handle applications. The image validation should then only cover the area, which is not modified by the program code dissemination of Contiki.

Heterogeneous Node Support

WSNs are seldom homogenous; nodes have different sensors, different roles, and different applications. PIDP separates heterogeneity only between hardware devices. If the hardware is not same, the image transfer is not started. To overcome this limitation, we have developed an autoconfigurator. The node is configured during the building process to its configuration: the connected

Figure 5. a) A typical use of PIDP, where operating system and WSN stack form the program image. b) If the operating system is reliable and will not require new features, it can be left out of the program image. c) With small modifications, PIDP can update the program image in two or more parts. This allows granular updating, but image headers increase overhead.

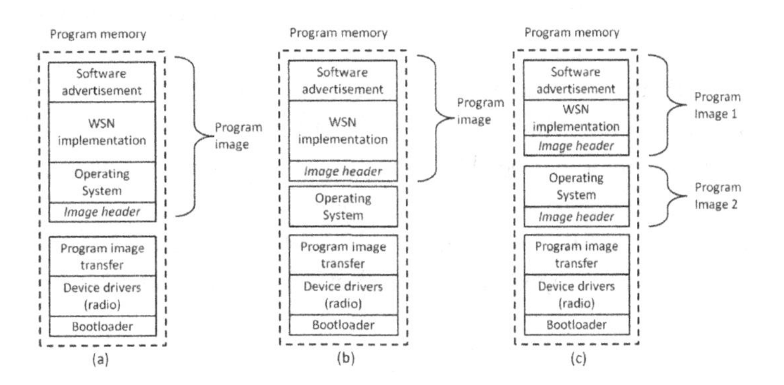

sensors, desired roles and required applications of the node are set to the EEPROM of the node. A program image is used, which contains all the necessary code for these configurable parts. Node selects used role and applications at the startup according to the configuration. This allows us to use one single program image for the whole WSN with various node configurations.

IMPLEMENTATION

PIDP and TUTWSN protocol stack are implemented using the C programming language and the Microchip MPLAB C compiler (Microchip Technology, 2009).

TUTWSN Protocol Stack

The TUTWSN MAC protocol forms a clustered tree topology (Kuorilehto, Kohvakka, Suhonen, Hämäläinen, Hännikäinen, & Hämäläinen, 2007). Each cluster contains a cluster head, a *headnode*, and several cluster members. Cluster members can be leaf nodes or headnodes of other clusters forming a tree of clusters. Each cluster within interference range operates on a separate *cluster channel* that is used for intra-cluster communications. Nodes share a common *network channel* that is used by the headnodes to advertise their clusters. Nodes scan the network channel at least once every hour to find new clusters. In addition, network scans occur when nodes lose their route to the network gateway.

The headnodes maintain a data exchange schedule. Time is divided into fixed length access cycles. Each access cycle begins with a superframe, which contains slots for data transfers at the cluster channel, and ends in an idle time. The length of the access cycle is set to two seconds. Cluster advertisements are sent on the network channel in the beginning of each superframe. Three additional cluster advertisements are also sent during the idle time with approximately 500 millisecond intervals between them.

A TUTWSN sensor node includes an 8 bit Microchip PIC18LF8722 microcontroller (Microchip Technology 2008) with 128 kilobytes of program memory and 3936 bytes of data memory. The microcontroller has an internal 1024 byte EEPROM memory. A Nordic Semiconductors nRF24L01 (Nordic Semiconductors, 2007) is used as the radio, which has a payload size of 32 bytes and a configured transmission rate of 1 megabit per second. The radio does support carrier sensing. It operates on the 2.4 gigahertz band and offers 126 channels and transmission powers of -18 dBm...0 dBm. TUTWSN node has a simple user interface, which consists of a push button and two light emitting diodes. TUTWSN sensor nodes can be equipped with multiple sensors, such as accelerometers, temperature sensors, and humidity sensors. Two 1.5 volt LR6-sized batteries are used as the power source. A TUTWSN sensor node circuit board is shown in Figure 8.

PIDP Implementation

Firmware version handshaking was embedded to the MAC layer of TUTWSN. Thus, nodes exchange version information when they perform association with each other. This allows rapid firmware dissemination within a TUTWSN cluster tree.

The periodic firmware advertisements are transmitted in the TUTWSN network channel, which allows nodes to receive advertisements while they are performing normal neighbor discovery. An advertisement is sent on each access cycle during the idle time. Thus, the interval T_a between the advertisements matches the length of the access cycle. In addition, advertisements on the TUTWSN network channel allow the program image to propagate between different cluster trees, but this method of dissemination is limited by the low frequency of network scans. The cluster channel is used for program image transfer to minimize collisions between concurrent program image transfers.

The bootloader and the program image transfer protocol are stored in a reserved segment in the beginning of the program memory. They are followed with the program image header and the main program image. The message authentication codes are implemented by using a modified 4 byte RC4 code similar to the code described in (Zhang, Yu, Huang, & Yang, 2008).

The program image transfer protocol uses a packet size of 32 bytes. Each packet has a 6 byte header followed by a payload with a length of 26 bytes.

Reliable image transfer is located on a memory section that cannot be updated with PIDP. It includes only the necessary modules to perform the program image transfer and the program memory rewrite. Modules are a radio driver, PIDP server, PIDP client, program memory writer, program image verification, and bootloader. This memory section has to be kept as small as possible since it reduces amount of available memory for the WSN implementation.

The reliable image transfer and the main program are never executed concurrently. Thus, the image transfer can utilize data memory segments that are normally reserved for the main program. Overlaying the data memory significantly reduces data memory requirements of the image transfer protocol. Despite the overlaying, a small amount of dedicated memory is needed for passing version information between the main program and the image transfer.

EVALUATION

Evaluation of PIDP was performed by analyzing the memory consumption, propagation speed and energy consumption impact.

Memory Consumption

Memory consumption was analyzed from the compiled program images for the TUTWSN platform.

From the results in Table 1 we can see that the memory consumption of the PIDP protocol is split in two parts. The first part contains the image transfer while the second part is stored within the program image and contains the necessary support for accessing the image transfer protocol and the implementation of the firmware advertisement scheme.

Although PIDP requires 815 bytes of data memory in total, the absolute increase in the data memory requirements stays at 22 bytes. The image transfer overlays data memory with the WSN stack.

Propagation Time

In order to give a reference point for the measured propagation times, the program image transmission and verification times between two nodes were first measured. Transferring a 123 kilobyte image between two nodes in optimal conditions was 51 seconds on average, thus achieving a transfer rate of 2.4 kilobytes per second. The program image verification time was a constant 23 seconds. Thus, the minimum time for updating a sensor node with this particular program image was 74 seconds.

Program image propagation experiments were performed in a typical office environment with various interference sources such as several

Table 1. Memory consumption of TUTWSN with PIDP in bytes. ©2010 IEEE. Used with permission.

Component	Program memory (B)	Data memory (B)
Reliable image transfer	3578	793 (overlayed)
Version handshaking	1386	15
Firmware advertisements	1425	7
Total	6389	22 + (793)

WLAN routers operating on the same frequency band. The first experiment included one gateway and 25 sensor nodes. The nodes were placed on a table in one group. Size of the table was less than one square meter. The purpose of this experiment was to see how PIDP performed in a situation where every node had multiple neighbors in close proximity and the amount of network activity was high. Update speed is presented in Figure 6. PIDP successfully reprogrammed the nodes in 12 minutes. 8 concurrent image transfers were observed during this period at the time of 500 and 600 seconds from the start. 28 updates were performed, which indicates that 3 updates failed and 3 nodes had to be updated again. Reason for these failures is unknown. T_a was approximated based on the measurements. It took 35 seconds from a node to be capable to disseminate received image. Figure 7. a) Tampere University of Technology campus, the coverage of the campus WSN, and the new program image injection point of the experiment in Computer Science building. b) A graph presenting the TUT Campus WSN experiment progress in percentage of updated nodes. The dissemination was stalled, because two neighboring nodes belonged to different clusters and had good routes to different gateways. Therefore, they did not associate until one hour periodic scan was due. Labels indicate the time when the update was completed for that building

Figure 6. The updating speed graph of PIDP on the 25 node experiment. The extra three updates were result of failed updates, which caused re-update. Ideal T_a presents how the network would be updated, if a node could start updating another one immediately after receiving the new image. Approximated T_a of the experiment indicates that one node disseminates 35 seconds after the update.

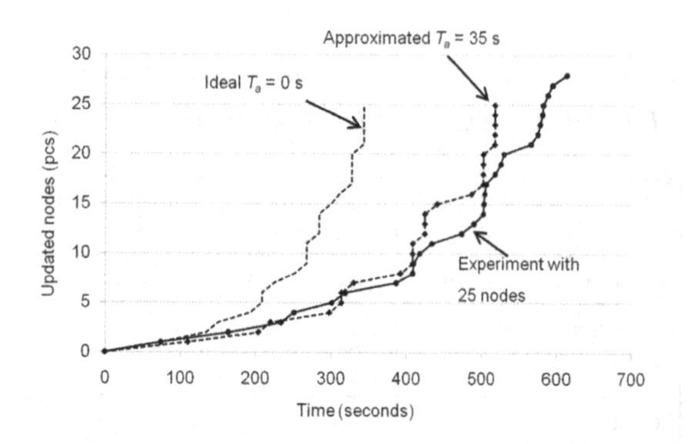

For the second experiment, the performance of PIDP was measured using the Tampere University of Technology campus WSN. This campus network has 178 sensor nodes and 13 gateways distributed in six buildings around the university campus. Figure 7a presents the campus and the coverage area of the campus WSN. The Computer Science building has sensor nodes in four floors while the others have nodes in only one floor. Distance between nodes ranges from 5 meters to 20 meters. The campus WSN is used as an application platform for students to implement their own applications on a WSN course. Measurement data of the campus WSN is provided for property maintenance. Figure 8 presents a humidity, luminance, and temperature measurement node at the campus of the Civil Engineering building. In addition, carbon dioxide and passive infrared based human activity are measured in the campus WSN. Students attending to the course may carry a node with them, which is tracked by the campus WSN.

The new program image was injected into the WSN by updating a single node on the 4th floor of the Computer Science building. Each node sent a report after a successful update procedure.

The results of the second experiment show that PIDP successfully propagated the program image through the campus WSN in five hours, as shown in Figure 7b. A delay was experienced between a pair of nodes located between the Civil Engineering building and the Main building. Once the image had spread to the Main building, it continued to propagate to the rest of the WSN. As a result, program image had to travel over 50 hops to achieve the last node in the Mechanical

Figure 7. a) Tampere University of Technology campus, the coverage of the campus WSN, and the new program image injection point of the experiment in Computer Science building. b) A graph presenting the TUT Campus WSN experiment progress in percentage of updated nodes

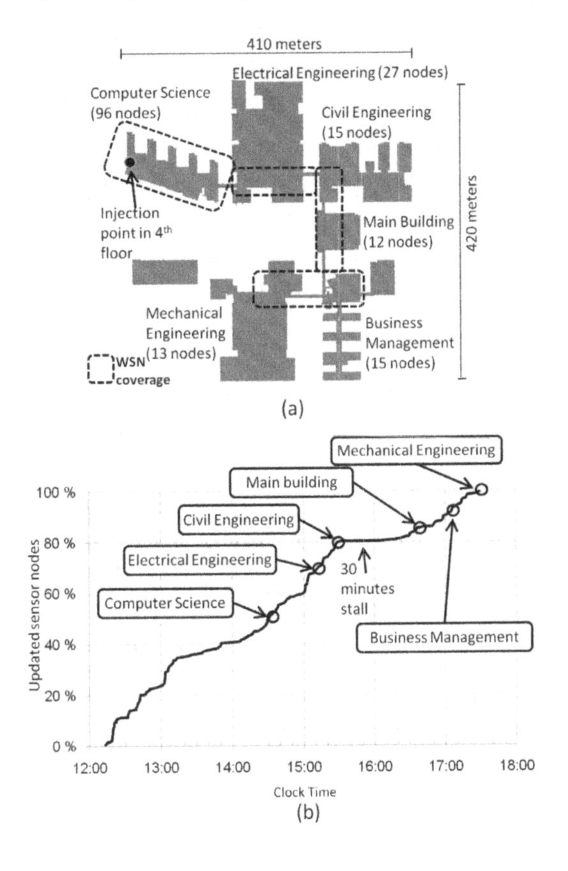

Engineering building (700 meters long path and average hop distance of 12.5 meters equals 56 hops).

The dissemination speed during the second test was mostly limited by the long interval between network scans. The new image propagated quickly within individual clusters e.g. inside one building, but spread slowly from one cluster to another. As most of the nodes had good routes to the nearest network gateway, they had no reason to perform additional network scans. This limitation can be avoided by disseminating the program image through the network gateways.

The transmission time between nodes varied from one minute to four minutes. This was caused by the differences in link reliability between different nodes. Due to the hardware restrictions, PIDP client chooses first received advertiser to be a PIDP server without considering link quality. This can lead to a situation where image transfer has to be attempted several times between nodes that are too far apart or suffer from low link reliability due to interference.

Energy Consumption Impact

PIDP energy consumption impact was measured on the TUTWSN hardware platform for the PIDP client and the PIDP server. Measurements were

Figure 8. A TUTWSN node platform, TUTWSN node in an enclosure, and two nodes installed to the Civil Engineering building of Tampere University of Technology campus in the TUT campus WSN

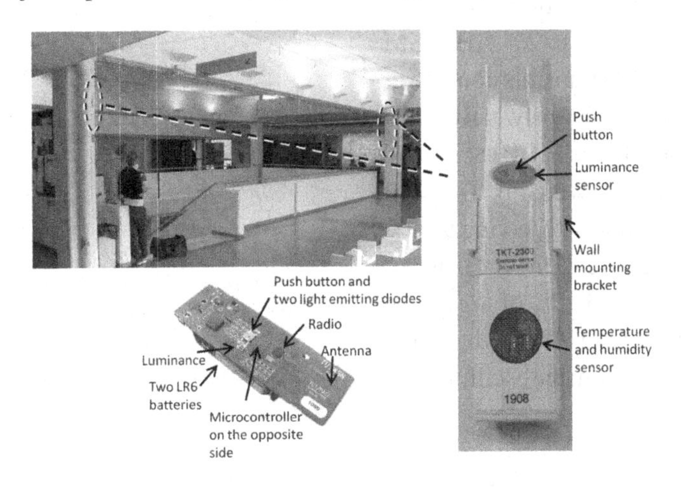

conducted with a stable power source and series resistor of known value. The voltage over the series resistor was measured with an oscilloscope to determine drawn root-mean-square current. Duration and current of the image transfer and verification operations were measured. Slight differences to the measurements in propagation time section are result of different updated image. Figure 9a and Figure 9b present screen captures of the oscilloscope, where the series resistor voltage is drawn over the time.

The PIDP client sends requests to the PIDP server, which are the narrow spikes in Figure 9a. Then the PIDP client listens for the image packets, which are the wide spikes in the capture. One packet has 26 B of payload and the program memory writing has to be done in 64 B blocks. Therefore, two to three packets are received before writing. The success of writing a block of the program image is always verified immediately. The whole image is verified after the image has been completely transferred. The PIDP server listens for requests of the PIDP client (wide spikes in Figure 9b), prepares a packet, writes it to the radio and sends it to the PIDP client (narrow spikes).

The TUTWSN platform was run on 4 MHz clock frequency and 2.5 V supply voltage. 4 MHz is the highest possible clock frequency with the specified supply voltage. The highest possible clock frequency is used to achieve the fastest possible dissemination time and the shortest affection time to the normal operation. The radio used a random channel from the 2.4 GHz ISM band and the highest possible transmission power of 0 dBm. The MCU is continuously active during the update operation. On the PIDP server, the radio is active for 35 second, which is 67.3% of *the image transfer time* and it is receiving 93.5% of that time. On the PIDP client, the radio is active for 17 seconds, which is 32.7% of *the image transfer time* and it is receiving 90% of that time. These values were obtained from the oscilloscope.

Energy consumption results of the PIDP client are presented in Table 2. Typical lithium AA batteries have approximately 20000 J of usable energy. Thus, one update consumes under 0.1‰ of the available energy of the PIDP client. If the expected node lifetime is two years, updating a node once a day would reduce lifetime approximately 10%. Impact is irrelevant on moderate amounts of updating. However, PIDP is not suitable for continuous application dissemination. This is due to the whole program image updating. If the image is sliced to smaller pieces as presented in Operating System Support section, the energy consumption impact will be reduced.

Figure 9. Screen captures of the oscilloscope drawing the series resistor voltage during the energy consumption measurements for the PIDP client and server. The scale is horizontally 10 ms/div and vertically 200 mV/div. a) The PIDP client sends a request packet and listens for the image packet. 2-3 image packets are received before writing to the program memory. b) The PIDP server listens for requests of the PIDP client, prepares an image packet and sends it. Note: a) is not in synchronization with b).

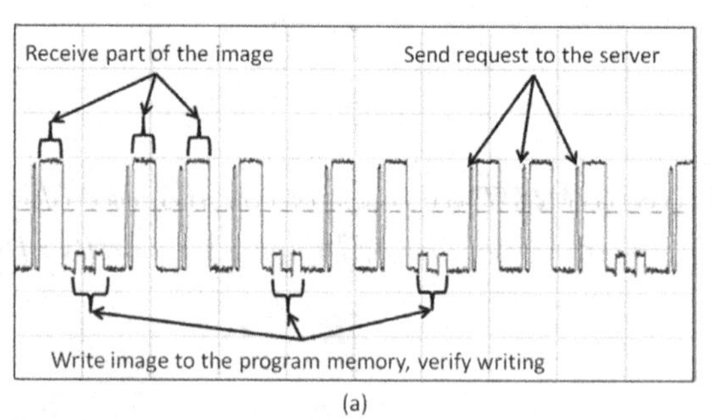

(a)

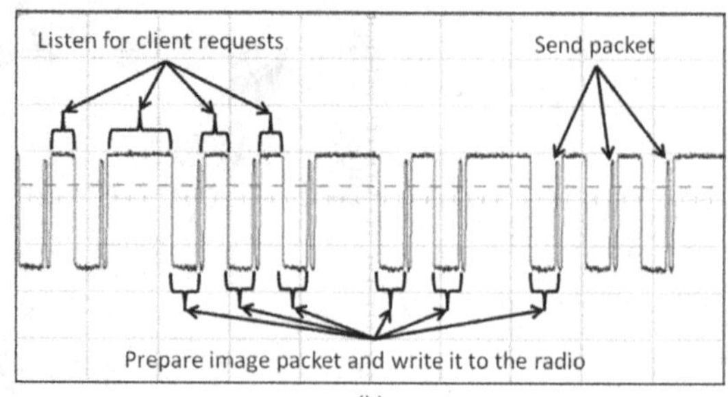

(b)

Table 2. Energy consumption measured of the PIDP client. Radio RX denotes for radio listening and receiving. Radio TX denotes for radio transmitting. MCU energy consumption during the image transfer includes the energy consumed in the program memory writing

Operation	Time consumed (s)	MCU (mJ)	Radio RX (mJ)	Radio TX (mJ)	Total (mJ)
Image transfer	49.62	711.99	436.14	48.46	1196.59
Image verification	24.44	272.44	0	0	272.44
Total:	74.06	984.43	436.14	48.46	1469.03

PIDP server energy consumption is presented in Table 3. The image transfer consumes more energy with the PIDP server, since it has to listen for longer periods. However, the PIDP server does not need to write or verify the image after the transfer and the total amount of consumed energy is similar to the PIDP client. In normal operation, one node acts once as a PIDP client and zero to multiple times as a PIDP server. Therefore, it is difficult to determine actual energy consumption impact of server duty in a network. In our experiments, one node acted as a PIDP server zero to three times. Thus, energy consumption impact varies between 1500 mJ – 6000 mJ, which is under 1‰ of the available energy.

PIDP is an energy efficient method to disseminate new program image to the network. The energy consumption impact increases significantly only if the network is updated often, e.g. once a day. Furthermore, the energy consumption impact is divided evenly across the whole network, since the image disseminates one hop a time and most of the nodes act once as the PIDP client and similar amounts as the PIDP server.

The energy efficiency could be improved by decreasing the radio transmission power or the radio listening time. If the radio transmission power is decreased that the total transmission energy consumption is half of the current consumption, it would reduce energy consumption 2% in total. Thus, the transmission power is not significant energy consumer. In optimum case, radio listening time would be the same as the transmission time. Therefore, reducing the radio listening time would reduce maximum of 26% of the PIDP client and 56% of the PIDP server energy consumption. Reducing the radio listening time is one major task in future work.

COMPARISON

Feature comparison of known multihop reprogramming methods for WSNs is presented in Table 4. PIDP manages to function without operating system. Also, it can function with any OS and update the OS as well. PIDP does not require external flash or reliable transport protocol, since it

Table 3. Energy consumption measured of the PIDP server. Radio RX denotes for radio listening and receiving. Radio TX denotes for radio transmitting

Operation	Time consumed (s)	MCU (mJ)	Radio RX (mJ)	Radio TX (mJ)	Total (mJ)
Image transfer	49.62	553.03	932.92	64.85	1550.80
Image verification	0	0	0	0	0
Total:	49.62	553.03	932.92	64.85	1550.80

Table 4. Feature comparison of known WSN reprogramming approaches

Protocol	Requires OS	Supports multiple OS	Requires Ext. Flash	Requires transport protocol	Update Scope
PIDP	No	Yes	No	No	Whole program image or parts
Deluge	Yes/TinyOS	No	Yes	Yes	Whole program image
Contiki	Yes	No	No	Yes	Application dissemination
Maté	Yes/TinyOS	No	No	Yes	Application dissemination
MOAP	Yes/TinyOS	No	Yes	Yes	Whole program image

has own transport protocol. Finally, PIDP usually updates the whole program image, but it can be modified to update it in parts.

CONCLUSION

This paper presents a lightweight, reliable and energy efficient program image dissemination protocol for WSNs. Unlike other dissemination protocols, PIDP does not require external memory storage, is independent of the WSN stack, offers a low overhead protocol for transferring program images, and can reprogram the whole WSN stack. PIDP is implemented using low-power WSN prototype nodes and tested in actual real-world conditions. The experimental results show that PIDP can reprogram 178 nodes in 5 hours and requires less than 7 kilobytes of ROM and 22 bytes of RAM and that it is possible to create a dissemination protocol that does not require external memory and yet achieves the epidemic dissemination capabilities of traditional dissemination protocols with low energy consumption.

Future work on PIDP will include new methods for inter-cluster advertisements, reliability improvements, energy consumption minimizing, and use of it for application dissemination with operating systems.

The new inter-cluster advertisement methods will speed up the propagation of the program image. This would remove stalls as seen in Figure 7b, where network was partitioned and both partitions considered their network situation satisfactory. In addition to multiple injection points, this can be solved with application level software advertisements, where nodes are informed in the application level that there might be newer program image available. Then the nodes could seek more eagerly for the new image.

To improve reliability, the PIDP protocol must select the transfer channel from non-interfering channels. Also, the PIDP client should start the image transfer with the best possible neighbor. These are difficult tasks to do and require novel designs and implementations to fit the PIDP design.

Energy consumption can be reduced significantly by reducing the radio listening time. This requires strictly synchronized protocol. The PIDP client and server could negotiate a timetable in every transmission for the next packet. This introduces research problems of what to do after unsuccessful transmissions and how to fit such a complex protocol on a restricted space.

For operating systems support, we will implement the two part program image dissemination to PIDP. SensorOS (Kuorilehto, Alho, Hännikäinen, & Hämäläinen, 2007) is used as operating system and WSN API (Juntunen, Kuorilehto, Kohvakka, Kaseva, Hännikäinen, & Hämäläinen, 2006) is used as an application layer. Also, we will experiment with Contiki to see, how the problems of dynamic application loading can be overcome with PIDP.

ACKNOWLEDGMENT

This paper is an updated and revised version of Määttä, Suhonen, Laukkarinen, Hämäläinen, and Hännikäinen, (2010), "Program Image Dissemination Protocol for Low-energy Multihop Wireless Sensor Networks", in Proceedings of the 2010 International Symposium on System on Chip (SoC). IEEE.

REFERENCES

Akyildiz, I., Weilian, S., Sankarasubramaniam, Y., & Cayirci, E. (2002). A survey on sensor networks. *IEEE Communications Magazine*, *40*(8), 102–114. doi:10.1109/MCOM.2002.1024422

Alliance, Z. (2010). *ZigBee specification*. Retrieved from http://www.zigbee.org/Standards/ZigBeeSmartEnergy/Specification.aspx

Crossbow Technologies. (2003). *Mote in-network programming user reference*. Retrieved from http://www.tinyos.net/tinyos-1.x/doc/Xnp.pdf

Deng, J., Han, R., & Mishra, S. (2006). Secure code distribution in dynamically programmable wireless sensor networks. In *Proceedings of the Fifth International Conference on Information Processing in Sensor Networks* (pp. 292-300).

Dunkels, A., Finne, N., Eriksson, J., & Voigt, T. (2006). Run-time dynamic linking for reprogramming wireless sensor networks. In *Proceedings of the Fourth ACM Conference on Embedded Networked Sensor Systems* (pp. 15-28).

Dunkels, A., Gronvall, B., & Voigt, T. (2004). Contiki - a lightweight and flexible operating system for tiny networked sensors. In *Proceedings of the 29th Annual IEEE International Conference on Local Computer Networks* (pp. 455-462).

Hill, J. W., Szewczyk, R., Woo, A., Hollar, S., Culler, D., & Pister, K. (2000). System architecture directions for networked sensors. *SIGPLAN Notes*, *35*(11), 93–104. doi:10.1145/356989.356998

Hui, J. W. (2005). *Deluge 2.0 - TinyOS network programming*. Retrieved from http://www.cs.berkeley.edu/~jwhui/deluge/deluge-manual.pdf

Hui, J. W., & Culler, D. (2004). The dynamic behavior of a data dissemination protocol for network programming at scale. In *Proceedings of the 2nd International Conference on Embedded Networked Sensor Systems* (pp. 81-94).

IEEE Standards Association. (2008). *Part 15.4: Wireless medium access control (MAC) and physical layer (PHY) specifications for low-rate wireless personal area networks (WPANs)*. Retrieved from http://standards.ieee.org/getieee802/download/802.15.4a-2007.pdf

Juntunen, J., Kuorilehto, M., Kohvakka, M., Kaseva, V., Hännikäinen, M., & Hämäläinen, T. (2006). WSN API: Application programming interface for wireless sensor networks. In *Proceedings of the IEEE 17th International Symposium on Personal, Indoor and Mobile Radio Communications* (pp. 1-5).

Kulkarni, S., & Wang, L. (2005). MNP: Multihop network reprogramming service for sensor networks. In *Proceedings of the 25th IEEE International Conference on Distributed Computing Systems* (pp. 7-16).

Kuorilehto, M., Alho, T., Hännikäinen, M., & Hämäläinen, T. D. (2007). SensorOS: A new operating system for time critical WSN applications. In *Proceedings of the 7th International Conference on Embedded Computer Systems: Architectures, Modeling, and Simulation* (pp. 431-442).

Kuorilehto, M., Kohvakka, M., Suhonen, J., Hämäläinen, P., Hännikäinen, M., & Hämäläinen, T. D. (2007). *Ultra-low energy wireless sensor networks in practice: Theory, realization and deployment*. New York, NY: John Wiley & Sons. doi:10.1002/9780470516805

Langendoen, K., Baggio, A., & Visser, O. (2006). Murphy loves potatoes: Experiences from a pilot sensor network deployment in precision agriculture. In *Proceedings of the 20th International Parallel and Distributed Processing Symposium* (p. 8).

Levis, P., & Culler, D. (2002). Maté: A tiny virtual machine for sensor networks. *SIGOPS Operating Systems Review*, *36*(5), 85–95. doi:10.1145/635508.605407

Levis, P., Patel, N., Culler, D., & Shenker, S. (2004). Trickle: A self-regulating algorithm for code propagation and maintenance in wireless sensor networks. In *Proceedings of the 1st Conference on Networked Systems Design and Implementation* (p. 2).

Microchip Technology. (2008). *PIC18F8722 product page*. Retrieved from http://www.microchip.com/

Microchip Technology. (2009). *MPLAB C compiler for PIC18 MCUs*. Retrieved from http://www.microchip.com/

Miller, C., & Poellabauer, C. (2008). PALER: A reliable transport protocol for code distribution in large sensor networks. In *Proceedings of the 5th Annual IEEE Communications Society Conference on Sensor, Mesh and Ad Hoc Communications and Network* (pp. 206-214).

Mukhtar, H., Kim, B. W., Kim, B. S., & Joo, S.-S. (2009). An efficient remote code update mechanism for wireless sensor networks. In *Proceedings of the IEEE Military Communications Conference* (pp. 1-7).

Mtt L., Suhonen, J., Laukkarinen, T., Hmlinen, T., & Hnnikinen, M. (2010). Program image dissemination protocol for low-energy multihop wireless sensor networks. In *Proceedings of the International Symposium on System on Chip* (pp. 133-138).

Nordic Semiconductors. (2007). *nRF24L01 product specification*. Retrieved from http://www.nordicsemi.com/

Reijers, N., & Langendoen, K. (2003). Efficient code distribution in wireless sensor networks. In *Proceedings of the 2nd ACM International Conference on Wireless Sensor Networks and Applications* (pp. 60-67).

Stathopoulos, T., Heidemann, J., Estrin, D., & SENSING, C. U. (2003). *A remote code update mechanism for wireless sensor networks*. Retrieved from http://www.isi.edu/~johnh/PAPERS/Stathopoulos03b.html

Suhonen, J., Kuorilehto, M., Hännikäinen, M., & Hämäläinen, T. (2006). Cost-aware dynamic routing protocol for wireless sensor networks - design and prototype experiments. In *Proceedings of the IEEE 17th International Symposium on Personal, Indoor and Mobile Radio Communications* (pp. 1-5).

Wang, Q., Zhu, Y., & Cheng, L. (2006). Reprogramming wireless sensor networks: Challenges and approaches. *IEEE Network*, *20*(3), 48–55. doi:10.1109/MNET.2006.1637932

Zhang, C., Yu, Q., Huang, X., & Yang, C. (2008). An RC4-based lightweight security protocol for resource-constrained communications. In *Proceedings of the 11th IEEE International Conference on Computational Science and Engineering Workshops* (pp. 133-140).

This work was previously published in the International Journal of Embedded and Real-Time Communication Systems, Volume 2, Issue 3, edited by Seppo Virtanen, pp. 50-68, copyright 2011 by IGI Publishing (an imprint of IGI Global).

Chapter 4
Asymmetric Geographic Forwarding:
Exploiting Link Asymmetry in Location Aware Routing

Pramita Mitra
University of Notre Dame, USA

Christian Poellabauer
University of Notre Dame, USA

ABSTRACT

Geographic Forwarding (GF) algorithms typically employ a neighbor discovery method to maintain a neighborhood table that works well only if all wireless links are symmetric. Recent experimental research has revealed that the link conditions in realistic wireless networks vary significantly from the ideal disk model and a substantial percentage of links are asymmetric. Existing GF algorithms fail to consider asymmetric links in neighbor discovery and thus discount a significant number of potentially stable routes with good one-way reliability. This paper introduces Asymmetric Geographic Forwarding (A-GF), which discovers asymmetric links in the network, evaluates them for stability (e.g., based on mobility), and uses them to obtain more efficient and shorter routes. A-GF also successfully identifies transient asymmetric links and ignores them to further improve the routing efficiency. Comparisons of A-GF to the original GF algorithm and another related symmetric routing algorithm indicate a decrease in hop count (and therefore latency) and an increase in successful route establishments, with only a small increase in overhead.

INTRODUCTION

Routing in wireless ad-hoc and sensor networks typically assumes that wireless links are bidirectional, i.e., wireless devices have identical transmission ranges. However, some recent empirical studies (Ganesan, Estrin, Woo, Culler, Krishnamachari, & Wicker, 2002; Woo, Tong, & Culler, 2003; Zamalloa & Krishnamachari, 2007; Zhao & Govidan, 2003) show that approximately 5-15% of the links in a low-power wireless network are asymmetric, and an increasing distance between

DOI: 10.4018/978-1-4666-2776-5.ch004

nodes also increases the likelihood of an asymmetric link. This trend is further exacerbated by the increasing use of power management techniques that may cause the nodes in a network to operate at different transmission ranges. Link asymmetry is also caused by node mobility, heterogeneous radio technologies, and irregularities in radio ranges and path and packet loss patterns. Based on these observations, it is expected that link asymmetry will become more common in future wireless ad-hoc and sensor networks.

With increasing use of Global Positioning System (GPS) and many other (possibly less accurate but more resource-efficient) localization schemes, Geographic Forwarding (GF) is becoming an attractive choice for widely scalable routing in wireless ad-hoc and sensor networks. GF incurs very low overhead since no prior route discovery is required before forwarding the data packets. Existing GF protocols are designed under the assumption of symmetric wireless links. That is, whenever a node receives a beacon packet from another node, it considers that node as its neighbor as it assumes the link is bi-directionally reachable. Such an assumption may not be realistic for practical wireless ad-hoc and sensor networks, since wireless links are often asymmetric. In Figure 1, node A discovers its neighbors B, C, and D by receiving beacons from them, which means that node A is within the wireless transmission ranges of all three nodes. Both node C and node D are also within the wireless transmission range of node A, so both the links A↔C and A↔D are symmetric. However, node B is outside node A's wireless transmission range and therefore, there is an asymmetric link B→A between these two nodes. Note that while all wireless links are to some extent asymmetric (i.e., the signal strength being stronger in one direction than the other but the wireless link still being connected in both directions), A-GF focuses on maximally asymmetric links (perfect connectivity one direction, zero in the other), as shown in Figure 1.

Figure 1. Concept of asymmetric links

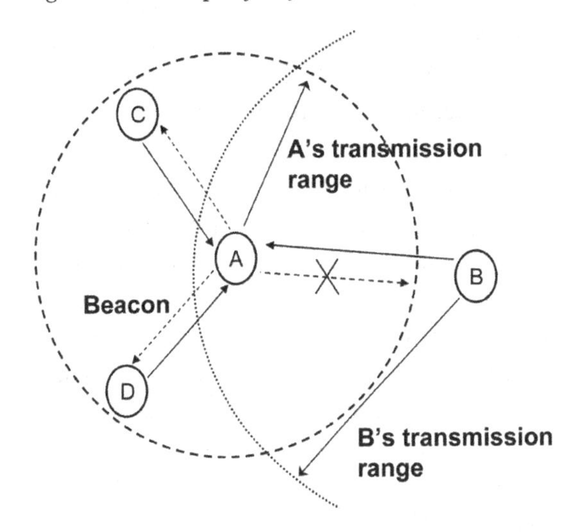

Similar to geographic location based protocols, many reactive routing protocols for ad-hoc and sensor networks, such as Dynamic Source Routing (Johnson & Maltz, 1996) and Ad Hoc On Demand Vector Routing (Perkins, Belding-Rower, & Das, 2003), assume that all links in a network are symmetric and, therefore, they can fail to find routes when this assumption does not hold true. These routing protocols often rely on a two-phase communication process, where the same path is used to communicate between the sender and the receiver. If a single link on this path is asymmetric, the route establishment will fail. Even if route discovery succeeds, protocols that avoid or ignore asymmetric links in their route establishment will often discover routes that are longer than necessary, thereby affecting the end-to-end communication latencies experienced between sender and receiver.

Some routing protocols (Couto, Aguayo, Bicket, & Morris, 2003; Zhou, He, Krishnamurthy, & Stankovic, 2004) detect asymmetric links and explicitly ignore them while making routing decisions. Eliminating asymmetric links discounts a substantial number of potentially stable routes with good one-way reliability and may lead to (1) routes that are longer than necessary and (2) no routes at all if there is not at least one fully symmetric path between the sender and receiver (Marina & Das, 2002; Sang, Arora, & Zhang, 2007). Therefore, it is difficult to achieve high network connectivity, high data transmission rates, and low transmission latencies if the network has many asymmetric links. In addition, some recent empirical studies (Chen, Hao, Zhang, Chan, & Ananda, 2009; Ganesan, Estrin, Woo, Culler, Krishnamachari, & Wicker, 2002) show that asymmetric links tend to span longer distances than symmetric links. As a result, inclusion of asymmetric links in routing decisions can further improve the network performance. On the other hand, in a highly mobile setting, communication among mobile nodes may only be temporary and short-lived. For example, while the wireless connection between two cars driving in the same direction on a highway may last for a long time, the connection between cars driving in opposite directions will break very quickly. Therefore, it may be necessary to not only identify the presence of an asymmetric link, but to also measure or predict the time-varying quality or stability of such a link.

This paper proposes *Asymmetric Geographic Forwarding (A-GF)*, which discovers and exploits asymmetric links in a network, i.e., instead of eliminating these links, A-GF uses them for routing to reduce the routing hop count (and latency) and to increase the routing reliability when there are no symmetric paths available. A-GF modifies the "Hello message" approach of GF to discover asymmetric links, proactively monitors the changes in conditions of discovered links, and ranks neighbors based on perceived link stability. During routing, a node considers both this ranking and the progress a neighbor can make towards the sink location in terms of geographical distance (thereby considering both reliability and performance). In order to avoid engaging in transient asymmetric relationships, A-GF closely watches the changes in condition of a newly detected asymmetric link and adds the node into its neighbor table only after the node's presence in the neighborhood for a certain amount of time.

The remainder of the paper is organized as follows: Section 2 provides a comparison of A-GF to related work. Section 3 describes a study of the origins and characteristics of link asymmetry in wireless sensor networks. Section 4 provides the details of A-GF, followed by experimental results in Section 5. Finally, Section 6 concludes the paper.

RELATED WORK

A number of existing routing protocols address the issue of non-uniform transmission ranges resulting in asymmetric links in wireless ad-hoc and sensor networks. This section compares the proposed A-GF protocol to related work.

Link Asymmetry in Location Based Routing

Roosta, Menzo, and Sastry (2005) propose a probabilistic method to select the forwarder nodes from a node's set of neighbors. They maintain forward and backward reliability values for a link in both directions, and a set of neighbors with good forward and backward values are chosen as forwarder nodes. Son, Helmy, and Krishnamachari (2004) assume that mobility can induce asymmetric links and address the asymmetric link problem by ensuring that during data forwarding, the distance to a neighbor node is less than the transmission range of a forwarding node. Chen, Hao, Zhang, Chan, and Ananda (2009) use unicast beacon packets for the detection of asymmetric links. When a node receives a MAC layer notification of failed delivery of these unicast beacon packets, the neighbor locally broadcasts a special beacon packet to announce the unidirectional link. Only symmetric links are used for calculating the planarization graph for perimeter routing.

Kim, Govidan, Karp, and Shenker (2004) show that planarization techniques fail in presence of asymmetric links, as the planarization techniques use the unit-graph assumption, which, in turn, assumes bi-directional communication. The authors propose the Cross-Link Detection Protocol (CLDP) to enable provably correct geographic routing on arbitrary graphs.

Most of these routing protocols employ techniques to discover asymmetric links in the network. However, they treat these asymmetric links as an anomaly and neglect them while making routing decisions. Such schemes scale well when the number of asymmetric links in the network is small. However, when the number of asymmetric links is large and the network still remains a connected graph, the routing protocols that do not utilize asymmetric links perceive the network as an unconnected graph. As a result, the set of asymmetric links remains unnoticed by these protocols. Therefore, the scalability and performance of such schemes decrease drastically with the increase in the number of asymmetric links in the network. A-GF does not only successfully find asymmetric links in the network but also takes advantage of these links, thereby improving the performance and minimizing data transfer latencies.

Link Asymmetry in On-Demand Ad-Hoc Routing

A few recent efforts have looked into link asymmetry in on-demand ad-hoc routing protocols. Sang, Arora, and Zhang (2007) propose a novel neighbor discovery technique based on a new one-way link metric to identify high reliability forward asymmetric links and present a local procedure for their estimation. Duros and Dabbous (1996) propose a modification to the well-known Internal Gateway Protocol OSPF to handle asymmetric satellite links by sharing neighbor tables in Hello packets. Wang, Ji, and Turgut (2004) discuss a routing protocol for power constrained networks with asymmetric links where they discover an asymmetric neighbor with the help of a mutual third-party proxy set. Kim, Toh, and Chou (2000) propose a routing protocol for asymmetric links on top of the popular on-demand DSR protocol with dual (forward and backward) paths for each source and sink node-pair. Most of these routing protocols employ asymmetric neighbor discovery techniques with the help of mutual witness neighbor nodes. However, due to node mobility or dynamic adjustments in transmission ranges, symmetric links can become asymmetric links, or asymmetric links can turn into broken links. Therefore, it is also important to monitor the condition of the discovered links to ensure successful data transmission. Toward that end, A-GF continuously monitors the discovered links allowing it to evaluate the reliability of each link.

The proposed asymmetric link discovery mechanism in A-GF is similar to the Discover and Exploit Asymmetric Links (DEAL) protocol proposed by Chen, Hao, Zhang, Chan, and Ananda (2009). However, DEAL uses data link layer beacon messages for finding and using asymmetric links in a wireless sensor network. Therefore, a cross-layer approach will be required in DEAL if the asymmetric links were to be used for efficient routing. On the other hand, A-GF uses routing layer Hello messages for asymmetric link discovery, which makes it possible to use asymmetric links for efficient routing without any assumptions about the underlying data link layer protocol. Toward that end, A-GF has two other major advantages over DEAL: (1) A-GF uses an on-demand method for asymmetric link notification and management which is suitable for a highly mobile network setting, whereas DEAL uses a common neighbor feedback based method for asymmetric link management. However, the common neighbor changes very frequently with changes in network topology in mobile environments, and that is why DEAL will result in many failed transmissions and large network traffic overheads. (2) In a highly mobile setting, asymmetric links can be transient and nodes may rapidly switch between discovering an asymmetric link and removing the same link from the neighbor table, thereby leading to significant control traffic overheads and energy consumptions. Unlike DEAL, A-GF is able to exploit the knowledge about the changing locations of mobile nodes (exchanged in periodic Hello messages) to detect such unreliable asymmetric links and to ignore them. Toward that end, an asymmetric link is announced and used only when the link is predicted to be available for some time. On the other hand, A-GF assumes unlimited memory available on the mobile nodes – however, in reality the mobile nodes have limited memory. This limitation poses a bound on the size the of neighbor table maintained at each node. In a densely populated network where a node has many neighbors, it may not be possible to store and maintain all of the neighbors in a small, fixed size neighbor table. Unless the neighbor table management scheme is aware of asymmetric links, such a limitation can cause symmetric links to be incorrectly identified as asymmetric links (false discovery) and asymmetric links to be evicted before they are discovered (early eviction). DEAL uses an asymmetry-aware caching scheme that eliminates false discovery and early eviction when memory size is limited.

CHARACTERISTICS OF ASYMMETRIC LINKS IN WIRELESS NETWORKS

In conventional routing protocols, whenever a node receives a beacon packet from another node, it considers that node as its neighbor and the link as bi-directional. Such an assumption may not be realistic for practical wireless sensor networks, since wireless links are often asymmetric. Link asymmetry is mainly caused by the differing transmission ranges of wireless devices.

Differing transmission ranges occur due to the following factors: (1) the increasing heterogeneity in wireless ad-hoc and sensor networks, e.g., networks consisting of different types of wireless devices, where these devices have different radio capabilities. (2) Transmission ranges are also affected by the transmit power used by a wireless device, which can be varied dynamically by resource management protocols in low-power networks. (3) Irregularities in path and packet loss patterns and varying noise floors of wireless devices (caused by distance, interferences and obstacles, and hardware variability, respectively) lead to differing transmission ranges (Cerpa, Wong, Kuang, Miodrag, & Estrin, 2005; Reijers, Halkes, & Langendoen, 2004; Srinivasan, Dutta, Tavakoli, & Levis, 2006). This section presents an empirical study of how some of these factors contribute to link asymmetry in wireless ad-hoc and sensor networks.

Heterogeneous Radios

To study the impact of heterogeneous radio technologies on link asymmetry, we performed simulations in the JiST/SWANS (JiST, 2005) simulator. The mobile nodes in the network have heterogeneous radio capabilities as a result of differing transmit powers, radio reception sensitivities, and radio reception thresholds of the nodes. Specifically, we used 4 different types of radios, each with different transmission range. The number of nodes was varied from 25 to 200 in steps of 25. The total area of the network is scaled in proportion so that the node density remains unchanged.

Figure 2 shows the percentage of symmetric and asymmetric links in the network with varying number of nodes. The percentage of asymmetric links in the network increases with increasing numbers of nodes with heterogeneous radio capabilities. For example, in a mobile network of size 75, the number of asymmetric links is 11.33% of the total number of links in the network. The number of asymmetric links increases to 24.79% of the total number of links when there are 200 nodes with heterogeneous radio capabilities in the network. Nodes with heterogeneous radio capabilities have different transmission ranges, which results in non-uniform wireless connectiv-

Figure 2. Percentage of symmetric and asymmetric links vs. network size

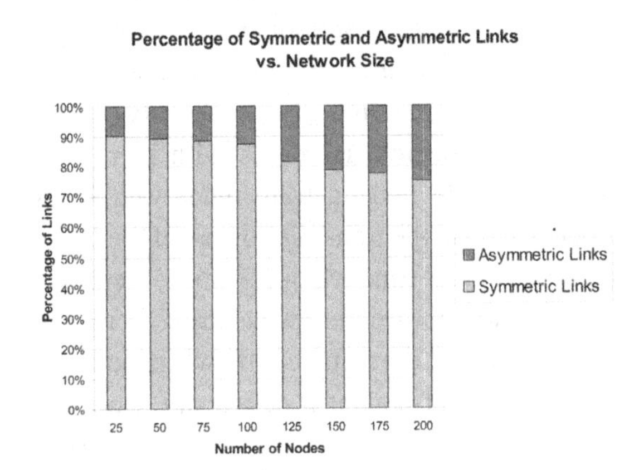

ity over both directions of a wireless link, thus causing asymmetric links in the network. Note that existing symmetric routing protocols, e.g., Greedy Perimeter Stateless Routing (GPSR) (Karp & Kung, 2000), do not consider asymmetric links, and therefore, discover only 75-90% of the total number of links in the network, which in turn may limit the data transmission success of the symmetric routing protocols.

Varying Transmit Power and Path Loss

Varying transmit power and path loss are two major reasons of varying transmission ranges leading to link asymmetry. To study the impact of transmit power and path loss variance on link asymmetry, we conducted outdoor field tests with 2 Crossbow Stargate (n. d.) wireless devices, placed 60 centimeters above the ground. The Stargate devices have an ARM processor running Linux, and each device is equipped with an 802.11b compact flash card, based on the PrismII chipset. The outdoor field tests were conducted in an open yard at the University of Notre Dame (Figure 3) during class hours. In Figure 3, the left-most dot represents the location of the static receiver device and the other dots represent the locations of the mobile sender device. The sender device was initially placed at a distance of 20 meters from the receiver device and then the distance was varied up to 140 meters along a straight line (in steps of 20 meters). At each step, the mobile sender device used 5 different transmit power levels, and broadcasted 200 beacon packets at the rate of 10 packets/second to the receiver device.

We studied the effects of varying transmit power and distance (path loss falls off by square of distance) on the successful packet transmission rate of the sender device. Figure 4 shows the corresponding successful packet transmission rate of the sender device with varying transmit power and distance. The figure shows that the successful packet transmission rate drops with increasing distance and decreasing transmit power levels. Table 1 provides the Received Signal Strength Indication (RSSI) and noise floor values measured

Figure 3. Outdoor field tests with two Crossbow Stargate devices

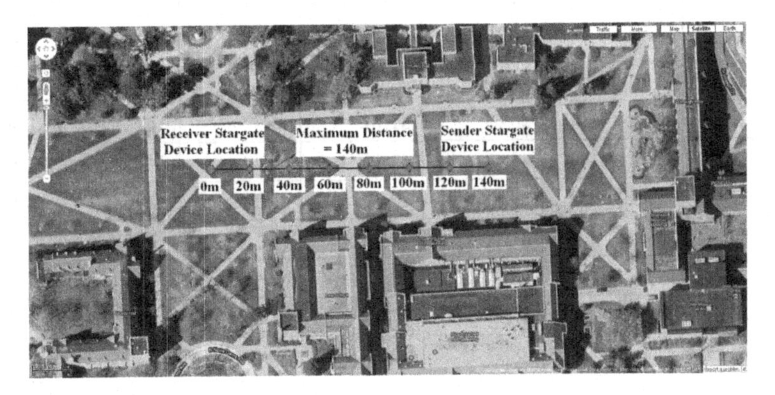

Figure 4. Successful packet transmission rate with varying distance and transmit power

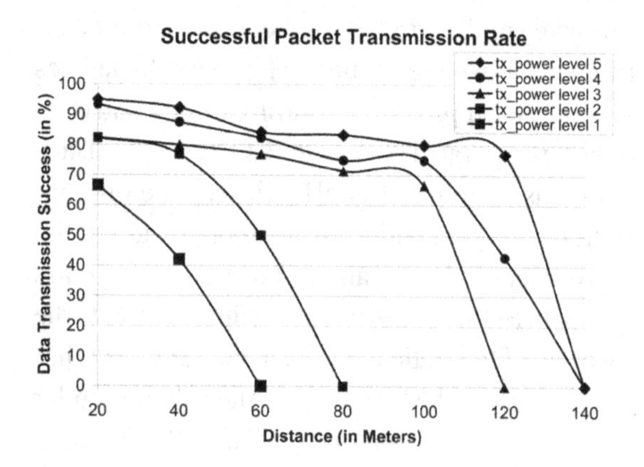

at the receiver device for the corresponding data transmission from the sender device. It can be observed from Table 1 that (1) the measured RSSI values drop with increasing distance, corroborating the previous observation that the successful packet transmission rate drops with increasing distance, and (2) when the successful packet transmission rate drops to zero, the measured noise floor is more than or very close to the measured RSSI value, i.e., no meaningful data reception was possible because the noise was stronger than the received signal.

The above outdoor experiment shows that the successful packet transmission rate drops with increasing distance and decreasing transmit power levels. In other words, the data transmission quality (a good measure of connectivity) over a certain wireless link decreases with increasing distance and decreasing transmit power. In an ad-hoc network, the transmit powers of the nodes may change over time when the nodes' resource managers attempt to preserve energy. In mobile ad-hoc networks, the distances between the nodes will also change frequently. As a result, there will be changes in wireless connectivity over a certain direction (e.g., when one node changes it's transmit power) or both directions (e.g., when a node moves) of a wireless link, thereby leading to asymmetric links.

ASYMMETRIC GEOGRAPHIC FORWARDING

In this section, we describe our assumptions and the protocol details of A-GF. We assume that each node knows its own geographic location and that every node in the network can be fully

Table 1. Measured RSSI and Noise Floor values with varying distances and transmit powers

Transmit power	Metric	Distance					
		20m	40m	60m	80m	100m	120m
Level 5	RSSI	45.63	40.87	40.73	40.63	36.32	33.48
	Noise	30.8	31.36	32.2	32.23	33.18	34.35
Level 4	RSSI	44.00	43.31	37.11	37.38	34.12	32.27
	Noise	30.69	31.14	31.3	31.53	31.69	32.35
Level 3	RSSI	43.66	42.43	39.44	34.42	33.56	30.17
	Noise	29.89	29.92	30.97	32.13	32.4	34.21
Level 2	RSSI	41.41	39.04	37.38	31.75	30.20	29.93
	Noise	29.86	29.92	30.6	30.67	30.78	32.5
Level 1	RSSI	37.78	35.44	33.86	32.67	32.13	32.08
	Noise	28.93	30.19	30.79	31.25	32.18	32.2

mobile. GF's *Perimeter Face Routing* technique to handle networks voids or dead ends has been left unchanged.

Detection of Asymmetric Links

In GF, each node broadcasts periodic Hello or beacon messages to its one-hop neighbors, containing the node's *node ID* and *location*. The exchange of periodic Hello messages helps to (1) discover new nodes that moved in the one-hop neighborhood; (2) update the neighbor list with the latest locations of known neighbors, and (3) identify and remove old neighbors that are have failed or moved out of range from the neighbor list.

A-GF modifies the Hello message approach of GF to discover asymmetric links in the network. Toward that end, in A-GF, each node augments the Hello messages with the list of neighbors it received beacons from previously. In Figure 5(a), when node B gets a periodic Hello packet from node A, B checks if it is recorded in the neighbor list in the received packet. If B is listed in A's neighbor list, then both A and B can hear each other and the link A↔B is symmetric. If B is not listed in A's neighbor list, B knows that the link A → B is asymmetric. We define A as the *up-link node* and B as the *down-link* node. In general, an asymmetric link is usable in the direction from the up-link node to the down-link node.

Notifying the Up-Link Node

At this point, only B is aware of the asymmetric link A→B, but A would not be able to exploit it. Sending a direct (single hop) report from B to A is not possible; therefore, A-GF finds a multi-hop *tunneling path* from the down-link node to the up-link node to notify the up-link node of the existence of the outgoing asymmetric link. It is reasonable to expect that the tunneling path will be built around the asymmetric link and thus would be short in length.

In Figure 5(b), the down-link node B routes a small control message, called *Asymmetric Notification (AN)*, towards the location of the up-link node A. In each step, the AN message is forwarded to the neighbor that minimizes the geographic distance to node A. If no neighbor is closer to the up-link node than the down-link node, then face routing helps to recover from that situation and find a path to another node, where greedy forwarding can be resumed (Bose, Morin, Stojmenovic & Urrutia, 1999). In Figure 5(b), the AN message is routed to node A along the tunneling path B→C→D→A. When node A receives the AN message, it realizes the existence of the asymmetric link A→B, and puts node B in its neighbor table while marking the link as asymmetric.

Proactive Link Condition Monitoring

Due to frequent changes in the topology of mobile ad-hoc/sensor networks, (1) symmetric links can become asymmetric links, and (2) asymmetric links can turn into broken links, as discussed below.

Case 1: A symmetric link connecting node A and node B (A↔B) turns into an asymmetric link if node B moves out of node A's wireless transmission range, but node A is still reachable from node B (Figure 6). In this case, the asymmetric path containing the link B→A is still usable in the forward direction.

Case 2: An asymmetric link connecting node A and node B in the forward direction (A→B) turns into a broken link if node B moves out of node A's wireless transmission range. Node A was already out of node B's wireless transmission range due to link asymmetry as shown in Figure 6.

Therefore, it is important to monitor the condition of the discovered links to account for these changes, and thereby ensure successful data transmission. Toward that end, A-GF *proactively*

Figure 5. Asymmetric link discovery in A-GF

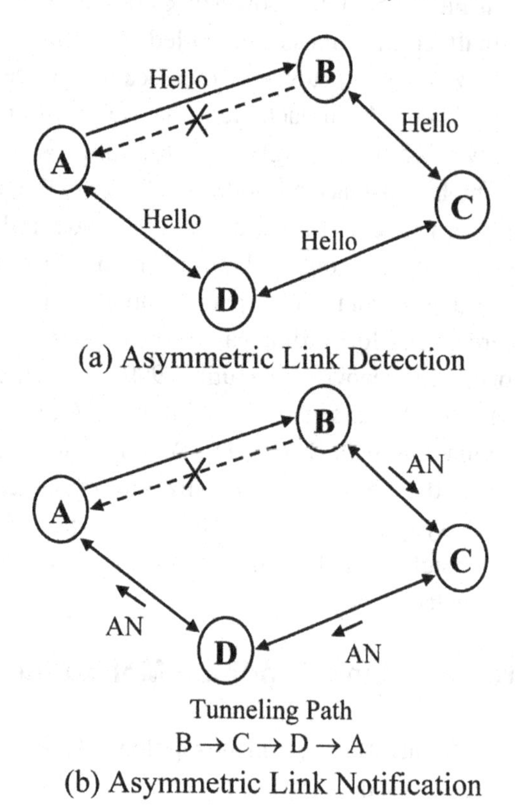

Figure 6. As node B moves away from node A, the symmetric link A↔B in (a) turns into an asymmetric link A←B in (b), and then into a broken link in (c)

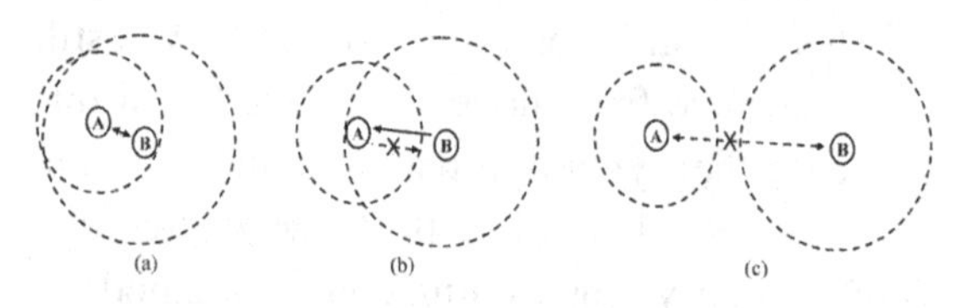

monitors the changes in conditions of discovered links and ranks neighbors based on perceived link stability. This proactive strategy, called *Proactive Link Condition Monitoring (PLCM)*, is discussed in detail below.

Propagation of Implicit Hello Messages

Neighbors connected via symmetric links receive periodic Hello messages from each other, which allow them to refresh their neighbor tables with the most recent locations of their known neighbors and to identify and remove "old" neighbors, i.e., neighbors that have failed or moved out of range. In case of an asymmetric link, the down-link node receives the periodic one-hop Hello messages from its up-link node. However, it is not possible for the up-link node to receive direct Hello messages from the down-link node because the asymmetric link is not usable from the down-link node to the up-link node. On the other hand, the down-link node could send periodic multi-hop Hello messages to the up-link node, along the same tunneling path that was used for routing the AN message in the asymmetric link notification phase (Section 4.2). However, sending periodic Hello messages over multiple hops for each asymmetric link in the network would not be a suitable solution for bandwidth constraint ad-hoc networks.

A-GF uses a bandwidth-efficient technique for the propagation of indirect Hello messages from the down-link node to the up-link node. During the periodic exchange of periodic one-hop Hello messages, an *Annotation* is inserted in the header of Hello messages sent by the down-link node to its neighbors. The Annotation contains the latest location of the down-link node and a confirmation that the down-link node received the latest periodic one-hop Hello message from the up-link node. The down-link node requests one of its one-hop neighbors to forward the Annotation to the up-link node. The neighbor which minimizes the geographic distance to the up-link node is requested for forwarding the Annotation. The requested neighbor inserts the Annotation in its next one-hop Hello message and repeats the same algorithm, and ultimately the Annotation is received by the up-link node. Thus the Annotation serves as an implicit multi-hop Hello message from the down-link node to the up-link node. The A-GF algorithm consumes very little extra bandwidth for propagating the Annotation from down-link node to up-link node because no actual messages are sent but only the size of the Hello messages increases a little due to the embedded Annotation. In this scheme, the maximum latency of the Annotation is bound by the upper limit of n $\times t_{HI}$, where n is the number of nodes forwarding the Annotation and t_{HI} is the periodic Hello message interval in the network. Note that during the forwarding of the Annotation from the down-link node to the up-link node, if no neighbor is closer to the up-link node, then a neighbor is randomly selected with the hope that a node would be found soon that would be closer to the up-link node. Such an assumption is reasonable because the up-link node is a neighbor of the down-link node and in a connected network the implicit path used for forwarding the Annotation from the down-link node to the up-link node should be short.

In Figure 7(a), the down-link node B puts an Annotation for node A and an *Annotation Forwarding Request (ANF-REQ)* for neighbor C in its periodic Hello messages. Similarly, node C inserts the Annotation in its next periodic Hello messages, with an ANF-REQ for neighbor D (Figure 7(b)). Finally, node A receives the Annotation from node D (Figure 7(c)), and updates the LCR value for the down-link node B in its neighbor table.

Monitoring of Link Conditions

A-GF uses the one-hop Hello messages to proactively monitor the changes in conditions of discovered symmetric links. Similarly, the implicit multi-hop Annotations are used by the up-link node, associated with an asymmetric link, to monitor the changes in condition of the down-link node. If subsequent Hello messages (or Annotations in case of asymmetric links) are received periodically from a neighbor node, the associated link is perceived to be stable. On the other hand, if a node misses subsequent Hello messages (or Annotations in case of asymmetric links) from a neighbor, the associated link is perceived to be less stable. Each node rates its one-hop neighbors based on this perceived link stability. This rating is called *Link Condition Rating (LCR)*. Note that this rating is personalized which means each node uses a self-policing evaluation of other nodes in its one-hop neighborhood, rather than a global evaluation for each node in the network. Thus it is possible for different nodes to have different LCR values for the same neighbor. We adopt a personalized evaluation scheme because (1) in a decentralized ad-hoc network, there is no centralized resource to maintain a global evaluation, and (2) a global evaluation is not required because the LCR values are only used for making local routing decisions at each node.

Figure 7. Annotation method for propagation of indirect Hello messages

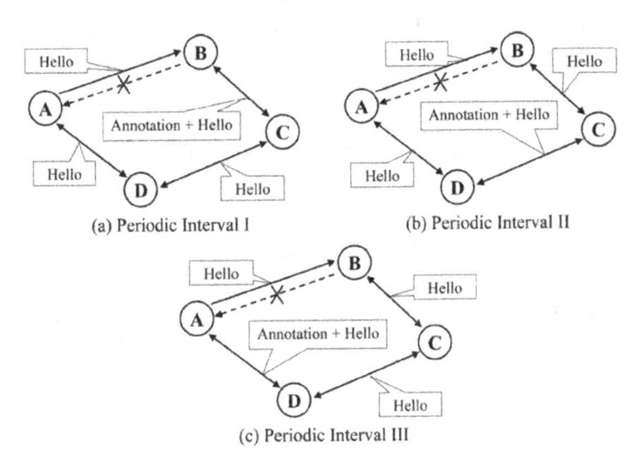

Each node runs a background daemon that periodically updates the LCR values in the neighbor table. Algorithm 1 (See Box 1) illustrates the *update_LCR_value* method. If subsequent Hello messages (or Annotations in case of asymmetric links) are received periodically from a neighbor node, its LCR value is incremented by a step size Δ_{lcr}. On the other hand, if a node misses subsequent Hello messages (or Annotations) from a neighbor, the LCR value is decremented by the same step size (i.e., Δ_{lcr}). The maximum possible LCR value for a neighbor is *MAX_LCR*. A neighbor entry is deleted from the neighbor table when its LCR value falls to or below zero. For a lossy connection, the LCR value is aggressively decreased so that it quickly falls down to zero for broken links. On the other hand, the LCR value is aggressively increased for a good connection and quickly reaches to MAX_LCR for stable links.

Box 1. Algorithm 1

Algorithm 1. update_LCR_value

Require: Neighbor Entry Variable *ne*
Ensure: the LCR value of the neighbor entry is made up-to-date

1. **if** (is_new_neighbor (*ne*)) **then**
2.   *ne.lcr* = MIN_LCR
3.   add_new_neighbor (*ne*)
4.   **return**
5. **else**
6.   **if** (is_timedout_update_interval (*ne*)) **then**
7. $ne.lcr = ne.lcr - \Delta_{lcr}$
8.   **elseif** ($ne.lcr <$ MAX_LCR) **then**
9. $ne.lcr = ne.lcr + \Delta_{lcr}$
10.   **endif**
11.   **if** ($ne.lcr \leq 0$) **then**
12.   delete_neighbor (*ne*)
13.   initiate_link_recovery
14.   **endif**
15. **endif**
16. **if** (*ne* != **NULL**) **then**
17.   reset_timeout_neighbor (*ne*)
18. **endif**

The timeouts are different for symmetric and asymmetric links – toward that end, the timeout for any type of link (i.e., symmetric or asymmetric) is equal to the interval at which an update message is expected to be received from the node on the other end of the link. However, the interval at which direct one-hop Hello messages are received over a symmetric link and the interval at which implicit Hello messages in form of Annotation are received over an implicit multi-hop path that abstracts an asymmetric link, are not equal. In fact, the implicit Hello messages in form of Annotation take much more time to travel the implicit multi-hop path than the direct one-hop Hello messages. Therefore, it is important to have different timeouts for symmetric and asymmetric links, in order to ensure fair rating of the link conditions. The timeout for a symmetric link is equal to ($t_{HI} + t_{GP}$) where t_{HI} is the periodic Hello message interval and t_{GP} is a certain grace period. Similarly, the timeout for an asymmetric link is equal to ($n \times t_{HI} + t_{GP}$) where n is the number of nodes that forwarded the implicit multi-hop Annotation from the down-link node to the up-link node Δ_{lcr}

A typical entry in the neighbor table maintained at each node has the format shown in Table 2. Each entry consists of five fields: (1) the node ID of the neighbor, (2) the latest location of the neighbor in terms of (x, y) coordinates, (3) the type of the link, i.e., symmetric or asymmetric, (4) the timestamp value stating when this entry was last updated, and (5) the LCR value assigned to the neighbor. Let us assume that the example shown in Table 2 is an entry in the neighbor table maintained at node A. The first field of the entry tells that the node ID of the neighbor associated with this entry is B. The second field tells that latest location of B is (5, 10). The third and fourth fields tell us that neighbor B is connected to A with an asymmetric link A→B (i.e., A is the up-link node and B is the down-link node) and the latest Annotation from B was received at time

Table 2. An example of neighbor table entry

Node ID	Location	Type of Link	Timestamp	LCR value
B	(5, 10)	Asymmetric	13:23:40	3

13:23:40 (expressed in hour-minute-second format). Lastly, the fifth field tells us that the current LCR value for node B is 3 which means at least 3 Annotations from down-link node B were received since this entry was created (i.e., the asymmetric link A→B was discovered).

When a neighbor is deleted from the neighbor table, a *Link Failure (LF)* message is routed to that neighbor, notifying it that the link is no longer usable. On receiving an LF message, the reaction by the receiver depends on the type of the associated link. If the link was originally an asymmetric link, then the neighbor sending the LF message is deleted from the neighbor table of the receiver of the LF message. However, if the link was originally a symmetric link then there can be on of the following two situations: (1) the receiver of the LF message is still receiving Hello messages from the sender of the LF message. In that case, the previously symmetric link has now turned into an asymmetric link in the direction from the sender of the LF message to the receiver of the LF message. Therefore, the receiver of the LF message responds by sending an AN message to notify the other node of the newly formed asymmetric link. (2) The receiver of the LF message is missing Hello messages from the sender of the LF message. In that case, the symmetric link is now broken, and the receiver of the LF message responds by sending a new LF message to sender of the LF message so that both parties are removed from each other's neighbor table. Figure 6 (which was previously discussed in Section 4.3) shows an example of how the link conditions change with node mobility and how the nodes respond to it, as follows:

In Figure 6(b), node B misses subsequent Hello messages from node A, as node B has moved out of node A's wireless transmission range. Therefore, node B sends an LF message to node A, which means that the symmetric link A↔B is no longer usable in the forward direction, (i.e., A→B). However, when node A receives this Hello message, it finds out that it is still receiving Hello messages from node B, which means that this link is still usable in the reverse direction, i.e, B→A. In other words, the symmetric link A↔B has turned into an asymmetric link B→A. In that case, node A sends an AN message to notify node B of the newly formed asymmetric link.

However, if the LF message from a down-link node is received at an up-link node, it means that the associated asymmetric link is no longer usable, and the neighbor entry is deleted from the neighbor table. In Figure 6(c), the down-link node B misses subsequent Hello messages from the up-link node A, so node B sends an LF message to node A. When node A receives the LF message, it realizes the asymmetric link A →B is no longer usable, and it deletes node B from its neighbor table. Thus, A-GF is able to detect the changes in condition of already discovered links, which helps the nodes to be aware of the latest topology changes in the neighborhood and ultimately results in increased reliability of data transmission.

Elimination of Transient Asymmetric Links

In A-GF, each node knows its own location through some positioning mechanism, and each node periodically advertises its location and one-hop neighbor table using Hello messages. While periodic Hello messages can be used to discover asymmetric links at the routing layer (similar techniques could be employed at the link layer), the knowledge of geographic locations of a node's one-hop neighbors can also assist in careful planning of bandwidth usage for such asymmetric link discovery and maintenance mechanisms. In A-GF, once an asymmetric link has been detected at a down-link node, the up-link node notification method is initiated so that the up-link node is informed about the existence of the asymmetric link. As described earlier in Section 4.2, a control packet called Asymmetric Notification (AN) is sent from the down-link node to the up-link node for asymmetric link notification. However, it may not be beneficial to initiate such notifications when the asymmetric link is transient, i.e., the cost of discovering, announcing, and exploiting an asymmetric link may be greater than the benefit of the asymmetric link (i.e., the potential availability of a shorter path).

As a consequence, A-GF initiates the up-link notification (Annotation), only after an initial *probation phase* to ensure that the discovered asymmetric link will be stable long enough to justify the cost of additional control packets. Algorithm 2 illustrates the *observe_new_neighbor* method that implements the probation phase. When an asymmetric link is detected for the first time at a down-link node, the up-link node associated with the asymmetric link enters a probation phase. During the probation phase, the changes in the distance between the up-link node and the down-link node are monitored. The duration of the probation phase is set equal to

$$\frac{DIST(uplink\_neighbor\_location, my\_location)}{2 \times uplink\_neighbor\_velocity}$$

i.e., until the current distance between the up-link node and the down-link node is at least half of the initial distance. The *Pythagorean Theorem with parallel meridians* is used to calculate the distance between the up-link node at $\Phi_{uplink}, \lambda_{uplink}$ and the down-link node at $\Phi_{downlink}, \lambda_{downlink}$ where the (Φ, λ) pairs indicate the location of the node in latitude and longitude. The formula being used to calculate the distance is as follows:

$$DIST(uplink\_node, downlink\_node) = R\sqrt{(\Phi_{uplink} - \Phi_{donwlink})^2 + (\lambda_{uplink} - \lambda_{donwlink})^2}$$

R is the radius of the Earth in the above formula.

At the end of the probation phase, the up-link node is notified about the asymmetric link only if it passes a test that includes the following three conditions:

1. More than one direct one-hop Hello messages from the up-link node were received at the down-link node during the probation phase;
2. The distance between the up-link node and the down-link node is either constant or decreasing over the probation phase;
3. The link is still asymmetric.

An asymmetric link has to meet all of these three conditions to pass this test. These three conditions, if all met; ensure that the asymmetric link will be stable for long enough to justify the cost of additional control packets cost for link notification. For example, condition 1 ensures that the up-link node is still in the neighborhood and the wireless connectivity from the up-link node to the down-link node is good. Condition 2 ensures that the up-link node and the down-link node are not moving away from each other, so the connection will most likely be available for some time. Condition 3 ensures that the link has not turned into a symmetric link, which is quite possible if conditions 1 and 2 are met and both nodes are within each other's transmission range. However, if one or more of these three conditions are not met, then the asymmetric link is not a stable one. In that case, the up-link node is not notified about the asymmetric link, and the corresponding cost of the asymmetric link notification and maintenance is avoided.

The method described in Algorithm 2 (See Box 2)returns an integer $i \in \{1, 0, -1\}$ (1) 1 is returned if at the end of the probation phase the asymmetric link meets all three conditions, and therefore, is predicted to be stable (i.e., non-transient); (2) 0 is returned if during or at the end of the probation phase, the asymmetric link does not meet all of these conditions, and therefore, is predicted to be transient; (3) -1 is returned if the asymmetric link meets all three conditions but the probation phase is not finished yet.

Box 2. Algorithm 2

```
Algorithm 2. observe_new_neighbor (ne, new_distance)

Require: Neighbor Entry Variable ne
Ensure: Transient links are eliminated

1.  if ((old_distance = look_up_old_distance (ne)) > 0) then
2.          if ((old_distance ≥ new_distance) &&
            (ne.link_type == ASYMMETRIC)) then
3.                  if ((is_timedout_observation_period(ne)) then
4.                          return 1
5.                  else
6.                          set_old_distance(ne, new_distance)
7.                          return -1
8.                  endif
9.          else
10.                 return 0
11.         endif
12. else
13.         set_old_distance(ne, new_distance)
14.         return -1
15. endif
```

The modified update_LCR_value algorithm that uses the *observe_new_neighbor* method is illustrated in Algorithm 3 (Box 3). When an asymmetric link is detected for the first time at a down-link node, the up-link node associated with the asymmetric link is entered into a probation phase. The action taken by the down-link node during the probation phase depends on the return value of the method observe_new_neighbor method. If the method returned 1 (i.e., the asymmetric link successfully passed the test at the end of probation phase), then the neighbor is entered into the neighbor table and the asymmetric link notification method is initiated. The return value is 0 when the asymmetric link failed the test at the end of probation phase, and in that case the neighbor is removed from the probation list. A return value of -1 means the asymmetric link is still in the probation phase, and in that case no action is taken. The LCR value is decremented either when a Hello message is lost or the distance between the previous and new Hello message has gotten bigger. Similarly, the LCR value is incremented when either a Hello message is received or the distance between two subsequent Hello messages decreases. The step size δ_{lcr}, that is used to modify the LCR value when the distance changes, may be different from the step size Δ_{lcr} that is used to modify the LCR value when a Hello message is received or lost.

$$\frac{new\_\,dis\tan ce}{2 \times ne.velocity}$$

Forwarding of Data Packets

When a node makes a data forwarding decision, it considers (1) the LCR rank of the neighbor and (2) the geographical progress that neighbor makes towards the destination location (thereby considering both reliability and performance). Toward that end, when a node forwards a data packet M towards a sink node S, it only considers next-hop neighbors with LCR values of at least *MAX_LCR/2*. The forwarding node calculates a *Next Hop Quality (NHQ)* value for each of its neighbors N_i. *NHQ* (N_i) consists of a weighted linear combination of the above mentioned parameters. The data packet will be forwarded to the neighbor with the highest NHQ value.

Box 3. Algorithm 3

```
Algorithm 3. update_LCR_value_modified (ne)

Require: Neighbor Entry Variable ne
Ensure: The LCR value of the neighbor entry is made up-to-date

1.  new_distance = DIST(ne_location, my_location)
2.  /*probation phase*/
3.  if (is_new_neighbor (ne)) then
4.          if (!(is_under_observation (ne))) then
5.                  ne.probation_period = new_distance / (2 × ne.velocity)
6.                  put_under_observation (ne)
7.                  return
8.          else
9.                  int result = observe_new_neighbor (ne, new_distance)
10.                 if (result == 1) then
11.                         ne.lcr = MIN_LCR
12.                         add_new_neighbor (ne)
13.                         initiate_asymmetric_notification (ne)
14.                         delete_from_observation (ne)
15.                 endif
16.                 if (result == 0) then
17.                         delete_from_observation (ne)
18.                         return
19.                 endif
20.                 if (result == -1) then
21.                         return
22.                 endif
23.         endif
24. else
25.         /*LCR update phase*/
26.         if (is_timeout_update_interval (ne)) then
27.                 ne.lcr = ne.lcr - Δ_lcr
28.         else
29.                 if (ne.lcr < MAX_LCR) then
30.                         ne.lcr = ne.lcr + Δ_lcr
31.                 endif
32.         endif
33.         if (((old_distance = look_up_old_distance (ne)) > new_distance)) then
34.                 ne.lcr = ne.lcr - δ_lcr
```

```
1.          else
2.                  if (ne.lcr < MAX_LCR) then
3.                          ne.lcr = ne.lcr + δ_lcr
4.                  endif
5.          endif
6.          if (ne.lcr == 0) then
7.                  delete_neighbor (ne)
8.                  initiate link recovery
9.          endif
10. endif
11. if (ne != NULL) then
12.         reset_timeout_neighbor (ne)
13. endif
```

$$NHQ(N_i) = (1-\alpha)\times(DIST(S,M) - DIST(S,N_i)) + \alpha \times LCR(N_i), i \in \{1,2\ldots n\}$$

The weight value (i.e., α) is the network designer's choice, i.e., if α is set to zero then A-GF works as the standard GF protocol. However, A-GF works as a proactive link condition based routing protocol if α is set to 1. Simulation results revealed that the protocol yields the desired performance (i.e., a combination of link reliability and geographic progress towards the sink location) when the value of α is set in the range 0.4-0.56.

EVALUATION

In order to evaluate the performance of A-GF, we performed simulations to study large and complex network topologies. In addition to simulation, we also implemented A-GF as a library on Linux and performed a small-scale experiment in order to see how accurately the simulation results match the results from a real system. As described later, the real experimental results exhibit similar trends as the simulation results, albeit on a smaller scale. The simulation and Linux implementation setups and results are described in detail.

Simulation

Setup

We used the JiST/SWANS simulator to simulate large and complex mobile ad-hoc networks. JiST (Java in Simulation Time) is a discrete event simulator designed to run over a standard Java Virtual Machine. SWANS (Scalable Wireless Ad-Hoc Network Simulator) is built on top of the JiST platform to provide the tools needed to construct a wireless mobile ad-hoc network.

The nodes in the simulated network have varying transmission ranges, as a result of differing transmit powers, radio reception sensitivities, and radio reception thresholds of the nodes. The default values of transmit power, radio reception sensitivity, and radio reception threshold are 16 dBm, -91 dBm, and -81 dBm, respectively. The radio transmission ranges of the nodes are chosen in the range *tx_radius ± ((tx_radius * x) / 100)*, where tx_radius is the default transmission range, and x is varied from 5 to 30 in steps of 5. It is reasonable to assume that the number of asymmetric links in the network will go up with an increasing variance of transmission ranges. There are 200 nodes in the network and they are placed randomly in a 1000 meters × 1000 meters field. Node mobility follows the Random Waypoint model, with a maximum velocity of 5 m/s. MAX_LCR is chosen as 4 and α is chosen as 0.4.

Twenty runs of experiments were conducted for each metric, and the final results were taken as the average of the twenty experiments. In each simulation, we generate n/2 CBR traffic flows where n is the number of nodes in the network, and each flow sends 100 packets of 1000 bytes each. Each simulation runs for 1000 seconds.

Simulation Results

We evaluate and compare the performance of A-GF to standard GPSR as well as *Symmetric Geographic Forwarding (S-GF)* (Zhou, He, Krishnamurthy, & Stankovic, 2004). S-GF allows a node to add the IDs of all its neighbors it has discovered into the periodic Hello messages. When a node receives a Hello message, it registers the sender as its neighbor in its local neighbor table, and then checks whether its own ID is in the Hello message. If the receiver finds its own ID in the neighbor list in the Hello message, then it marks the communication link connecting it to the sender as symmetric. Otherwise, it marks the communication link between them as asymmetric. Whenever a node needs to forward a packet, it selects only those neighboring nodes connected via symmetric links. Thus, S-GF uses a similar approach as in A-GF to discover the asymmetric links in the network. However, S-GF ignores asymmetric links whereas A-GF exploits them while making routing decisions, and this results in an increase in data transmission success and a decrease in latency (or route length), as shown in the simulation results. Note that we did not compare the performance of A-GF to DEAL (Chen, Hao, Zhang, Chan, & Ananda, 2009), even though both protocols are similar in that they discover the presence of an asymmetric link by exchanging each node's knowledge about their one-hop neighborhood in the periodic messages. However, DEAL uses data link layer beacon messages for finding asymmetric links, whereas A-GF uses routing layer Hello messages for the same purpose. Therefore, DEAL improves network connectivity at the data link layer but a cross-layer approach will be required if the routing layer protocol wanted to use the asymmetric links for efficient routing. Without having the cross-layer support in DEAL, it is not possible to compare its performance with A-GF in terms of metrics (e.g., data transmission success, average latency or hop count) which are commonly used to evaluate routing layer protocols.

Figure 8 shows the *data transmission success* for these three protocols. We define the data transmission success as the ratio of the number of data packets received at the destinations to the number of data packets sent by the source node. Figure 8 shows that A-GF performs as well as S-GF and GPSR when the nodes in the network have only 5% variance in their transmission ranges. However, as the variance of transmission ranges of nodes in the network increases (i.e., number of asymmetric links increases), the data transmission success of both S-GF and GPSR decreases. Both A-GF and S-GF always result in better performance than GPSR because GPSR ignores asymmetric links and consequently suffers from low network connectivity and data transmission failures. S-GF discovers the asymmetric links in the network and avoids them during routing, thereby reducing the transmission failures incurred by GPSR in an asymmetric network. A-GF not only discovers the asymmetric links in the network, but also exploits the asymmetric links in routing, thereby offering better performance than S-GF.

Figure 9 shows the *average route length* for these three protocols, which is the average of total number of hops traveled by all data packets in the network. Figure 9 shows that when the variance of transmission range of the nodes in the network increases (i.e., number of asymmetric links increases), the average route lengths for all three protocols go up. However, the increase in the average route length in A-GF is less substantial than in S-GF and GPSR, because A-GF explicitly exploits asymmetric links for routing. As indicated earlier in Section 1, asymmetric links often tend to span longer distances than symmetric links. Therefore, using asymmetric links in routing decisions often results in shorter routes, whereas ignoring them results in routes longer than necessary (as in S-GF and GPSR).

Figures 10, 11, and 12 compare the overhead of the A-GF, S-GF, and GPSR protocols. The total overhead in any routing protocol in wireless ad-hoc networks consists of the following main components: (1) control traffic generated by the protocol, (2) overheads from data traffic that is forwarded over sub-optimal routes, and data transmission failures due to node movement or failure, and (3) link maintenance overheads expressed in form of the average size of Hello messages. Figure 10 shows the *control packet overhead* in all three protocols. We define control packet overhead as the total number of control packets

Figure 8. Data transmission success

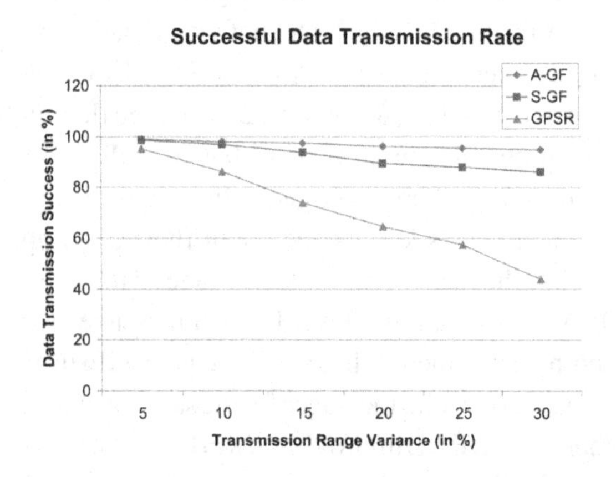

Figure 9. Average route length

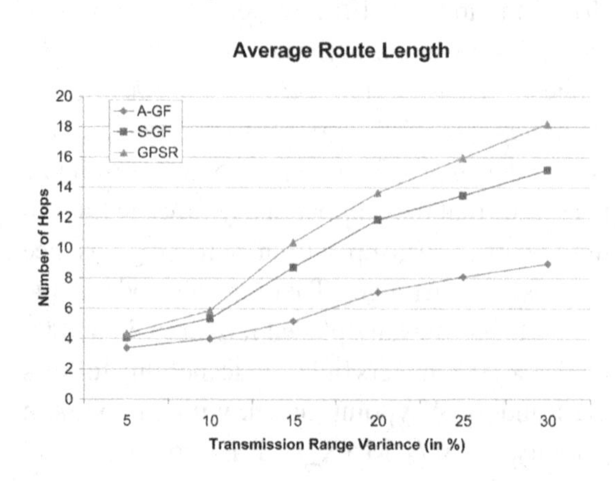

Figure 10. Control packet overhead

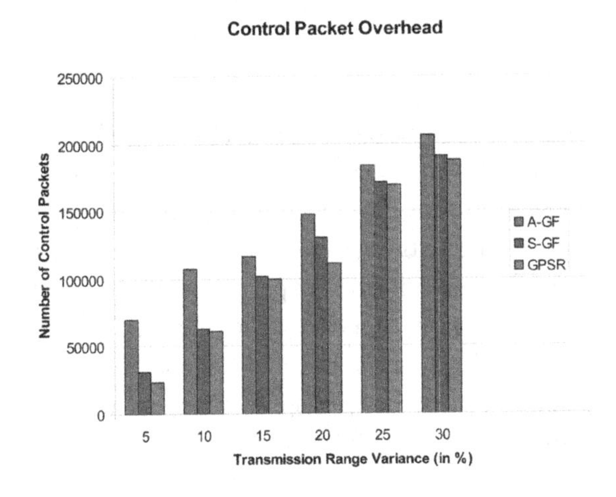

Figure 11. Total data bandwidth consumption

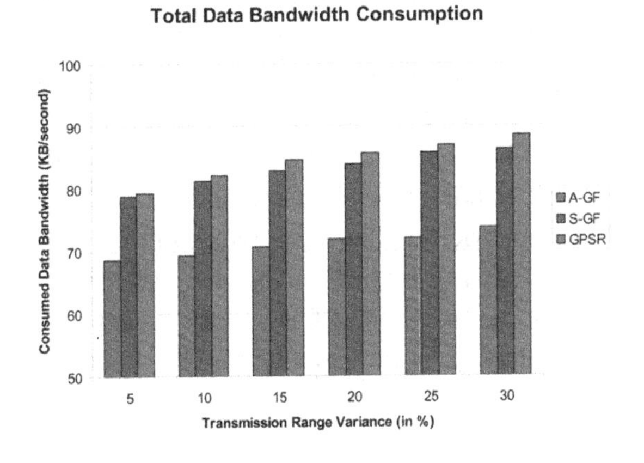

Figure 12. Link maintenance overhead

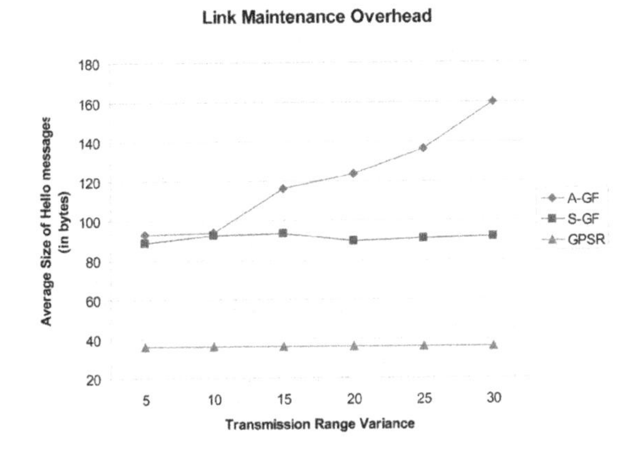

in the network. Figure 10 shows that the control packet overhead in all three protocols increases when the transmission range variance of the nodes in the network increases (i.e., number of asymmetric links increases). However, A-GF always results in higher control packet overhead than S-GF and GPSR, because A-GF adds more control packets in the network for asymmetric link discovery and maintenance. Toward that end, when the nodes in the network have 30% variance in their transmission ranges, A-GF results in 8.16% and 9.67% higher control packet overhead than S-GF and GPSR, respectively.

Figure 11 shows the *total data bandwidth consumption* of these three protocols. We define total data bandwidth consumption as the total number of bytes per second in the network that are needed to successfully deliver all data packets to their destinations. Figure 11 shows that A-GF consumes less data bandwidth than S-GF or GPSR. Toward that end, when the nodes in the network have 30% variance in their transmission ranges, A-GF consumes 17.15% and 17.53% less data bandwidth than S-GF and GPSR, respectively. A-GF needs fewer retransmissions of data packets to reliably deliver the data packets to all destinations, whereas S-GF and GPSR result in higher retransmissions of data packets due to transmission failures over asymmetric links. Furthermore, A-GF eliminates transient asymmetric links, thereby increasing data transmission success and further reducing retransmissions.

Figure 12 shows the *link maintenance overhead* of these three protocols. We define link maintenance overhead as the average size of the Hello messages in bytes. Figure 12 shows that GPSR always results in fixed sized Hello messages, and that is because in GPSR the one-hop Hello messages always contain a fixed number of fields that include the node ID and current location of the node. However, the S-GF Hello messages have the neighbor table inserted in their header, and therefore, are bigger in size than GPSR Hello messages. Note that the average size of the S-GF Hello messages do not vary much over varying transmission ranges. Given the network density remains constant, the average number of neighbors a node in the network does not vary much, and as a consequence, the average size of Hello messages does not very much either. On the other hand, the average size of A-GF Hello message increases when the transmission range variance of the nodes in the network increases (i.e., number of asymmetric links increases). A-GF Hello messages contain the neighbor table and the Annotations used for asymmetric link maintenance. The number of embedded Annotations increases as the number of asymmetric links in the network increases, thereby leading to the increase in average size of A-GF Hello messages. When the nodes in the network have 30% variance in their transmission ranges, the A-GF Hello messages are 74.15% and 343.94% bigger in size than the S-GF Hello messages and GPSR Hello messages, respectively.

Figure 13 shows the *asymmetric link utility* of A-GF. We define asymmetric link utility as the lifetime of asymmetric links vs. the number of times these links were used by A-GF in making routing decisions. The average lifetime of an asymmetric link in the simulations was 17 seconds, with the least and most available links being active for 1 second and 25 seconds, respectively. Figure 13 shows that A-GF chooses the asymmetric links with longer lifetime than the asymmetric links with shorter lifetime, thereby successfully identifying and ignoring transient asymmetric links for efficient routing. Figure 13 also shows that the overall usage of asymmetric links in routing goes up when the variance of transmission range of the mobile nodes increases; this is because increasing variability in the radio capabilities of the mobile nodes results in an increase in the total number of asymmetric links in the network.

Linux Implementation

Setup

We implemented the A-GF protocol in C on a Linux platform using an event based group communication tool developed at the University of Notre Dame. In order to evaluate the performance of A-GF, we performed experiments in a testbed of 20 Crossbow Stargate devices and a Lenovo ThinkPad X60 sink, connected to the same ad-hoc network. One of the Stargate devices was selected as the source node. The nodes were initially arranged in a 5×4 grid, and a random offset was applied to the initial location coordinates in order to create a random topology. Some of the nodes used a lower transmission range than other nodes in the network, thereby leading to irregular radio capabilities in the network. The locations of the nodes were changed at the beginning of each periodic beaconing interval to create a mobile network. In our experiments, we vary the number of asymmetric links in the network in the range of 5%-20%. The mobile nodes have a maximum speed of 5 m/s. MAX_LCR and α are set to the same values as in the simulation, i.e., 4 and 0.4 respectively.

Figure 13. Asymmetric link utility

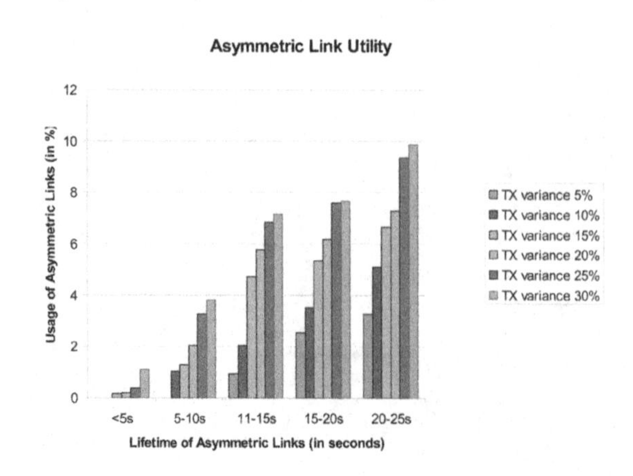

Experimental Results

We evaluate and compare the performance of A-GF to standard GF in terms of three metrics as described below. Ten runs of experiments were conducted for each metric and the final results were taken as the average of the ten experiments. During each run, the source node sent packets to the sink node at a rate of 1 packet/second for 200 seconds.

Figures 14, 15, and 16 exhibit similar trends as shown in Figures 8, 9, and 10, respectively; albeit at a smaller scale. Figure 14 shows that the *data transmission success* of A-GF and GF are comparable when only 5% of the links in the network are asymmetric. However, in case of 20% link asymmetry in the network, Figure 14 shows that A-GF results in better performance than standard GF (96% over 90%). Figure 15 shows that for 5% asymmetric links in the network, the *average route length* in A-GF and standard GF is comparable. However, with an increasing percentage of asymmetric links in the network, the average route length in A-GF is almost constant, while it increases for standard GF. Figure 16 shows that for 5% asymmetric links in the network, the *total network traffic* in A-GF and standard GF is comparable. However, the total network traffic in A-GF increases when then percentage of asymmetric links in the network increases, due to the additional control packets generated by A-GF for asymmetric link discovery and proactive link condition monitoring. Toward that end, the network traffic is 8% higher in A-GF than standard GF when 20% of the links in the network are asymmetric.

Figure 14. Data transmission success

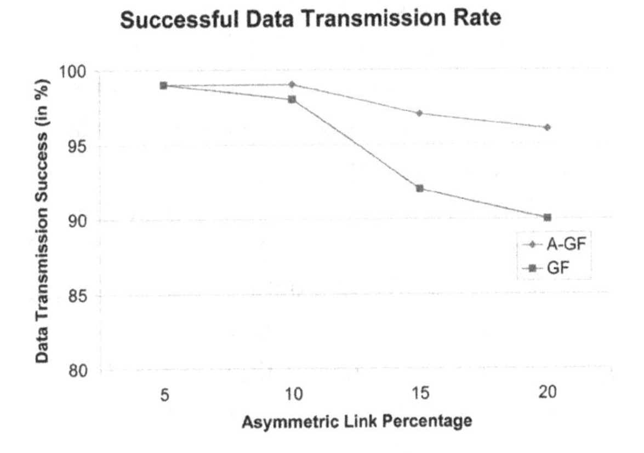

Figure 15. Average route length

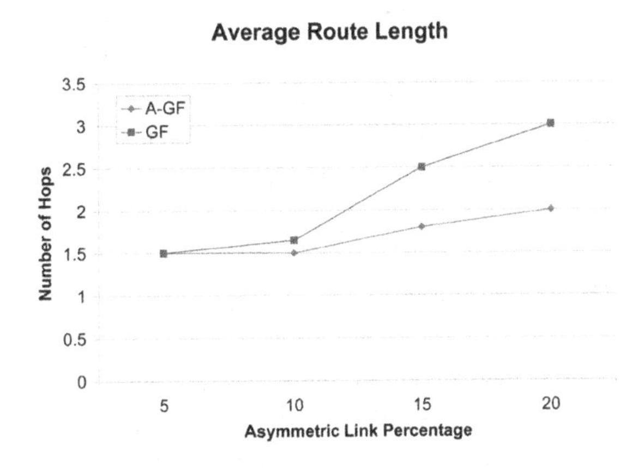

Figure 16. Total network traffic

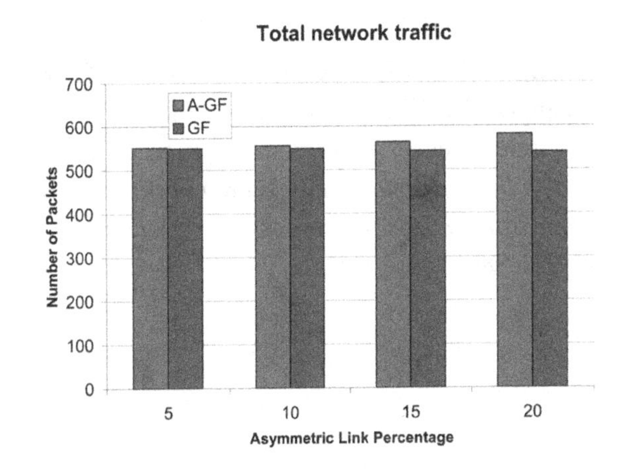

CONCLUSION AND FUTURE WORK

This paper introduces A-GF, a location aware routing approach that actively exploits asymmetric links to increase the reliability and performance of ad-hoc routing. Our results indicate that A-GF succeeds in reducing latencies and increasing successful route discoveries, while keeping the overheads small. The increase in routing success also indicates that utilizing asymmetric links is overall beneficial, even when occasional asymmetric links fail due to dynamic adjustments in transmission range and node mobility. A-GF combines the two metrics stability and minimum latency while making routing decisions. The stability metric in A-GF considers the number of Hello messages received during a certain time window and the changes in distance between the up-link and down-link nodes caused by node mobility. Toward that end, A-GF eliminates transient asymmetric links for efficient resource management schemes in wireless ad-hoc and sensor networks. Our future work will further investigate the possibility of combining stability and energy efficiency in asymmetric wireless networks.

REFERENCES

Bose, P., Morin, P., Stojmenovic, I., & Urrutia, J. (1999). Routing with guaranteed delivery in ad-hoc wireless networks. In *Proceedings of the 3rd International Workshop on Discrete Algorithms and Methods for Mobile Computing and Communications*, Seattle, WA (pp. 48-55).

Cerpa, A., Wong, J. L., Kuang, L., Miodrag, P., & Estrin, D. (2005). Statistical model of lossy links in wireless sensor networks. In *Proceedings of the 4th International Symposium on Information Processing in Sensor Networks*, Los Angeles, CA (pp. 81-88).

Chen, B. B., Hao, S., Zhang, M., Chan, M. C., & Ananda, A. I. (2009). DEAL: Discover and exploit asymmetric links in dense wireless sensor networks. In *Proceedings of the 6th Annual IEEE Communications Society Conference on Sensor, Mesh and Ad Hoc Communications and Networks*, Rome, Italy (pp. 297-305).

De Couto, D., Aguayo, D., Bicket, J., & Morris, R. (2003). A high-throughput path metric for multi-hop wireless routing. In *Proceedings of the 9th Annual International Conference on Mobile Computing and Networking*, San Diego, CA (pp. 134-146).

Duros, E., & Dabbous, W. (1996). *Handling of unidirectional links with OSPF*. Retrieved from http://tools.ietf.org/html/draft-ietf-ospf-unidirectional-link-00

Ganesan, D., Estrin, D., Woo, A., Culler, D., Krishnamachari, B., & Wicker, S. (2002). *Complex behavior at scale: An experimental study of low-power wireless sensor networks* (Tech. Rep. No. CSD-TR 02-0013). Los Angeles, CA: University of California at Los Angeles.

JiST. (2005). *Java in simulation time/scalable wireless ad hoc network simulator.* Retrieved from http://jist.ece.cornell.edu/

Johnson, D. B., & Maltz, D. A. (1996). Dynamic source routing in ad hoc wireless networks. *Mobile Computing*, *5*, 153–181. doi:10.1007/978-0-585-29603-6_5

Karp, B., & Kung, H. T. (2000). Greedy perimeter stateless routing for wireless networks. In *Proceedings of the 6th Annual International Conference on Mobile Computing and Networking*, Boston, MA.

Kim, D., Toh, C. K., & Chou, Y. (2000). RODA: A new dynamic routing protocol using dual paths to support asymmetric links in mobile ad hoc networks. In *Proceedings of the 9th International Conference on Computer Communications and Networks*, Las Vegas, NV (pp. 4-8).

Kim, Y. J., Govidan, R., Karp, B., & Shenker, S. (2004). Practical and robust geographic routing in wireless networks. In *Proceedings of the 2nd International Conference on Embedded Networked Sensor Systems*, Baltimore, MD (pp. 295-296).

Liu, R. P., Rosberg, Z., Collings, I. B., Wilson, C., Dong, A., & Jha, S. (2008). Overcoming radio link asymmetry in wireless sensor networks. In *Proceedings of the International Symposium on Personal, Indoor and Mobile Radio Communications*, Cannes, France (pp. 1-5).

MacDonald, J. T., & Roberson, D. A. (2007). Spectrum occupancy estimation in wireless channels with asymmetric transmitter powers. In *Proceedings of the 2nd International Conference on Cognitive Radio Oriented Wireless Networks and Communications*, Orlando, FL (pp. 245-249).

Marina, M. K., & Das, S. K. (2002). Routing performance in the presence of unidirectional links in multihop wireless networks. In *Proceedings of the 3rd International Symposium on Mobile Ad Hoc Networking and Computing*, Lausanne, Switzerland (pp. 12-23).

Misra, R., & Mandal, C. R. (2005). Performance comparison of AODV/DSR on-demand routing protocols for ad-hoc networks in constrained situation. In *Proceedings of the IEEE International Conference on Personal Wireless Communications*, New Delhi, India (pp. 86-89).

Perkins, C., Belding-Rower, E. M., & Das, S. (2003). *Ad-hoc on demand vector (AODV) routing*. Retrieved from http://www.ietf.org/rfc/rfc3561.txt

Reijers, N., Halkes, G., & Langendoen, K. (2004). Link layer measurements in sensor networks. In *Proceedings of the 1st IEEE International Conference on Mobile Ad-hoc and Sensor Systems*, Fort Lauderdale, FL (pp. 224-234).

Roosta, T., Menzo, M., & Sastry, S. (2005). Probabilistic geographic routing in ad hoc and sensor networks. In *Proceedings of the Wireless Networks and Emerging Technologies*, Banff, AB, Canada.

Sang, L., Arora, A., & Zhang, H. (2007). On exploiting asymmetric wireless links via one-way estimation. In *Proceedings of the 8th International Symposium on Mobile Ad Hoc Networking and Computing*, Montreal, QC, Canada (pp. 11-21).

Son, D., Helmy, A., & Krishnamachari, B. (2004). The effect of mobility-induced location errors on geographic routing in ad hoc networks: Analysis and improvement using mobility prediction. *IEEE Transactions on Mobile Computing*, 233–245. doi:10.1109/TMC.2004.28

Srinivasan, K., Dutta, P., Tavakoli, A., & Levis, P. (2006). Understanding the causes of packet delivery success and failure in dense wireless sensor networks. In *Proceedings of the 4th International Conference on Embedded Networked Sensor Systems*, Boulder, CO (pp. 419-420).

Srinivasan, K., & Levis, P. (2006). RSSI is under appreciated. In *Proceedings of the 3rd Workshop on Embedded Networked Sensors*, Cambridge, MA.

Stargate. (n. d.). *Resource links*. Retrieved from http://platformx.sourceforge.net/Links/resource.html

Wang, G., Ji, Y., & Turgut, D. (2004). A routing protocol for power constrained networks with asymmetric links. In *Proceedings of the International Workshop on Performance Evaluation of Wireless Ad Hoc, Sensor, and Ubiquitous Networks*, Venice, Italy (pp. 69-76).

Woo, A., Tong, T., & Culler, D. (2003). Taming the underlying challenges of reliable multi-hop routing in wireless networks. In *Proceedings of the 1st International Conference on Embedded Networked Sensor Systems,* Los Angeles, CA (pp. 14-27).

Zamalloa, M. Z., & Krishnamachari, B. (2007). An analysis of unreliability and asymmetry in low-power wireless links. *ACM Transactions on Sensor Networks, 3*(2).

Zhao, Y. J., & Govidan, R. (2003). Understanding packet delivery performance in dense wireless sensor network. In *Proceedings of the 1st International Conference on Embedded Networked Sensor Systems*, Los Angeles, CA (pp. 1-13).

Zhou, G., He, T., Krishnamurthy, S., & Stankovic, J. A. (2004). Impact of radio irregularity on wireless sensor networks. In *Proceedings of the 2nd International Conference on Mobile Systems, Applications, and Services*, Boston, MA (pp. 125-138).

This work was previously published in the International Journal of Embedded and Real-Time Communication Systems, Volume 2, Issue 4, edited by Seppo Virtanen, pp. 46-70, copyright 2011 by IGI Publishing (an imprint of IGI Global).

Section 2
Model-Based Testing of Embedded and Real-Time Communication Systems

Chapter 5
Requirements Traceability within Model-Based Testing:
Applying Path Fragments and Temporal Logic

Vanessa Grosch
University of Ulm, Germany

ABSTRACT

Requirements traceability enables the linkage between all development artifacts during the development process. Within model-based testing, requirements traceability links the original requirements with test model elements and generated test cases. Current approaches are either not practical or lack the necessary formal foundation for generating requirements-based test cases using model-checking techniques involving the requirements trace. This paper describes a practical and formal approach to ensure requirements traceability. The descriptions of the requirements are defined on path fragments of timed automata or timed state charts. The graphical representation of these paths is called a computation sequence chart (CSC). CSCs are automatically transformed into temporal logic formulae. A model-checking algorithm considers these formulae when generating test cases.

INTRODUCTION

In practice, model-based testing still lacks the confidence of the actual users. Usually only structural or similar coverage criteria (Ryser et al., 1998; Gaston & Seifert, 2005) are used to select the tests. This type of test selection is not satisfying, even if the test engineers understand the algorithms behind test case generation. For example, the structure-oriented criteria do not answer the question whether a test case should be run during a new software release or whether it might be left on hold until the next release. Requirements traceability with an adequate depth of information contributes to requirements-based test case generation and could help solve this type of issues. The test cases should be connected to dependant test model fragments and original system

DOI: 10.4018/978-1-4666-2776-5.ch005

requirements. When this is done, the context of a test case is then linked to the original requirements as well. Requirements traceability leads to a better understanding of test cases by introducing a powerful requirements description and a more efficient reuse and maintenance of test models. The need to test according to the requirements is demanded by quality and safety norms.

There are several different approaches to requirements traceability. They range from simple state annotation to formal requirements specification. In the former case, the content of a requirement cannot be captured by labeling a single transition; a timed sequence of transition changes is needed.

Three examples give an idea of the problems occurring. (1) A requirement states "The maximum delay between a turn indication request and the turn indicator flashes is 600ms". In a model containing the whole turn indication functionality, many transitions have to be taken between the turn indication request and the turn indicator flashing. The transitions may have different timing restrictions. Only annotating all the transitions, which is a lot of effort and error-prone task, would reflect the statement of a maximum delay of 600ms. (2) Another requirement states "A turn indication request that got interrupted by an emergency flashing request will continue when withdrawing the emergency request". This requirement needs the parallel visit of the two states *turn indication requested* and *emergency flashing requested*, ensuring that *turn indication requested* was visited first. Then, the *emergency flashing requested* state has to be left while the *turn indication requested* state is still visited. These timing constraints cannot be reflected by just annotating the requirements transitions. (3) The last example addresses the tip indication of turn indicator lights which results in three flashes. The requirement states that "Repeating tip indication requests result in the last three flashes". As a consequence summing up the flashes is forbidden. To reflect the requirement, at least one repetition of the tip indication request is needed. This results in a cycle and revisit of already visited test model elements. Cycles in a path are usually avoided by test case generators and cycles cannot be described by simple annotation of transitions with the requirements identifier.

However, a completely new notation to achieve a formal and powerful requirements specification that is able to handle the examples above, is not efficient, because the test engineer defining the requirements trace needs to specify the same requirement in two different ways. Some examples of formal graphical requirements specifications will be given in chapter *Related Work*.

This paper offers an approach to requirements traceability through application of state-based test models in a balanced manner. Balanced means that the approach uses the states of the test model to reduce the effort of composing the requirements trace, it adapts the modeling paradigm of the test model and therefore does not request another interpretation of the requirement and it adds new information of the sequential flow of a requirement as a cluster of paths through the test model. It aims at narrowing the gap between the operational specification of a system designed as state charts and the declarative specification expressed in temporal logic. As result a formal but at the same time easy to create approach is described.

The requirements descriptions, called computation sequence charts, offer enough dynamic range to address most behavioral aspects (e.g., hard real time, distribution, etc.). The creation of these requirements descriptions is quick because they are composed of pre-existing test model elements. As an additional advantage, these same requirements descriptions can be formalized and therefore used in automatic test case generation using model-checking techniques.

BACKGROUND

In this paper, model-based testing is understood in the following fashion. Assuming that the test engineer models system behavior using state-transition techniques (e.g., a composition of parallel state charts) based on system requirements described using natural language. This model is called a test model. In general, this test model is an additional model to an already available implementation model. The test model is built to include the technical constraints of the target test bench and therefore allows for advances such as fault simulation modeling. Based on the test model, a test case generator calculates paths through the state charts. The generated test cases are run against the implementation of the system. Thus, it can be shown whether the system implementation is conforming to the test model or not (Utting & Legeard, 2007).

The state-based test models are used as basis of the requirements traceability presented in this paper because the method is part of research and development in the automotive domain, for example at the Daimler, Inc. (http://www.daimler.com) (Schmid, 2008). Most of the functionalities in vehicles are realized using the paradigms for development of embedded systems. These compose of (hard) real-time requirements distributed over electronic control units and are realized as reactive and parallel processes. There is no global clock synchronizing those control processes. A vehicle system behavior, like the stop/start functionality of a car can therefore be described as a composition of timed automata (Alur & Henzinger, 1992). The complexity of these kinds of systems makes manual test case selection and implementation unreliable and expensive. Being able to specify the systems in a test model and generating test cases based on them gives the advantage of reproducible, measurable and reliable test coverage. Being able to generate test cases especially for atomic requirements, as requested by internal and external process directives was not possible in an efficient manner at the beginning of this research.

Today, there are several methods for tracing requirements throughout the model-based testing process (Aizenbud-Reshef et al., 2006; Egyed & Gruenbacher, 2002; Zisman et al., 2003). Considering a state-based test model, the simplest is to annotate states, transitions, or model parts with requirement identifiers. This though easy to achieve does not offer enough information for properly separating single requirements and does not give information about specific system behavior that is required to address the content of a requirement accurately, as can be seen in the three examples of the chapter *Introduction*. Some other approaches capture tracing information during the modeling process (Pfaller, 2008). The opposite procedure specifies the requirement in a completely separate manner by introducing a new notation (Naslavsky et al., 2005; Naslavsky et al., 2007). For example, one can model the whole system behavior in state charts and a single requirement as a use case diagram. This leads to a better understanding of the requirement content, but demands much more effort because of introducing a completely new notation. Chapter *Related work* comprises the closest approaches to the one presented in this paper.

There are several model-checking theories using system properties specified in temporal logics to verify the system model (Alur et al., 1990; Tan et al., 2004; Fraser & Wotawa, 2008). The system properties are then used to trace the corresponding requirements. The model-checking algorithms identify whether a system model satisfies a system property or not. The algorithm generates a witness as a proof if the model rejects the property. A witness is a path through the model showing the property does not hold. The model-checking techniques can be used to generate test cases. The negation of a property should be rejected, given the system model is correct and it satisfies the property. The witness calculated by the model-checking algorithm can then be interpreted as a test case and run against the system implementation. Thus, modeling system requirements in temporal logic is a good basis for test case generation. The model-checking algorithms are very advanced and efficient and still an important research area in the field of model-based testing (Peleska, 2009; Clarke et al., 2001). In the presented research model-checking based on timed automata and binary decision-diagrams as well as model-checking based on timed state charts and bounded model-checking theories was considered. The aspects reused and the effect of the computation sequence charts on the techniques will be presented in chapter *Approach*.

Difficulties arise during the creation of system requirement specifications in temporal logic. The syntax of a temporal logic formula is learnable; however, finding the correct formula to express a system property is a challenging and error-prone task. There are efforts making this process easier to fulfill by introducing a graphical notation of the temporal logic (Schloer, 2001; Braberman et al., 2005). This supports the understanding of temporal logic formulae, but demands a completely different notation in comparison to the operational specification of the system as a state chart.

The definition of temporal logic in general and especially of the branching-time logic (Computation Tree Logic) shall be introduced briefly in the following section. The selected variant of time logic is called Timed CTL (TCTL) and will be discussed in more detail in the proceedings. The formalization of the requirements description presented in this paper is based on TCTL formula.

Temporal logic is an extension of the propositional logic and provides a way to specify properties of system calculations. The *Kripke structure* is a structure which properties can be defined in temporal logic (Clarke et al., 2000).

Definition 1.1. Kripke Structure

Let AP be a set of atomic propositions. A Kripke structure K is a tuple $K = (S, S_0, T, L)$ where

- S is the finite set of *states*,
- S_0 is a set of *initial states*, $S_0 \subset S$,
- $T \subseteq S \times S$ is the set of *complete transition relations*, i.e., $\forall s \in S\ \exists s' \in S : (s, s') \in T$,
- $L : S \rightarrow 2^{AP}$ is the *labeling function*, labeling each state with the set of atomic propositions with *true* in that state. 2^{AP} is the powerset of AP.

A Kripke structure specifies the system behavior as a sequence of pre- and post-states (Figure 1). A state change is controlled by the transition relation. A state itself is not observable. Instead, all atomic propositions can be observed. The state is visited if all the atomic propositions that need to be true in the state evaluate to true. The state graph is a graphical representation of the Kripke structure.

Focusing on the computations of a system, one can draw the state graph with no specific entry state and no terminal state. All multiple transitions are grouped together to form a single transition. By selecting an initial state in the state graph a computation tree can be generated by unwinding the state graph. Each row inside the computation tree corresponds to one computation step and its possible results expressed as circles. Computation steps 1 to 3 are already labeled.

1. COMPUTATION TREE

The infinite paths of the computation tree (e.g., some are marked as a sequence of black circles in Figure 1) correspond to one computation of the system. With temporal logic one can define properties of states or paths. They allow for specification of the system change over time by introducing a non-explicit time notion. Temporal logic addresses either one linear computation

Figure 1. State graph (l.) and computation tree (r.) of a Kripke structure

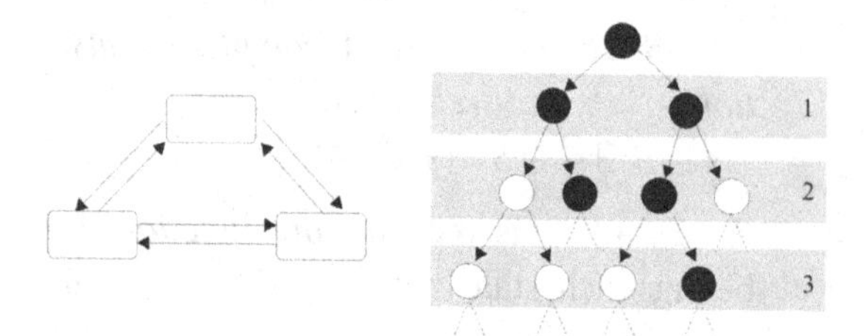

of the system (i.e., linear time logic – LTL) or computations with alternative outcomes (i.e., branching time logic – CTL).

To define CTL we need to distinguish between state and path formulas. A state formula specifies the rules to be satisfied by the atomic propositions in a single state, while path formulae express temporal logic of computational paths, in other words of a sequence of states (Baier & Katoen, 2008).

Definition 1.2. Syntax of CTL

The grammar of CTL state formulae is formed as follows:

$$\Phi ::= true \mid a \mid \Phi_1 \wedge \Phi_2 \mid \neg\Phi \mid \exists\phi \mid \forall\phi$$

with $a \in AP$ and ϕ a path formula.

The grammar of CTL path formulae is defined as follows:

$$\phi ::= Next\ \Phi \mid \Phi_1\ Until\ \Phi_2$$

The CTL path formula $Next\ \Phi$ requires that the property Φ holds in the second state of path ϕ.

Φ_1 *Until* Φ_2 holds if there is a state on path ϕ where property Φ_2 holds and in all preceding states property Φ_1 holds.

Definition 1.3. Satisfaction Relation of CTL

Let K be a Kripke structure, $a \in AP$ an atomic proposition, Φ, Ψ be CTL state formulae and ϕ a CTL path formula. $s \in S$ is a state and π a path of K. The satisfaction relation $\models$ is defined:

$$s \models true$$

$$s \models a \text{ iff } a \in L(s)$$

$$s \models \neg\Phi \text{ iff not } s \models \Phi$$

$$s \models \Phi \wedge \Psi \text{ iff } (s \models \Phi) \text{ and } (s \models \Psi)$$

$$s \models \exists\phi \text{ iff } \pi \models \phi \text{ for some } \pi \in Paths(s)$$

$$s \models \forall\phi \text{ iff } \pi \models \phi \text{ for all } \pi \in Paths(s)$$

$$\pi \models Next\,\Phi \text{ iff } \pi[1] \models \Phi$$

$$\pi \models \Phi\, Until\, \Psi \text{ iff}$$
$$\exists j \geq 0.(\pi[j] \models \Psi \wedge (\forall 0 \leq k < j.\pi[k] \models \Phi))$$

$$\pi \models Finally\,\Phi \text{ iff } s_j \models \Phi \text{ for some } j \geq 0$$

$$\pi \models Globally\,\Phi \text{ iff } s_j \models \Phi \text{ for all } j \geq 0$$

where $\pi = s_0 s_1 s_2 \ldots$ and $\pi[i] = s_i$.

The *Implies* operator shall be introduced as one-pointed arrow. $\Phi \rightarrow \Psi$ equals $\neg\Phi \vee \Psi$ saying that if Φ becomes true, Ψ has to become true either. In this case, we do not know anything about the situation when Φ does not become true.

The aim of model-checking is to decide whether a transition system TS satisfies a CTL formula Φ ; in short whether $TS \models \Phi$ is a tautology.

APPROACH

The approach to requirements tracing based on state charts is composed of two steps. Firstly, a graphical notation, called a *computation sequence chart*, is introduced. Secondly, the formalization of a requirements description as TCTL formula is presented. For this purpose transformation rules are defined and the generation algorithm is sketched.

As the main industrial application of the computation sequence charts are time-critical reactive systems as needed in the automotive domain, timed automata or alternatively timed state charts are selected as a basic structure of the temporal logic (i.e., Timed CTL). A timed automaton / timed state chart is a finite automaton on infinite words and allows a finite set of independent real-valued clocks. Real-valued clocks allow convenient modeling of hard real-time requirements. Computation sequence charts are not only restricted to timing requirements but requirements including timing constraints demand the possibility of synchronizing events. Therefore mainly timing requirements will be presented although non-timing requirements can be described by computation sequence charts as well.

As presenting the timed state chart as well as the timed automaton alternative would be too comprehensive, only timed automata and generating test cases by means of binary decision diagrams and computation sequence charts will be explained in more detail. The formal definition of a timed automaton is as follows (Alur, 1999):

Definition 2.1. Timed Automaton

Atimedautomatonisatuple $A = (\Sigma, L, L_0, X, I, T)$ where

- Σ is the finite *alphabet*;
- L is the finite set of *locations*;
- L_0 is a set of *initial locations*, $L_0 \subset L$;
- X is a finite set of *clocks*;
- $I : L \rightarrow C(X)$ is a mapping that labels each location with a set of clocks, called the *location invariant*;
- $L \times \Sigma \times C(X) \times 2^X \times L$ is a set of *switches*. A switch is a 5-tupel $\langle l, a, \phi, \lambda, l' \rangle$ from location l to location l' labeled with a, a clock constraint ϕ, assigning when a switch can be taken and set of clocks $\lambda \subseteq X$, that are reset to zero when the switch is taken.

2^X is the powerset of X.

The model of a timed automaton is a finite state-transition graph $T(A) = (\Sigma, Q, Q_0, R)$ (Clarke et al., 2000). Each state in Q is a pair $\langle l, \nu \rangle$ with $l \in L$ a location and $\nu : X \rightarrow R^+$ mapping each clock to a non-negative real number. The transition relation R is a composition of two types of transitions, the so called *action* and *delay* transitions. An action transition is taken instantaneously once the switch labeled with a operates. Along with the action transition some of the clocks may be reset to zero. A delay transition can be taken within a location. It represents the ongoing delay of time while staying in a location. Time delays are only allowed in the boundaries of the assigned clock invariants.

A test case based on a timed automaton is a finite path fragment of the corresponding state-transition graph. State pairs within the path fragment belong to the transition relation R.

2. AUTOMATON

Figure 2. contains one way of drawing a state-transition graph of a timed automaton. The boxes labeled a, b and c are locations of the timed automaton. y and x are variables. The black dot is the entry point of the graph, saying that initially, any execution starts in location a. The only clock t is reset to zero nearly every time the graph changes its location. Inequalities of the form $t >= x$ are clock invariants stating the duration the graph can spend time in a location. The state graph may either rest in location b for 300 time units and then change to location a or once the variable x increases its value to higher than 5, swap to location c.

An *execution* or *run* of a transition system is an alternating sequence of states and actions ending with a state (Baier & Katoen, 2008). The execution represents one precise behavior of the modeled system. Concerning an original requirement of the system behavior, a single execution would not be powerful enough to express the whole content. Usually system requirements represent a cluster of executions. Thus, the first assumption of this approach states that an abstract view on state changes will be exact enough to express a requirement. As a result, one does not need to describe several executions and group them as a requirements description. Instead, a single abstract description is sufficient.

The second assumption claims that actions do not need to be redrawn within a requirements description. Action labels of transitions are only important for modeling communication. This is already captured within the test model. Deleting the actions within the execution leaves us with a sequence of states, called a *path fragment*.

Strictly speaking, a state of a transition system is not observable. Instead, the atomic propositions true in the state combined with the valuation of the clock variables express the state the system is in. The states *observed* result in the so called *trace*. The trace of a path fragment is thus the induced finite or infinite word over the alphabet 2^{AP} as well as the clock valuation of the state changes. The wording *trace* leads directly to the requirements trace wanted. A single requirement can be associated with a class of traces of a transition system. Thus, the test engineer needs a convenient way of selecting traces of path fragments that correspond to his understanding best with the original requirement he is interested in. The graphical representation of these traces of path fragments is called a computation sequence chart and will be presented in chapter *Graphical notation*.

A computation sequence chart is a cluster of witnesses assigned to a requirement. A single witness is one trace through the test model satisfying the requirement. To be able to generate test cases considering the computation sequence chart as a property to hold along the witness, the computation sequence chart needs to be translated into temporal logic formulae. Temporal logic formulae are chosen to be able to reflect timing requirements.

Figure 2. State graph of a timed automaton

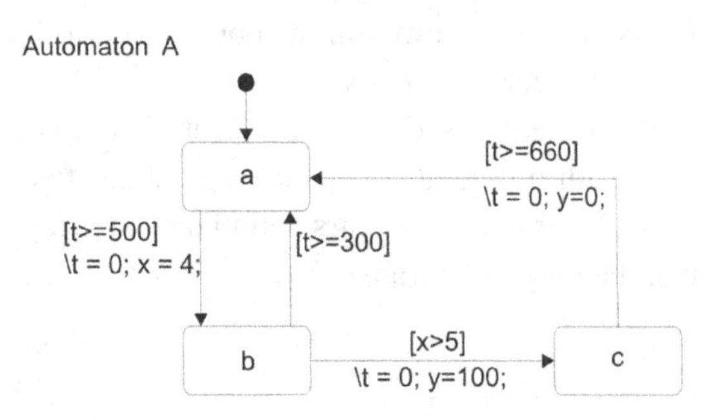

Graphical Notation

On the manual side of the requirements traceability approach, the test engineer describes fragments of the system behavior of a single requirement by choosing path fragments within the state-based test model. These path fragments consist of state sequences or rows of a decision table. The test engineer is supported by dedicated software ensuring consistency and fast completion of the computation sequence chart. In the case study, Enterprise Architect (http://sparxsystems.eu/) was chosen as the case tool to draw the test models and the software for generating the computation sequence charts was realized as a plug-in solution.

The process consists of the following steps. The test engineer selects a sequence of not necessarily directly connected locations of a state-transition graph. The only presumption is that each location transition can be reached within a finite number of steps. This helps shortening the process of creating the requirements description. Example (1) in chapter *Introduction* can therefore be traced easily. Two sequential relations reflect the difference between directly and non-directly connected locations.

Example (2) needs the possibility to synchronize locations, because the requirement results in parallel as well as sequentially ordered visits of locations. Therefore two timelines can be connected via parallel relations. There are different ways of synchronizing locations between two different timelines and state charts, respectively. At this point, we can see one difference between using timed automata or timed state charts. To be more precise, the underlying semantics of these modeling paradigms can differ, depending on the semantic used. Timed automata usually are related to interleaving semantics, timed state charts to true parallelism semantics. In interleaving semantics two locations can never be visited in parallel. They get interleaved and temporal logic properties make a difference between interleaved sequences. This results in a parallel relation stating that two locations only need to be visited during a specific period of time but they do not need to be visited at exactly the same instance of time.

Example (3) demands the possibility of cycles in a computation sequence chart, which is therefore allowed.

Once the trace is finished the plug-in software generates a timeline and inserts the locations in the given order.

3. CSC

Figure 3 shows a requirements description composed of two state charts A and B. The variable t marks time constraints so they can be distinguished from other constraints like signal evaluations. There is no need of timing constraints. A relation without a constraint allows any point in time for the relation to become true. The different types of relations and the meaning of the time constraints connected are subsumed in Figure 4 and in Definitions 2.4 and 2.5. The timelines have to be read from top to bottom. A single location within a timeline is directly linked with the same location in the original state chart.

The computation sequence charts are automatically transformed into temporal logic formulae which can be used to generate test cases based on model-checking algorithms or can be compared with already generated test cases. The translation of the computation sequence chart and there contribution to the model-checking techniques is described later.

The plug-in software supports name modifications and deletion of original locations by pointing out all requirement descriptions that are under investigation. Name modifications are resolved automatically. Deleted locations of state-transition graphs need to be interpreted. Deleting the location

Figure 3. Computation Sequence Chart (CSC)

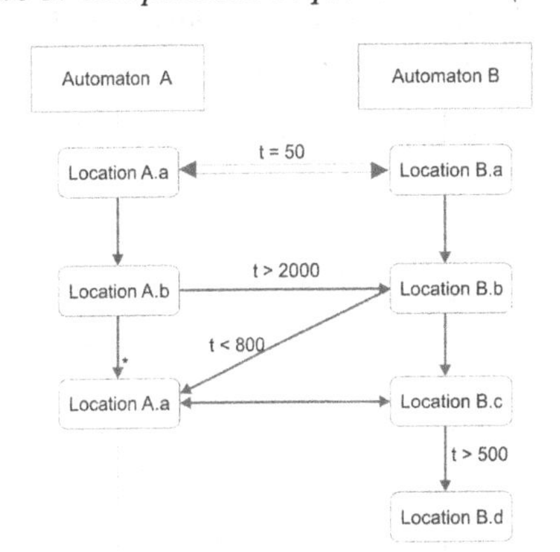

in all CSCs referring to it automatically leaves relations having either no target or no source location. Additionally, a deleted location might change the content of a trace and the requirements description should be adapted manually.

4. ELEMENTS

Since sequential as well as parallel relations allow time intervals as boundaries, there might be a potential timing problem. Two or more relations might exclude each other, as can be seen in Figure 4. The sequential relation with guard [t < 800] states that Location A.a must be left before 800 time units have passed. The parallel relation with guard [t > 2000] on the other hand forces automaton A to stay in Location A.a for at least 2000 time units.

Therefore a decision about solvability should be made. The decision is based on a system of inequalities. The inequalities express all timing conditions of pairwise locations.

5. CONFLICTS

Selecting the parallel relation of *A.a* and *B.a* in Figure 5, we get *timetick(B.a) – timetick(A.a) > 2000* with *timetick(location)* giving the point

Figure 4. Elements of the graphical notation of a computation sequence chart

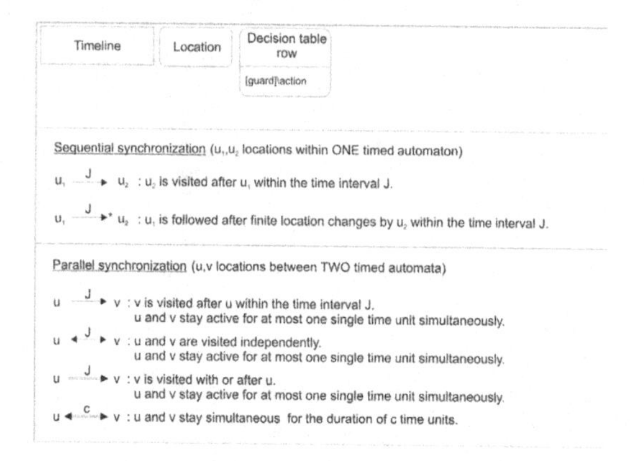

Figure 5. Computation sequence chart with conflicting relations

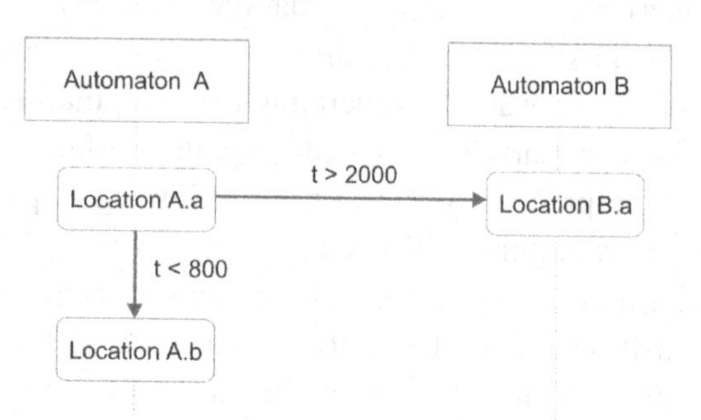

of time the location is entered. The sequential relation leads to *timetick(A.b) – timetick(A.a) < 800.* Additionally *timetick(B.a) < timetick(A.b)* is given. Generating all such relation inequalities, a system of inequalities is obtained, whose existence of a solution can be decided using the Fourier-Motzkin elimination (Bik & Wijshoff, 1995). This is handled within the transformation algorithm of computation sequence chart.

Advantages of the graphical requirement descriptions are the readability, the easier way of creation, and the supported maintenance. Content and changes of requirements lead directly to the affected test model elements via requirements descriptions. Content and changes inside the model lead directly to the requirements addressed.

An Example

The different types of flashing of a vehicle have been specified in a test model. There is normal flashing, tip flashing, emergency flashing, theft flashing and many others. The test model consists of approximately 50 parallel statecharts.

6. NORMALFLASHING

The statecharts are two examples of this test model. They describe the logic behind normal flashing combined with tip flashing (Figure 6) und emergency flashing (Figure 7).

Figure 6. Statechart of the normal flashing behavior

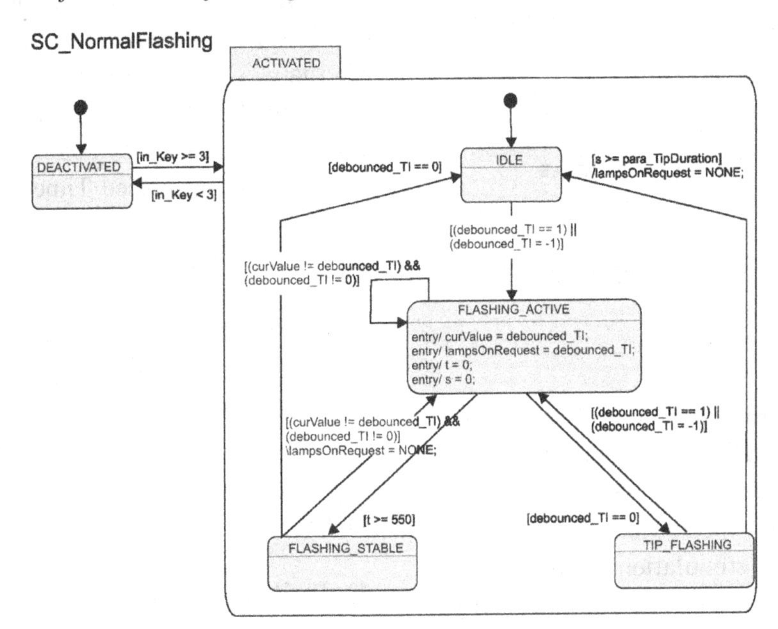

Figure 7. Statechart of the normal flashing behavior

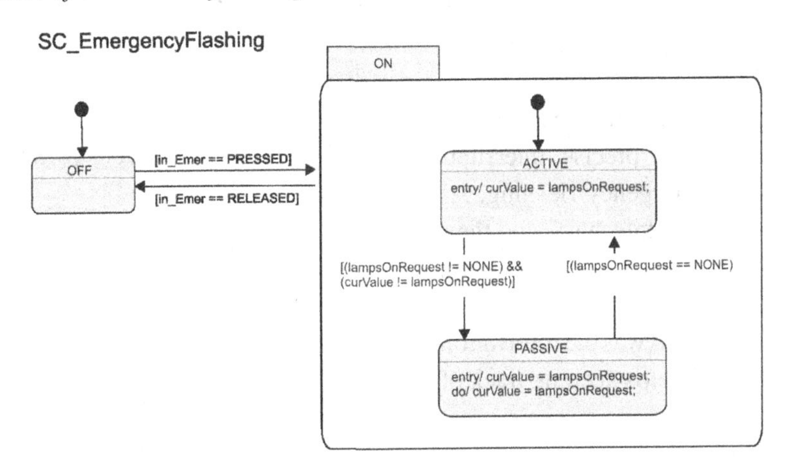

7. EMERGENCYFLASHING

One requirement of the flashing system says "Normal flashing can be interrupted by emergency flashing". The two statecharts SC_NormalFlashing and SC_EmergencyFlashing are therefore related with each other. To be able to generate a test case testing exactly the behavior specified in the requirement, the following CSC was created.

8. INTERRUPTION

The CSC in Figure 8 describes the situation where the system is in the location SC_NormalFlashing. FLASHING_STABLE saying the system behaves like normal flashing. At this point the start of emergency flashing is needed. So after at least one flashing cycle of 550ms, emergency flashing starts by changing to SC_EmergencyFlashing. ACTIVE. Whether the system behaves correctly

Figure 8. CSC of the requirement "Normal flashing can be interrupted by emergency flashing"

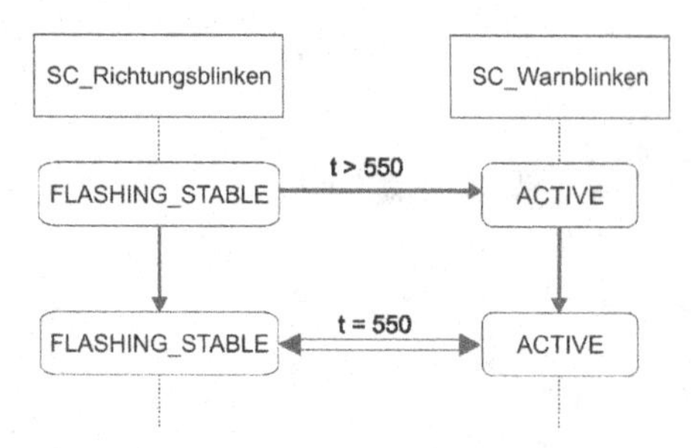

by interrupting normal flashing is not part of a CSC as the expected behavior is specified in the test model. A test case generated based on the test model has the expected behavior included automatically. The stimulation needed to see normal flashing or emergency flashing will be calculated by the test case generator. That is the reason why a CSC does not need to describe the stimulation of a system. The possibilities to stimulate the system are also part of the test model. The only content of a CSC is one path fragment enabling the test engineer to observe the behavior of a single requirement, to be precise, interrupting normal flashing by emergency flashing. As the test engineer knows the statecharts and their dependencies, usually because he defined them, he can tell which path fragments relate to which requirement. The requirement was the motivation for the statecharts. A CSC helps keeping track of this information.

Temporal Logic Formulae

The CSC helps understanding the intention of a single requirement as a sequence of test model elements. To associate test cases with requirements or run test case generation based on requirements, we need a formal representation of a computation sequence chart. Thus, to benefit from all advantages of model-checking algorithms in generating test cases, the CSC needs to be translated to characteristics of paths in the computation tree. Best practice is to translate the CSC to a temporal logic. Because a requirement not only describes one linear computation but most of the time also alternatives, the branching time logic would be best (Pnueli, 1977). In CTL notation an explicit time condition is not available. This is the main reason Alur introduced Timed CTL (TCTL) in 1990 (Alur, 1990). TCTL extends CTL by allowing time intervals for specifying hard real-time properties.

TCTL with the operators *Until*, *Finally,* and *Globally* has the following syntax and semantic. Let X be a set of clocks and $AC(X)$ the set of clock constraints true in either a state or a transition of the timed automaton A with respect to the clocks in X (Baier & Katoen, 2008).

Definition 2.2. Syntax of Timed CTL

The grammar of TCTL state formulae is formed as follows:

$$\Phi ::= true \mid a \mid g \mid \Phi_1 \wedge \Phi_2 \mid \neg\Phi \mid \exists\phi \mid \forall\phi$$

with $a \in AP$, $g \in AC(X)$ and ϕ a path formula.

The grammar of TCTL path formulae is defined as follows:

$$\phi ::= \Phi_1 \; Until^J \; \Phi_2$$

$J \subseteq R_{\geq 0}$ is a natural bounded time interval.

Definition 2.3. Satisfaction Relation of TCTL

Let $A = (\Sigma, L, L_0, X, I, T)$ be a timed automaton, $a \in AP$ an atomic proposition, $g \in AC(X)$ and $J \subseteq R_{\geq 0}$ a natural bounded time interval. For state s = $\langle l, \eta \rangle$ in the state graph T(A), let Φ, Ψ be TCTL state formulae and ϕ a TCTL path formula. The satisfaction relation $\models$ is defined as follows:

$s \models true$

$s \models a$ iff $a \in L(l)$

$s \models g$ iff $\eta \in g$

$s \models \neg\Phi$ iff not $s \models \Phi$

$s \models \Phi \wedge \Psi$ iff $(s \models \Phi)$ and $(s \models \Psi)$

$s \models \exists\phi$ iff $\pi \models \phi$ for some $\pi \in Paths_{div}(s)$

$s \models \forall\phi$ iff $\pi \models \phi$ for all $\pi \in Paths_{div}(s)$

$\pi \models \Phi\, Until^{J}\, \Psi$ iff $\exists i \geq 0. s_i + d \models \Psi$ for some $d \in [0, d_i]$ with

$\sum_{k=0}^{i-1} d_k + d \in J$ and

$\exists j \leq i. s_j + d' \models \Phi \vee \Psi$ for any $d' \in [0, d_j]$ with

$\sum_{k=0}^{j-1} d_k + d' \leq \sum_{k=0}^{i-1} d_k + d$

$\pi \models Finally^{J}\Phi$ iff $\exists i \geq 0. s_i + d \models \Phi$ for some $d \in [0, d_i]$ with $\sum_{k=0}^{i-1} d_k + d \in J$.

$\pi \models Globally^{J}\Phi$ iff $\forall i \geq 0. s_i + d \models \Phi$ for any $d \in [0, d_i]$ with $\sum_{k=0}^{i-1} d_k + d \in J$.

where $\pi \in s_0 \overset{d_0}{\Rightarrow} s_1 \overset{d_1}{\Rightarrow} \ldots$ is a time-divergent path and for $s_i = \langle l_i, \eta_i \rangle$ and $d \geq 0$ we have $s_i + d = \langle l_i, \eta_i + d \rangle$.

As *time-convergent* paths are not realistic the satisfaction relation of TCTL is defined only on *time-divergent* paths. Time-convergent paths would assume time progresses up to a certain value and not until infinity.

In the case of $J = [0, \infty)$, we set $\Phi\, Until^{[0,\infty)}\, \Psi = \Phi\, Until\, \Psi$ and $Finally^{[0,\infty)}\, \Phi = Finally\, \Phi$ and $Globally^{[0,\infty)}\, \Phi = Globally\, \Phi$.

Remark 2.1. The Until Operator in TCTL

Comparing the Until operator of CTL and TCTL highlights one important difference. In CTL the Until operator $\Phi\, Until\, \Psi$ holds when $\neg\Psi \wedge \Phi$ holds at all previous steps of the first occurrence of Ψ. In opposition to CTL, the Until operator in TCTL holds, when $\Phi \vee \Psi$ holds in all previous time instants of the first occurrence of Ψ at some point in J.. For example, the requirement "stay in location A for at least 500 time units until changing to location B" cannot be written as $A\ Until^{\geq 500} B$. In this case, B could already be visited before 500 time units.

Example

"Once a train is *far*, within 1 minute the gate is *up* for at least 1 minute". This requirement can be expressed by the TCTL formula:

$$\forall Globally\,(far \rightarrow \forall Finally^{\leq 1}\ \forall Globally^{\leq 1}\ up)$$

Definition 2.4. TCTL Formulae for CSCs (One Timed Automaton)

Let A $= (\Sigma, L, L_0, X, I, T)$ be a timed automaton, $u_1, u_2 \in L$ locations. $c \in N$ a clock. We define the sequential dependencies of locations within a requirements description as the following state formulae:

$$u_1 \overset{>c}{\longrightarrow} u_2 : \Phi = \forall Globally$$
$$(u_1 \rightarrow ((\exists Globally^{\leq c} u_1) \wedge \exists (u_1\ Until^{>c} u_2)))$$

$$u_1 \overset{\geq c}{\rightarrow} u_2 : \Phi = \forall Globally\ (u_1 \rightarrow ((\exists Globally^{<c} u_1) \wedge \exists(u_1\ Until^{\geq c} u_2)))$$

$$u_1 \overset{<c}{\rightarrow} u_2 : \Phi = \forall Globally(u_1 \rightarrow \exists(u_1 Until^{<c} u_2))$$

$$u_1 \overset{\leq c}{\rightarrow} u_2 : \Phi = \forall Globally(u_1 \rightarrow \exists(u_1 Until^{\leq c} u_2))$$

$$u_1 \rightarrow u_2 : \Phi = \forall Globally\ (u_1 \rightarrow \exists(u_1 Until\ u_2))$$

$$u_1 \overset{>c}{\rightarrow}\!* u_2 : \Phi = \forall Globally\ (u_1 \rightarrow ((\exists Globally^{\leq c} \neg u_2) \wedge (\exists(Finally^{>c} u_2)))$$

$$u_1 \overset{\geq c}{\rightarrow}\!* u_2 : \Phi = \forall Globally\ (u_1 \rightarrow ((\exists Globally^{<c} \neg u_2) \wedge \exists(Finally^{\geq c} u_2)))$$

$$u_1 \overset{<c}{\rightarrow}\!* u_2 : \Phi = \forall Globally(u_1 \rightarrow \exists Finally^{<c} u_2)$$

$$u_1 \overset{\leq c}{\rightarrow}\!* u_2 : \Phi = \forall Globally(u_1 \rightarrow \exists Finally^{\leq c} u_2)$$

$$u_1 \rightarrow\!* u_2 : \Phi = \forall Globally\ (u_1 \rightarrow \exists Finally\ u_2)$$

In words, $u_1 \overset{\geq 500}{\rightarrow} u_2$ expresses that location u_1 will hold at least 500 time units until u_2 becomes true. There is no other location allowed during the location change. Considering remark 2.1, the TCTL formula had to ensure that location u_2 cannot become true before location u_1, which can not be achieved by using the *Until* operator alone. Therefore a subformula had to be constructed, that separates the time interval into two parts, the one $<$ ($\leq$) 500 time units, where u_1 has to stay true *Globally* and the time interval $\geq$ ($>$) where u_1 *until* u_2 can be applied. If intermediate location changes are allowed, the one-pointed arrow with * should be used. In this case, we do not need to observe location u_1. It does not matter whether u_1 will be left immediately after its first occurrence or whether the automaton stays in u_1 until u_2.

Definition 2.5. TCTL Formulae for CSCs (Two Timed Automata)

Let $A_1 = (\Sigma_1, L_1, L_0^1, X_1, I_1, T_1)$ and $A_2 = (\Sigma_2, L_2, L_0^2, X_2, I_2, T_2)$ be two parallel timed automata, $u \in L_1, v \in L_2$ locations. $c \in N$ a clock. We define the parallel dependencies of locations within a requirements description as the following state formulae:

$$u \overset{>c}{\rightarrow} v : \Phi = \forall Globally\ ((u \wedge \neg v) \rightarrow ((\exists Globally^{\leq c}(u \wedge \neg v)) \wedge \exists(u\ Until^{>c}\ (u \wedge v))))$$

$$u \overset{\geq c}{\rightarrow} v : \Phi = \forall Globally\ ((u \wedge \neg v) \rightarrow ((\exists Globally^{<c}(u \wedge \neg v)) \wedge \exists(u\ Until^{\geq c}\ (u \wedge v))))$$

$$u \overset{<c}{\rightarrow} v : \Phi = \forall Globally\ ((u \wedge \neg v) \rightarrow \exists(u\ Until^{<c}\ (u \wedge v)))$$

$$u \overset{\leq c}{\rightarrow} v : \Phi = \forall Globally \\ ((u \wedge \neg v) \rightarrow \exists(u\ Until^{\leq c}\ (u \wedge v)))$$

$$u \rightarrow v : \Phi = \forall Globally \\ ((u \wedge \neg v) \rightarrow \exists(u\ Until\ (u \wedge v)))$$

$$u \overset{>c}{\leftrightarrow} v : \Phi = \forall Globally \\ ((u \rightarrow \exists(u\ Until^{>c}\ (u \wedge v))) \\ \vee(v \rightarrow \exists(u\ Until^{>c}\ (u \wedge v))))$$

$$u \overset{\geq c}{\leftrightarrow} v : \Phi = \forall Globally \\ ((u \rightarrow \exists(u\ Until^{\geq c}\ (u \wedge v))) \\ \vee(v \rightarrow \exists(u\ Until^{\geq c}\ (u \wedge v))))$$

$$u \overset{<c}{\leftrightarrow} v : \Phi = \forall Globally \\ ((u \rightarrow \exists(u\ Until^{<c}\ (u \wedge v))) \\ \vee(v \rightarrow \exists(u\ Until^{<c}\ (u \wedge v))))$$

$$u \overset{\leq c}{\leftrightarrow} v : \Phi = \forall Globally \\ ((u \rightarrow \exists(u\ Until^{\leq c}\ (u \wedge v))) \\ \vee(v \rightarrow \exists(u\ Until^{\leq c}\ (u \wedge v))))$$

$$u \leftrightarrow v : \Phi = \forall Globally\ (\exists Finally\ (u \wedge v))$$

$$u \overset{>c}{\Rightarrow} v : \Phi = \forall Globally \\ (u \rightarrow ((\exists Globally^{\leq c}(u \wedge \neg v)) \\ \wedge(\exists(u\ Until^{>c}(u \wedge v))))$$

$$u \overset{\geq c}{\Rightarrow} v : \Phi = \forall Globally \\ (u \rightarrow ((\exists Globally^{<c}(u \wedge \neg v)) \\ \wedge(\exists(u\ Until^{\geq c}(u \wedge v)))$$

$$u \overset{<c}{\Rightarrow} v : \Phi = \forall Globally(u \rightarrow \exists(u\ Until^{<c}(u \wedge v)))$$

$$u \overset{\leq c}{\Rightarrow} v : \Phi = \forall Globally(u \rightarrow \exists(u\ Until^{\leq c}(u \wedge v)))$$

$$u \Rightarrow v : \Phi = \forall Globally \\ (u \rightarrow \exists(u\ Until\ (u \wedge v)))$$

$$u_1 \overset{c}{\Leftrightarrow} u_2 : \Phi = \forall Globally \\ ((u \wedge v) \rightarrow \exists Globally^{\leq c}\ (u \wedge v))$$

Sometimes, it is important that location changes correspond to specific valuations of variables and signals, for example when there are two possible transitions between two locations that need to be differentiated. Therefore we included the possibility of adding a simple logical expression on atomic propositions to any kind of dependency, e.g.

$$u \underset{(signal==0)\&\&(x>=5)}{\overset{\leq 500}{\rightarrow}} v : \Phi = \forall Globally \\ ((u \wedge \neg v) \rightarrow \exists(u\ Until^{\leq 500} \\ (u \wedge v \wedge (signal == 0) \wedge (x \geq 5))))$$

Based on the subformula templates defined in Definitions 2.4 and 2.5 the graphical representation of a requirements description can be transformed into a TCTL formula. An algorithm starting with the first two locations inside two timelines (called InterTimelines) traverses the requirements description stepwise until the end of both timelines and the formula is composed recursively. The transformation algorithm involves the steps shown in Table 1.

The Main() function is the entry point of the transformation algorithm. The first two locations are selected. Depending on whether there is a

Table 1. Main

for all *InterTimelines* **do**
GetFirstLocations()
if *ExistsParallelTransition()* **then**
if *SourceEquals(LocationOne)* **then**
RecursiveGeneration(locationOne, locationTwo)
else
RecursiveGeneration(locationTwo, locationOne)
end if
else
RecursiveGeneration(locationOne, locationTwo)
end if
end for
if *tctlFormula != null* **then**
$\{\exists Finally((startFormula) \wedge (tctlFormula))\}$
end if

Table 2. RecursiveGeneration

if *ExistsParallelRelation ()* **then**
if *!parRel.AlreadyProcessed()* **then**
Φ_1 = ComputeParRelation()
end if
else
$\{(\exists Finally\ (locationOne)) \wedge (\exists Finally\ (locationTwo))\}$
end if
if *locationOne.HasMultipleParallelRelations()* **then**
Ψ_1 = ComputeMultRelations()
end if
if *locationTwo.HasMultipleParallelRelations()* **then**
Ψ_2 = ComputeMultRelations()
end if
Φ_1 = CombineMultRelations(Ψ_1, Ψ_2)
if ! *locationOne.seqRel.AlreadyProcessed()* **then**
Φ_2 = ComputeSeqRelation()
end if
if ! *locationTwo.seqRel.AlreadyProcessed()* **then**
Φ_3 = ComputeSeqRelation()
end if
CombineSubformulae(Φ_1, Φ_2, Φ_3)
GetNextLocation()
if *!locationOne.IsLast() && !locationTwo.IsLast()* **then**
if *ExistsParallelRelation()* **then**
if *SourceEquals(LocationOne)* **then**
RecursiveGeneration(locationOne, locationTwo)
else
RecursiveGeneration(locationTwo, locationOne)
end if
else
RecursiveGeneration(locationOne,location)
end if
end if

parallel synchronization between them and which location is the source of the relation, the function RecursiveGeneration() gets called. At the end, the recursively generated tctlFormula is composed with the startFormula of the first two locations. The reason behind that is discussed after the function RecursiveGeneration() shown in Table 2.

The functions *ComputeParRelation()* and *ComputeSeqRelation()* build the TCTL formulae. The function to resolve multiple parallel relations *ComputeMultRelations()* shall be skipped in this paper.

CombineSubformulae (Φ_1, Φ_2, Φ_3) extends the tctlFormula. Depending on the existence of Φ_1, Φ_2 or Φ_3 we get $tctlFormula = tctlFormula \rightarrow (\Phi_1 \rightarrow (\Phi_2 \wedge \Phi_3))$. The *Implies* operator is a link between the parallel and sequential dependencies and forms the sequence of location changes during a requirements description.

Note that in a formula like $\Phi \rightarrow \Psi$, Φ might never become true and still the formula is satisfied. We do need Φ to become true to actually see the requirement during a test case. The first two locations of a parallel system of two timelines are required. From there the sequence of location changes is given. That is why at the end of Main() we define $\exists Finally\ ((startFormula) \wedge (tctlFormula))$ stating that at some point of time the startFormula composed of the first two states and at the same time the implication starting with startFormula have to become true. This makes the formula flexible. It is not important when during a test run the requirement has to be satisfied, nevertheless is has to be done at some point.

Test Case Generation Using Computation Sequence Charts

The size of a test model specifying an industrial system allows two different ways of model-checking to generate test cases; model-checking on binary decision diagrams and bounded model-checking. The following section shows, how the resulting temporal logic formulae of a computation sequence chart are considered in the model-checking techniques.

For model-checking on binary decision diagrams the process is the following (Baier & Katoen, 2008):

- The TCTL formula of the CSC will be rewritten in existential normal form.
- The syntax tree of the formula will be expanded.
- Starting at the leaves of the tree, the satisfiability sets $Sat(\phi_i)$ will be constructed. To eliminate the timing parameters of the TCTL formula, the timed automaton is extended with a new clock z. As soon as $Sat(\phi_i)$ is calculated, the clock z can be reused to generate the other satisfiability sets. Therefore, the state space is only extended by one additional clock.
- If the initial set of states of the timed automaton is a subset of the satisfiability set $Sat(\phi)$, the state-transition system satisfies the TCTL specification of the CSC.

Depending on the size of the system, explicit or symbolic model-checking can be used to calculate the satisfiability sets. For small industrial model sizes, symbolic model-checking on binary decision diagrams proves sufficient (Clarke et al., 2001).

A case study in the automotive domain showed, that these industrial test model sizes could only be handled with advanced bounded model-checking techniques. There are two alternatives representing the CSC when choosing bounded model-checking as test case generation algorithm. The first approach uses the TCTL representation of the CSC, transforms it to a CTL formula which then gets rewritten in first order logic. The second approach directly represents the CSC as first order logic formula. If one would want to be flexible on the model-checking algorithms, an intermediate representation as TCTL formula would be useful. Otherwise directly representing the TCTL formula in first order logic is more efficient. The transformation from TCTL to first order logic is part of the dissertation of the author appearing in 2011. Having the first order representation P of the CSC in conjunctive form, it becomes part of the bounded model-checking instance (BMC-instance).

$$[[M,P]] = \bigwedge_{j=0}^{c-1} R(i(k_0+j), s(k_0+j), s(k_0+j+1))$$
$$\wedge P(i(k_0), s(k_0), o(k_0), \ldots, i(k_0+c), s(k_0+c), o(k_0+c)).$$

M is the test model. R is its transition relation, i is the input vector, s is the state vector and o is the output vector of the test model M.

For keeping the complexity of the BMC-instance as low as possible, a stepwise generation should be followed. How this can be done is described in Peleska (2010).

RELATED WORK

The graphical notation of the requirements description is comparable to scenario-based visual languages like Live Sequence Charts (Harel & Marelly, 2002), Message Sequence Charts (Recommendation Z.120), Symbolic Timing Diagrams (Schloer, 2001) or Visual Timed event Scenarios (Braberman et al., 2005). Other approaches follow the same aim of supporting the specification of system requirements by introducing a TimeLine Editor helping to draw events and constraints (Smith et al., 2001) or defining a set of commonly used temporal logic patterns. Some of those solutions can be reused easily (Dwyer et al., 1999). All approaches suggest graphical notations that include underlying formal representations so that model-checking techniques can be performed. They differ in the way the system executions are regarded (e.g., interleaving or partial-order semantics) and in the power and amount of logics supported (e.g., CTL, LTL, etc.). Additionally the original motivation behind the definition of the graphical notation reflects their complexity. Some approaches start with the needs of engineers, others with the formal expressiveness that shall be realizable.

Message Sequence Charts (MSC) and the closely related UML Sequence Charts are widely accepted. Their motivation is driven by the need of usability and efficiency in the daily routine of engineers. The result is an easy to learn visual notation. Applying temporal logics such as TLC (Alur et al., 1995) to MSCs is possible (Muscholl, Peled, & Su, 2005). Defining a new visual notation and not adopting MSCs is motivated by the excessive expressiveness of MSCs that is not required for requirements traceability.

Live Sequence Charts (LSC) are closely related to Message Sequence Charts (MSC) and Computation Sequence Charts (CSC). LSCs extend MSCs by adding the possibilities to differentiate between existential and universal behavior as well as to define the progress of LSC instances as *hot*, saying an instance has to leave its location or as *cold*, stating an instance may remain in its location. The instances of LSCs or MSCs are activities or objects. They belong to the so called *inter-object* specifications which extend the *intra-object* specifications, e.g. statecharts, of a system. Inter-object specifications describe the relations and communication between instances of a system. Intra-object specifications define the behavior of the entire system without referring to instances implementing the behavior. CSCs are comparable with LSCs in the sense that they define a sequential order of locations of a system but they are no inter-object specification as they do not describe relations or communication between instances. An instance of a CSC is a single statechart. A CSC is no additional specification of the system. It defines a single computation of the system to achieve traceability of a single requirement and requirement-based test cases. Thus, a CSC is a sequential intra-object specification. The possibility of LSCs to distinct between universal und existential behavior as well as hot and cold locations is not yet supported in CSCs but is part of the future work.

SUMMARY AND CONCLUSION

The requirements traceability approach presented in this paper offers an easy notation to trace a requirement throughout the model based-testing process. The notation only consists of already existing test model elements and six relations for synchronization. The requirements descriptions are easy to create. They help understand the test cases that were generated. They link the requirements in natural language with test model elements and vice versa. By that, reuse and maintenance is supported. They build a good basis to discuss functionality with other testers and developers.

The transformation of the Computation Sequence Chart to TCTL formulae, which is also discussed in this paper, offers the whole power of generating requirements-based test cases using model-checking algorithm. The interpretation of each synchronization relation as TCTL template and transformation rules on CSCs result in an algorithm to generate a TCTL formula representing the requirement. As test case generation is always only a selection of the amount of possible test cases, a guided reference to the original requirements helps ensuring that at least some realistic test cases are created. Requirements-based test cases help building up the confidence in model-based testing techniques as the test engineer recognizes the test cases that otherwise would have been derived manually.

At the beginning of 2010, Daimler, Inc. started a case study investigating not only the efficiency of model-based testing but also the improvement of fault finding by using Computation Sequence Charts additionally to existing structural model coverage like transition or state coverage. The first attempt of the study annotated transitions to achieve requirements traceability. It showed that not all requirements could be traced (examples can be found in chapter *Introduction*). The second approach tried to use temporal logic to specify single requirements directly. The results showed, that this work is very time consuming and error-prone. Different test engineers specified requirements differently so all formulas had to be reviewed by a specialist on temporal logic. As creating the test model already takes a lot of time this second time-consuming task would be too inefficient. The graphical notations that can be transformed into temporal logics as described in chapter *Related work* did not reuse the already existing elements and paths of the test model and therefore would have generated additional effort of specifying them. These arguments lead to the definition of computation sequence charts. They reuse the test model elements and the motivation behind the test model creation. As the composition of the CSC can be performed inside the same case tool the test model was created in (in the case study this is a realized via a plug-in solution of Enterprise Architect, see chapter *Graphical Notation*), the work-flow of the CSC creation can be performed in parallel to the test model creation. Still, creating the CSC takes more time than just annotating transitions with requirement labels. The time investigated can only be argued if more faults can be found using CSCs for test case generation. Anyway, safety and quality norm demand at least one test case per requirement. So, if the requirement can not be traced only by annotating single transitions, this kind of traceability is not good enough. The generation of test cases based on CSC is implemented by a German company developing test case generators. The measurement of the efficiency of the test case generator is part of the case study.

Future work investigates three main issues. First, it focuses on the combination of the pairwise formulae of more than two timelines via the *and* operator and checks whether it leads to the intended system behavior. Second, as one test case per requirement is not a satisfying definition of requirements coverage the methods of defining coverage criteria on requirements descriptions to achieve meaningful requirements coverage are investigated. The third issue comprises the efficiency of TCTL language, as now with the predefined formula templates its entire power is not leveraged yet. In particular, such universal properties as, for example, "there is never a critical situation" cannot be specified at present. The best way of handling universal requirements within the graphical notation is still under development.

REFERENCES

Aizenbud-Reshef, N., Nolan, B., Rubin, J., & Shaham-Gafni, Y. (2006). Model traceability. *IBM Systems Journal*, *45*(3), 515–526. doi:10.1147/sj.453.0515

Alur, R. (1999). Timed automata. In N. Halbwachs & D. Peled (Eds.), *Proceedings of the 11th International Conference on Computer Aided Verification* (LNCS 1633, pp. 8-22).

Alur, R., Courcoubetis, C., & Dill, D. (1990). Model-checking for real-time systems. In *Proceedings of the 5th IEEE Annual Symposium on Logic in Computer Science*, Philadelphia, PA (pp. 414-425). Washington, DC: IEEE Computer Society.

Alur, R., & Henzinger, T. A. (1992). Logics and models of real time: A survey. In J. W. de Bakker, C. Huizing, W. P. de Roever, & G. Rozenberg (Eds.), *Proceedings of the Workshop on Real-Time: Theory in Practice* (LNCS 600, pp. 74-106).

Alur, R., Peled, D., & Penczek, W. (1995). Model-checking of causality properties. In *Proceedings of the 10th IEEE Symposium on Logic in Computer Science*, San Diego, CA (pp. 90-100). Washington, DC: IEEE Computer Society.

Baier, C., & Katoen, J. P. (2008). *Principles of model checking*. Cambridge, MA: MIT Press.

Bik, A. J. C., & Wijshoff, H. A. G. (1995). Implementation of Fourier-Motzkin elimination. In *Proceedings of the 1st Annual Conference of the Advanced School for Computing and Imaging* (pp. 377-386).

Braberman, V., Kicillof, N., & Olivero, A. (2005). A scenario-matching approach to the description and model checking of real-time properties. *IEEE Transactions on Software Engineering, 31*, 1028–1041. doi:10.1109/TSE.2005.131

Clarke, E., Biere, A., Raimi, R., & Zhu, Y. (2001). Bounded model checking using satisfiability solving. *Formal Methods in System Design, 19*(1), 7–34. doi:10.1023/A:1011276507260

Clarke, E. M., Grumberg, O., & Peled, D. A. (2000). *Model checking*. Cambridge, MA: MIT Press.

Dwyer, M. B., Avrunin, G. S., & Corbett, J. C. (1999). Patterns in property specifications for finite-state verification. In *Proceedings of the 21st International Conference on Software Engineering* (pp. 411-420). New York, NY: ACM Press.

Egyed, A., & Gruenbacher, P. (2002). Automating requirements traceability: Beyond the record & replay paradigm. In *Proceedings of the 17th IEEE International Conference on Automated Software Engineering* (pp. 163-171). Washington, DC: IEEE Computer Society.

Fraser, G., & Wotawa, F. (2008). Using model-checkers to generate and analyze property relevant test-cases. *Software Quality Journal, 16*, 161–183. doi:10.1007/s11219-007-9031-6

Gaston, C., & Seifert, D. (2005). Evaluating coverage based testing. In M. Broy, B. Jonsson, J.-P. Katoen, M. Leucker, & A. Pretschner (Eds.), *Proceedings of the International Conference on Model-Based Testing of Reactive Systems* (LNCS 3472, pp. 293-322).

Harel, D., & Marelly, R. (2002). Playing with time: On the specification and execution of time-enriched LSCs. In *Proceedings of the 10th IEEE International Symposium on Modeling, Analysis, and Simulation of Computer and Telecommunications Systems* (pp. 193-202). Washington, DC: IEEE Computer Society.

ITU. (1999). *Series Z: Languages and general software aspects for telecommunication systems: Formal Description Techniques (FDT) - Message Sequence Chart.* Retrieved from http://www.itu.int/ITU-T/studygroups/com17/languages/Z120.pdf

Muscholl, A., Peled, D., & Su, Z. (2005). Deciding properties of message sequence charts. In M. Nivat (Ed.), *Proceedings of the 1st International Conference on Foundations of Software Science and Computation Structure* (LNCS 1378, pp. 226-242).

Naslavsky, L., Alspaugh, T. A., Richardson, D. J., & Ziv, H. (2005). Using scenarios to support traceability. In *Proceedings of the 3rd International Workshop on Traceability in Emerging Forms of Software Engineering* (pp. 25-30). New York, NY: ACM Press.

Naslavsky, L., Ziv, H., & Richardson, D. J. (2007). Towards traceability of model-based testing artefacts. In *Proceedings of the 3rd International Workshop on Advances in Model-Based Testing* (pp. 105-114). New York, NY: ACM Press.

Peleska, J. (2009). The automation problems and their reduction to bounded model checking. In Drechsler, R. (Ed.), *Model-based testing of embedded control systems in the railway, avionic and automotive domain*. [Proceedings of the TuZ, 21 Workshop für Testmethoden und Zuverlässigkeit von Schaltungen und Systemen]

Pfaller, C. (2008). Requirements-based test case specification by using information from model construction. In *Proceedings of the 3rd International Workshop on Automation of Software Test* (pp. 7-16). New York, NY: ACM Press.

Pnueli, A. (1977). The temporal logic of programs. In *Proceedings of the 18th IEEE Annual Symposium on Foundations of Computer Science* (pp. 46-57). Washington, DC: IEEE Computer Society.

Ryser, J., Berner, S., & Glinz, M. (1998). *On the state of the art in requirements-based validation and test of software* (Tech. Rep. No. 12). Zurich, Switzerland: University of Zurich.

Schloer, R. (2001). *Symbolic timing diagrams: A visual formalism for model verification*. Unpublished doctoral dissertation, University of Oldenburg, Germany.

Schmid, H. (2008). *Hardware-in-the-loop Technologie: Quo Vadis?* [Tagungsband Simulation und Test in der Funktions- und Softwareentwicklung für die Automobilelektronik]. (pp. 195–202). Stuttgart, Germany: Verlag.

Smith, M. H., Holzmann, G. J., & Etessami, K. (2001). Events and constraints: A graphical editor for capturing logic requirements of programs. In *Proceedings of the 5th IEEE International Symposium on Requirements Engineering* (pp. 14-22). Washington, DC: IEEE Computer Society.

Tan, L., Sokolsky, O., & Lee, I. (2004). Specification-based testing with linear temporal logic. In *Proceedings of the IEEE International Conference on Information Reuse and Integration* (pp. 483-498). Washington, DC: IEEE Computer Society.

Utting, M., & Legeard, B. (2007). *Practical model-based testing*. San Francisco, CA: Morgan Kaufmann.

Zisman, A., Spanoudakis, G., Perez-Miana, E., & Krause, P. (2003). Tracing software requirements artefacts. In *Proceedings of the International Conference on Software Engineering Research and Practise*, *1*, 448–455.

This work was previously published in the International Journal of Embedded and Real-Time Communication Systems, Volume 2, Issue 2, edited by Seppo Virtanen, pp. 1-21, copyright 2011 by IGI Publishing (an imprint of IGI Global).

Chapter 6
Model-Based Testing of Highly Configurable Embedded Systems in the Automation Domain

Detlef Streitferdt
Ilmenau University of Technology, Germany

Florian Kantz
ABB Corporate Research, Germany

Philipp Nenninger
ABB Corporate Research, Germany

Thomas Ruschival
ABB Corporate Research, Germany

Holger Kaul
ABB Corporate Research, Germany

Thomas Bauer
Fraunhofer IESE, Germany

Tanvir Hussain
Fraunhofer IESE, Germany

Robert Eschbach
Fraunhofer IESE, Germany

ABSTRACT

This article reports the results of an industrial case study demonstrating the efficacy of a model-based testing process in assuring the quality of highly configurable systems from the automation domain. Escalating demand for flexibility has made modern embedded software systems highly configurable. This configurability is often realized through parameters and a highly configurable system possesses a handful of those. Small changes in parameter values can account for significant changes in the system's behavior, whereas in other cases, changed parameters may not result in any perceivable reaction. This case study addresses the challenge of applying model-based testing to configurable embedded software systems to reduce development effort. As a result of the case study, a model-based testing process was developed and tailored toward the needs of the automation domain. This process integrates existing model-based testing methods and tools, such as combinatorial design and constraint processing. The testing process was applied as part of the case study and analyzed in terms of its actual saving potentials, which reduced the testing effort by more than a third.

DOI: 10.4018/978-1-4666-2776-5.ch006

INTRODUCTION

In the automation domain, large and complex systems like chemical or power plants are common practice. The products of these plants are part of our daily lives, and our living standard depends directly on their reliable supply. This dependency accounts for the *high quality* requirements for these plants, which adds to the burden of voluminous costs for engineering and operation. Of course, such *high quality* is required for almost all of the components of a plant in order to ensure the proper functioning up to the point that even certain failures should not lead to unbearable consequences. On the upper level, *control systems* based on workstation platforms (e.g. Microsoft Windows®) are used to control the overall function of the plant, for example the generation of energy in a power plant. Between the control system layer and the lowest sensor and actuator level, several layers of embedded systems of varying complexity are used to collect and pass on sensor data (like temperature or pressure values), monitor the proper function of plant sub modules and actuate upon requests from the upper level *control system* (e.g., close a valve).

The main challenge in the application of model-based testing for embedded systems is their simple behavior visible from the outside, which internally gets dramatically complex due to configurable parameters. Each system has many parameters and within this system, a *configuration* is a set of parameters with concrete values selected for each parameter. Such configurations are intended for various purposes, for example for dealing with different modes of operation, different types of user interactions, error and exception handling etc. Different kinds of system behavior are directly related to configurations and as a result, the verification of the system is cumbersome and difficult as the number of available configurations rises.

This article presents the results of the industrial automation domain case study of the ITEA2-project D-MINT (http://www.d-mint.org), driven by ABB. The case study aimed at answering questions regarding the most promising model-based testing methods and tools as a way of addressing the goal of reduced testing efforts. In addition, the questions of how and when to apply model-based testing within this domain were answered and ultimately led to a new and holistic view on model-based testing for embedded systems in the automation domain, based on (Bauer, Eschbach, Groessl, Hussain, Streitferdt, & Kantz, 2009). Finally, the case study delivers an analysis and precise numbers of the actual savings as a result of applying the developed model-based testing process.

In the section "Softstarter Example", the softstarter is introduced as an example of an embedded device in the automation domain and a basis for the case study. In the section "Model-Based Testing Process for Embedded Devices in the Automation Domain", the integrated testing process is discussed as a key concept of the case study. In the section "Evaluation of the Approach", the results of applying the process in the case study are presented. In section "Related Work", an overview of the relevant testing technologies and methods is given. Finally, this article concludes with a brief summary and topics for further research.

SOFTSTARTER EXAMPLE

Electric motors are common actuators in process automation. For this article, the starting and the stopping of an electric motor is taken as example. The device used in this article is a softstarter as shown in Figure 1, which is used to smoothly ramp up/down a motor. This functionality is needed for large motors where the peak current consumption from the power grid may cause a breakdown in voltage or for conveyor belts where sudden steep acceleration ramps may damage the transported goods.

Besides ramping the motor up and down, a softstarter monitors the motor to detect, e.g., a locked or overheated motor or disturbances in the power supply network that might damage the motor.

The softstarter is able to control a wide range of electric motors in different scenarios (e.g., stone crushers, conveyor belts, fans or water pumps). Besides the mechanical installation of the softstarter, its behavior can be configured by a set of user changeable parameters for different scenarios. During commissioning, the softstarter parameters like the ramp up time need to be set according to the desired usage scenario – the softstarter is configured.

This kind of configuration is done by setting the values of 150 parameters; a selection of these parameters is shown in Table 1. This complexity poses a challenge for testing the softstarter. To ensure complete coverage, the $1.0{\cdot}10^{110}$ possible combinations of parameter values (see section "Parameter Modeling") would have to be tested. Without a doubt, this is not testable within an acceptable time frame. The only way to handle this complexity is the structured selection of a subset of parameter configurations.

Figure 1. Softstarter for different motor sizes

TEST ENVIRONMENT

The model-based testing approach described in this article relies on a test environment needed to control and execute the test cases for the softstarter. In Figure 2, the test environment is visualized. The *Developer PC* is used to implement the test process, generate the test cases, and finally analyze the test results. The *Test Execution Hardware* is important for the test environment since the softstarter is an embedded system and the analysis of the test results depends on timing constraints, which cannot be met on an office PC platform with a general-purpose operating system. The generated test cases are transferred to the *Test Execution Hardware*, which in turn operates the *Softstarter*, e.g., it emulates the press of a button on the *External Keypad*, as prescribed by the test cases. Besides measuring voltage and current, the *Test Execution Hardware* tracks the menu content of the *External Keypad* via a tapped serial connection. With this information, it is possible to monitor any press of a button.

On the *DeveloperPC*, test cases are assembled out of simple test steps, which are methods of a test interface specifically developed for testing the softstarter. The test interface offers methods for operating the softstarter as well as the test environment, like Start_Motor(), Read_Temp() or Set_int_Parameter().

MODEL-BASED TESTING PROCESS FOR EMBEDDED DEVICES IN THE AUTOMATION DOMAIN

The development of embedded devices in the automation domain is aligned along the V-Model (Reinhold, 2003). As shown in Figure 3, the model-based testing activities discussed here focus on the upper levels "System Requirements Definition" and "System Validation".

In the case study, the inclusion of behavioral changes connected to the parameters in the device led to an explosion of states in the test model.

Table 1. Softstarter Parameters, adaptable by the user based on the Softstarter User Manual, page 81

Param. number	Description	Display text	Setting range	Default value
1	Setting current	Setting Ie	9,0...1207A	Individual
2	Start ramp	Start Ramp	1...30s, 1...120s	10s
3	Stop ramp	Stop Ramp	0...30s, 0...120s	0s
4	Initial voltage	Init Volt	30...70%	30%
5	End voltage	End volt	30...70%	30%
6	Step down voltage	Step down	30...100%	100%
7	Current limit	Current Lim	2,0...7,0xle	4,0xle
8	Kick start	Kick Start	Yes, No	No
9	Kick start level	Kick Level	50...100%	50%
10	Kick start time	Kick Time	0,1...1,5s	0,2s
11	Start ramp range	Start Range	1-30s, 1-120s	1-30s
12	Stop ramp range	Stop Range	0-30s, 0-120s	0-30s
13	Overload protection type	Overload	No, Normal, Dual	Normal
14	Overload protection class	OL Class	10A, 10, 20, 30	10
15	Overload class, dual type, start class	OL Class S	10A, 10, 20, 30	10
16	Overload class, dual type, run class	OL Class R	10A, 10, 20, 30	10
17	Overload protection, type of operation	OL Op	Stop-M, Stop-A, Ind	Stop-M
...	...	...	...	...

Thus, the test model was not manageable any more. To keep the test model at a manageable level of complexity and address parameter-related behavioral changes, existing approaches were composed into a development process that is new to the automation domain. This test process is shown in Figure 4. It is a black-box test process addressing the user-visible behavior of the system.

Figure 2. Test Environment for the automated generation and execution of test cases for the softstarter

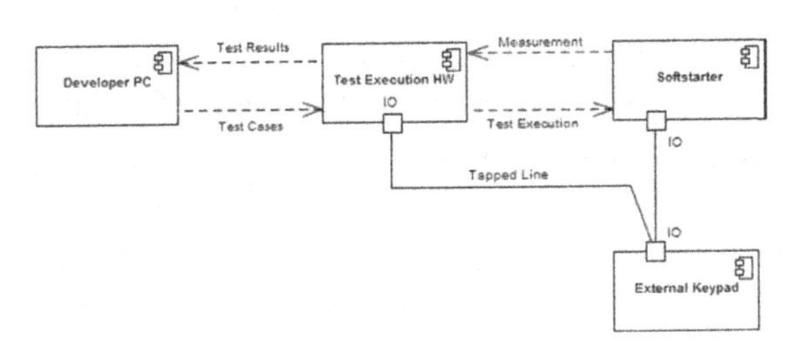

Within the domain analysis phase, many sources of information are analyzed, such as requirements documents, use-case models, or the user manual. This information is structured so as to be ready for referencing in the following phases. In the early development stage of system requirements definition, a *parameter model* and a *generic test model* are added to the development process. The *generic test model* formally describes the valid system usage scenarios independent of the system configurations.

Based on the type of development, the *parameter model* and the *generic test model* are roughly sketched and further refined in the subsequent development phases of the V-Model for developments starting from scratch. Or, the model can be fully reused for some development such as an update of existing products.

Figure 3. Integration of this model-based testing approach into the V-Model

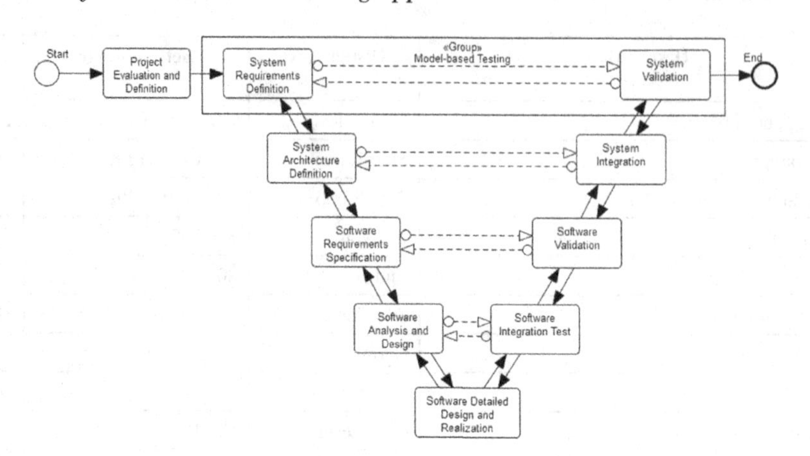

Figure 4. Detailed model-based testing process in the automation domain

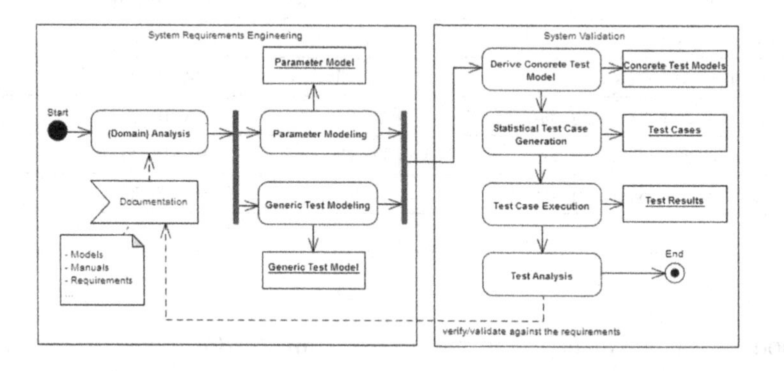

In the last stage of the development, the system validation, concrete test models are derived from the generic test model and the parameter model. The concrete test models serve as input for the statistical test case generation, which results in executable test cases. While the test cases are executed, several measurements are made and stored. They are used in the final step, the test analysis.

Although the documents are concise and detailed, the task of developing a parameter model and a test model constitutes advanced engineering work, where experience and expert knowledge are an advantage. The following sections correspond to the development steps in Figure 4, starting with parameter modeling.

PARAMETER MODELING

Many parameters (in our case 150), partly shown in Table 1, are available inside the softstarter to adapt it to a specific application, such as operating a fan, a stone crusher, or a conveyor belt. The behavior visible from the outside is simple. An electric motor can be started and stopped again (after a while), which would be one important test case. To achieve complete coverage, this test case would have to be executed with all possible permutations of the parameters. In real life, this is not possible due to the huge number of parameter permutations. The following sections briefly explain the method of clustering, the application of constraints, and pairwise testing. Finally, the approaches are integrated.

Table 2 describes the parameter types typically used in devices in the automation domain together with the number of permutations for a single parameter of the given type.

Based on the number of possible values for each parameter, it is easy to calculate $n_{complete}$ as the mathematically complete permutations of all parameters in the softstarter using Eq. 1.

$$n_{complete} = \prod_{i=1}^{N_{params}} n_i \quad (1)$$

N_{param}: Number of parameters
n_i: Number of possible values of parameter i

The softstarter uses 150 parameters: 65 Boolean parameters, 55 integer values and 30 enumerated values. Assuming about 100 values for integers using allowed ranges and 7 for enumerations, $n_{complete}$ is reduced to $n_{reduced}$.

$$n_{reduced} = 2^{65} \cdot 100^{55} \cdot 7^{30} = 8.3 \cdot 10^{154}$$

It takes between five seconds and five minutes (our assumption is an average of 1 minute of testing time, $t_{testcase}$) to execute a single test case for the softstarter. The combination of this execution time with the set of parameter permutations results in the testing time t_{test} of:

$$t_{test} = n_{reduced} \cdot t_{testcase} \approx 1.6 \cdot 10^{149} a$$

a: Years

The testing time t_{test} exceeds the age of the universe of about $14 \cdot 10^9$ years dramatically. Thus, further reduction of the parameter permutations to be tested is required, which is accomplished using methods described in the following sections, which are based on (Kantz, Ruschival, Nenninger, & Streitferdt, 2009). These methods as well as the corresponding tools have been chosen or developed based on the state of the art and on experiences of the testing domain experts in the D-MINT project, our own assessments of the usability and acceptance of the methods in our industrial domain, and our own expert knowledge in applying and integrating scientific methods towards industrial usage.

First, the parameters will be organized into sub-sets referred to as *clustering*. A cluster is a set of parameters according to the following rules:

1. All parameters within a cluster must be independent of all parameters outside the cluster.
2. Parameters inside a cluster may be dependent on each other.

The calculation of the overall permutations is the sum of the permutations of all individual clusters. In order to group the parameters into the clusters, expert knowledge is necessary to ensure the requirement of independence (see first rule above). This knowledge is highly dependent on the device functionality and on the specific device design.

Applying the clustering approach to the softstarter, we obtained 15 functionally independent groups with about 10 parameters each. For the running example and a simplified calculation, the clusters *Protections* (containing parameters like locked rotor protection, phase imbalance protection, or high current protection), *Warnings* (containing parameters like high current warning, overload warning, or thyristor overload

Table 2. Parameter types

Type	Permutations	Comment
Boolean	2	Standard type
Integer	2^8, 2^{16} or 2^{32}	Standard type
String	$\approx 50^{20}$	Assuming a character set with 50 chars and strings with an average of 20 chars
Enumeration	≈ 7	Assuming an average of 7 values

warning), and *Faults* (containing parameters like phase loss fault, fieldbus fault, or frequency fault) were selected. All three clusters together contain 50 parameters. By clustering, the permutations for these parameters can be reduced from $n_{simplified}$ as calculated above with Eq.1 and the parameter types, to $n_{cluster}$.

$$n_{simplified} = 4.92 \cdot 10^{37} \quad n_{cluster} = 4.67 \cdot 10^{17}$$

Taking this reduction rate as an average for the whole parameter set, it is possible to reduce the number of permutations to

$$\frac{n_{cluster}}{n_{simplified}} = \frac{4.67 \cdot 10^{17}}{4.92 \cdot 10^{37}} \approx 9.49 \cdot 10^{-19}\%$$

of the entire configuration space. Applying this huge reduction to the initially calculated $1.0 \cdot 10^{110}$ parameter permutations $n_{reduced}$, $1 \cdot 10^{90}$ permutations still remain.

A second approach, *parameter constraints*, has been used to further reduce the parameter permutations. The dependencies of parameters inside a group can be used for further reduction of the number of permutations needed to test the device. After computing the possible permutations (brute force) per parameter group, each parameter set needs to fulfill the constraints; otherwise, it is discarded. The following three constraints were analyzed.

The first constraint, *Mutual Exclusion*, is present if the selected value of a single parameter switches between parameter sets. For the parameter "Start Mode", two values "Volt" or "Torque" are possible. Either all Start/Stop cycles will be controlled by the voltage passed on to the motor or all Start/Stop cycles will be controlled by the torque the electric motor delivers. Each mode is further parameterized by a set of sub-parameters (e.g., ramp-up time or initial start voltage level if the motor is voltage-controlled) belonging only to this mode and therefore the parameters of other modes are mutually excluded (e.g., torque limit if the motor is torque-controlled). As soon as a mode is selected, other parameters are not taken into account for test parameter permutations. In most cases, expert knowledge is needed to identify mutual exclusions and all affected parameter sets.

The second constraint is a specialized form of mutual exclusion, referred to as *Function Switching Parameters*. In this case, a parameter switches a mode or functionality *on* or *off* whereas further parameters are used for a detailed configuration of the mode or functionality. The parameter "Locked rotor protection" can be switched *on* or *off*. This type of protection is triggered by the current flow through the motor and can be parameterized by the maximum duration of this current flow and the desired action (e.g., complete stop or automatic restart). Without this protection, all the permutations for the sub-parameters are irrelevant.

The third constraint is the *Selection of Ranges*. It directly influences the possible permutations of a parameter. While calculating the permutations, a parameter that rules over the range of another parameter forms the trigger for the reduction of the parameter space depending on the range selected. Of course, it is useful to create values slightly outside the defined range while testing the limits on the selected range.

The application of parameter constraints in each cluster led to a reduction of the permutations to $1.45 \cdot 10^{17}$. In the case study, scripting and spreadsheets were used to realize this reduction.

Finally, the *Pairwise Testing* approach assumes (based on empirical data of 329 error reports) that a majority of faults occur when changing values in a pair of parameters independently of further parameters. Thus, the testing effort can again be reduced significantly (Kuhn et al., 2004).

Due to the number of parameters in the clusters, the following example is presented only with three parameters, A: *Kick-Start* (before ramping up, kick-start the motor if, e.g., stones are blocking a stone crusher), B: *External By-Pass* (by-pass the softstarter once the motor is running at full speed), and C: *Fieldbus Control* (remote control the softstarter). All parameters can have two values, *yes* or *no*. Considering such a system, eight parameter sets are possible, see Table 3 leftmost part. Each of the eight parameter sets would have to be tested. By building pairs as in the three middle parts of Table 3, the possible permutations of the pairs A-B, A-C and B-C need to be present in the final testing parameter set. The gray parameter sets, rightmost part in Table 3, are enough to cover all individual permutations of the three middle pairs. Thus, five instead of eight parameter sets will be tested.

Complex errors depending on the combination of three or more parameter values cannot be found systematically by using the approach of pairwise testing. In order to enhance the coverage and detect these faults, the application of 3-wise testing with the softstarter parameter set results in a reduction of the parameter permutations to $3.7 \cdot 10^{9}$.

In the sections above, possible techniques for reducing the parameter permutations are presented. Although the resulting number of permutations still remains large, the *combination of the different approaches* resulted in a reasonable reduction.

The initial step for the optimization will be clustering. In the second step, a decision between two possibilities needs to be made. Either parameter constraints or the pairwise testing approach can be applied. Figure 5 gives an overview of the number of combinations that could be obtained with the different approaches.

$$n_{simplified} = 4.92 \cdot 10^{37}$$

$$n_{clust} = 4.67 \cdot 10^{17}$$

$$n_{clust+constr} = 1.45 \cdot 10^{17}$$

$$n_{clust+pair} \approx 3.7 \cdot 10^{9}$$

Starting with the simplified subset of parameters $n_{simplified}$, the clustering approach results in a reduced configuration space containing n_{clust} permutations. By applying constraints to the clustered parameters, the number of permutations can be further reduced to $n_{clust+constr}$. Finally, using the pairwise testing approach on the clustered parameters, the number of permutations can be reduced to $n_{clust+pair}$. The reduction for the paths Clustering-Constraints or Clustering-Pairwise in Figure 6 is between 10^{-27}% and 10^{-19}%. Despite this huge reduction, the remaining configuration space is still large. Thus, future efforts will be spent on the combination of clustering, constraints, and the pairwise testing approach (the dotted line in Figure 5).

GENERIC TEST MODELING

Model-based statistical testing (MBST) is a *black-box* testing technique that enables the generation of representative or failure-sensitive test cases from the tester's or user's perspective (Prowell et al., 1999; Prowell, 2005). The central element of the approach is a state-based test model, which describes the relevant system inputs and usages, and the expected system responses. Test models can be annotated with probabilistic weights to express frequency of use, costs, or criticality of inputs, outputs, and usages. Models that incorporate frequency of use are also called *usage models*.

Table 3. n-Way testing

A	B	C
n	n	n
n	n	y
n	y	n
n	y	y
y	n	n
y	n	y
y	y	n
y	y	y

A	B
n	n
n	y
y	n
y	y

A	C
n	n
n	y
y	n
y	y

B	C
n	n
n	y
y	n
y	y

A	B	C
n	n	n
n	n	y
n	y	n
n	y	y
y	n	n
y	n	y
y	y	n
y	y	y

Parameters
A : Kick-Start
B : External By-Pass
C : Fieldbus Control

Figure 5. Combination with resulting permutations of parameter permutation reduction approaches

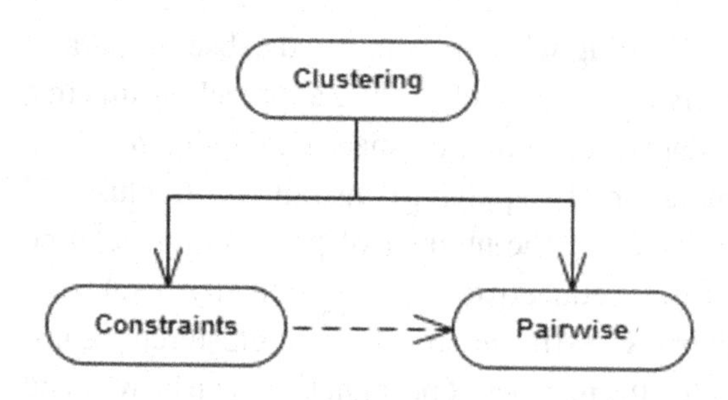

MBST allows the estimation of the system's reliability considering the given usage profile. The approach has been extended for risk-based testing (Zimmermann et al., 2009; Bauer et al., 2008) and applied to safety-critical embedded systems (Kloos et al., 2009; Bauer, Böhr, Landmann, Beletski, Eschbach, & Poore, 2007). MBST was used to construct generic test models and to automatically generate test cases from the configuration-specific test models.

Figure 6 shows the steps of the model-based statistical testing approach. In the first stage on the left side, *Test Model Construction*, a test model is built from the system requirements, which represents relevant system inputs, usages, and the expected system responses. The second stage comprises the *Automated Testing*, which is divided into automated generation of test cases from the test model, automated execution of test cases in the test environment, and automated evaluation of test results. The derivation of configuration-specific test models from the generic model is an additional step, which is not part of the original MBST approach.

The technique for the systematic construction of the model is called sequence-based specification (SBS) (Prowell, & Poore, 2003). SBS is a systematic approach for formalizing textual requirements and transforming them into a consistent and complete black-box specification represented as a finite state machine. The black-box specification is a mapping, which associates a stimulus history with its associated response. It defines the required external behavior of a system. During the SBS, the original system requirements are inspected with respect to system stimulation, usages, and responses. Every modeling decision in the SBS has to be justified and linked to the original system requirements. This assures the construction of a consistent, complete, and traceable black-box specification.

The first step of the SBS the system boundary is identified, i.e., interfaces and associated stimuli and responses. Consequently, a sequence enu-

Figure 6. Process steps of the model-based statistical testing approach

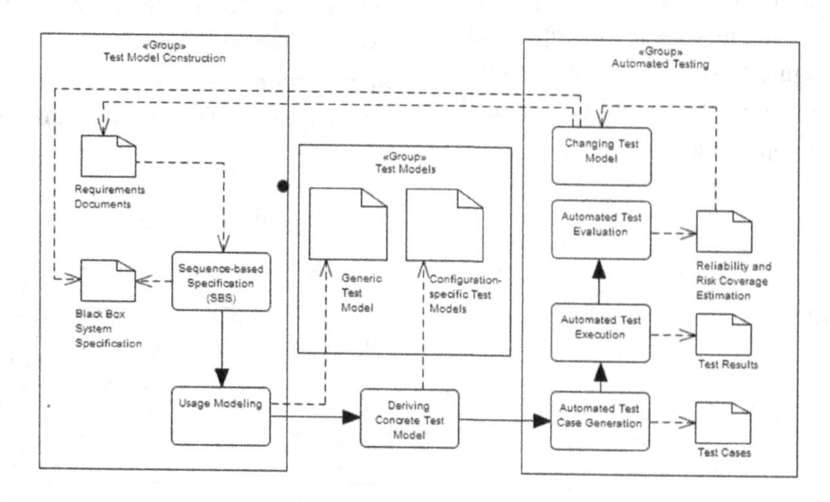

meration is performed, leading to a well-defined blackbox specification mapping each stimulus sequence to a sequence of responses.

The process of sequence enumeration aims at systematically writing down each possible stimulus sequence, starting with sequences of length one. A sequence of stimuli represents one history of the usage of the system under test. For every sequence, it is determined whether the sequence is valid or invalid related to the system requirements and whether corresponding expected responses are given. Empty responses are also possible. Only valid enumerated sequences will be extended and analyzed again in the next step. Every sequence is compared to previously analyzed sequences in terms of equivalence. Two sequences are equivalent if their responses to future stimuli are identical. Equivalent sequences are reduced to the shorter equivalent sequence and not extended.

A valid sequence that cannot be reduced to a shorter equivalent sequence is called a *canonical sequence*. A canonical sequence corresponds to the shortest sequence of stimuli, which leads to a state in the test model. The set of canonical sequences corresponds to the set of usage states identified for the test object. Usage states are named according to their meaning in the test model, see Figure 7. Alternatively, state variables can be introduced to label the states of the model. The enumeration stops if all sequences are invalid or reduced to equivalent sequences. All construction decisions for validity, equivalence, responses, and requirement coverage are traced back to the original system requirements.

The enumeration step provides a finite state machine that can be used further for usage modeling. The enumeration helps in revealing inconsistencies, missing specifications, unclear requirements, and vague formulations in the original requirements document.

Like most of the models used in any model-based testing approach, the models constructed using the abovementioned technique also contain certain abstractions. The abstractions help in restricting the size of the model and thus help in understanding, analyzing as well as in test case generation.

For each selected product configuration of the test object, a particular configuration specific test model is required which is later used for automated test case generation. In our approach, a generic test model is first built using the above mentioned steps, as depicted in Figure 7. The transitions in the model use a guard expression for the "FailureMode" parameter. The values of the parameter "FailureMode" are "manual" and "automated". In the configuration specific test models, the variability of parameters and multiple

Figure 7. Generic test model for the softstarter

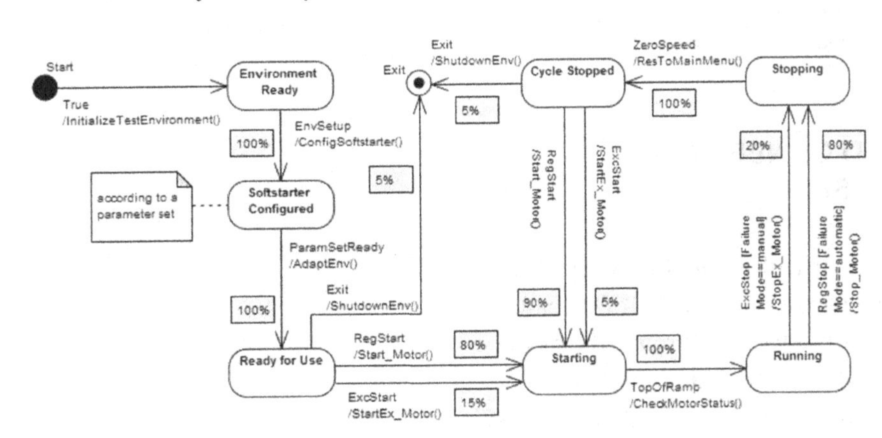

transitions are resolved. The resulting models do not contain variables and guards, which facilitates automated test case generation.

The generic model describes the configuration-independent functional aspects of the device, i.e., the mapping from stimuli sequences to responses. Specific models related to each configuration are then derived from that model using guards that exclude certain transitions and unreachable states from the generic model. A simple example of this concept is illustrated in Figure 8. Here, the configuration is meant to define the behavior of the software with regard to failure handling. When configured for manual handling of a failure, the regular stop transition labeled "RegStop" (which is the automatic handling) is not valid any more. The selection of a value for the "FailureMode" parameter removes the "RegStop" transition, as shown in the right part of Figure 8.

The resulting state machine represents the structure of the usage model and describes all possible usages identified from the system requirements. By adding probabilities to the transitions, the state machine becomes a discrete-time Markov chain. The probabilities reflect the frequency or criticality of the system usage. Data for probability distributions is sometimes available from domain experts or monitoring data from similar systems, partially or completely, for usage environments. Without such data, uniform probabilities are assumed. This ultimately implies that all outcomes are equally likely. In our case, we used risk analysis data to define the usage profile of the test object, as shown in Figure 7.

AUTOMATED GENERATION OF TEST CASES

Test models have particular states for the initialization (START) and finalization (EXIT) of test cases. A test case is an arbitrary path through the model from START to EXIT traversing a sequence of states and transitions. The state START describes the system state at the beginning of a test case. The state EXIT marks the end of a test case and can be reached from all states where a test case can end. Different strategies for automated test case generation exist, e.g.:

- **Model Coverage Tests:** Make sure that the whole model is covered by test cases. This means that each transition and each state of the test model is tested.
- **Random Tests:** Are randomly generated paths through the test model based on an operational profile, e.g., frequency or criticality.

From every configuration-specific test model, a set of test cases was generated: one set of test cases to cover all model elements and one test set of test cases with a number of random tests. The tool-supported generation of test cases takes just seconds, but the execution of the test cases is dependent on the planned and available test effort.

TEST CASE EXECUTION AND ANALYSIS OF RESULTS

The transitions of the test model are annotated with scripts for the test runner, which is part of the execution hardware. Hence, during the gen-

Figure 8. Creation of a configuration-specific test model from a generic model

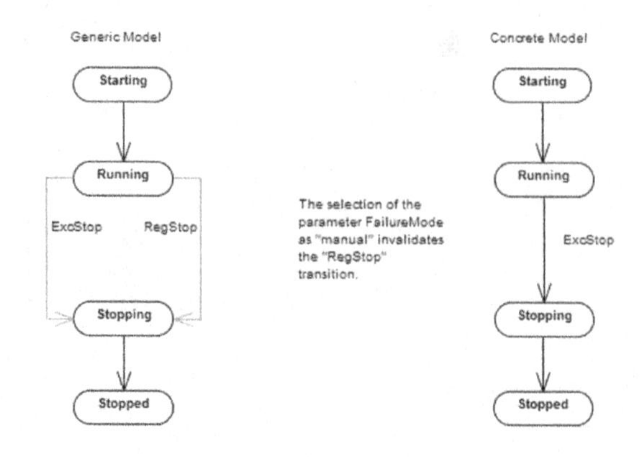

eration of a specific test case, the scripts of the transitions on the path will be aggregated one after another in order to build a concrete test case that is executable in the selected test environment as shown in Figure 2.

As an example, Figure 9 shows a test case generated from a concrete test model with manual acknowledgement for failures due to an exceptional stop of the motor. The sequence includes the testing hardware, the environment, and the test interface towards the softstarter itself. The test cases and the corresponding parameter set are transferred to the testing hardware. The test case will be executed for each parameter set of the reduced parameter permutations according to the section "Parameter Modeling".

First, the test execution hardware is initialized and the parameter set is prepared for use with this test case. The current parameter configuration is transferred to the softstarter via the external keypad, after which the test execution hardware will set up the environment according to the parameters (e.g., surrounding temperature). The measurements of voltage and current, for example, are started before the actual test sequence is started. After issuing the start command to the softstarter, the test execution hardware checks (with a time limit) whether the connected motor has reached the running state and whether the test sequence can continue. The concrete test model now only allows for an exceptional stop, for which an exception needs to be triggered. This forces the softstarter to stop the motor, which is checked again by the test execution hardware. Finally, the test execution hardware stops and stores the measurements for the upcoming analysis of the test case. This sequence is repeated for all parameter sets in the current list of parameter permutations.

Figure 9. Test case execution sequence with the test environment elements

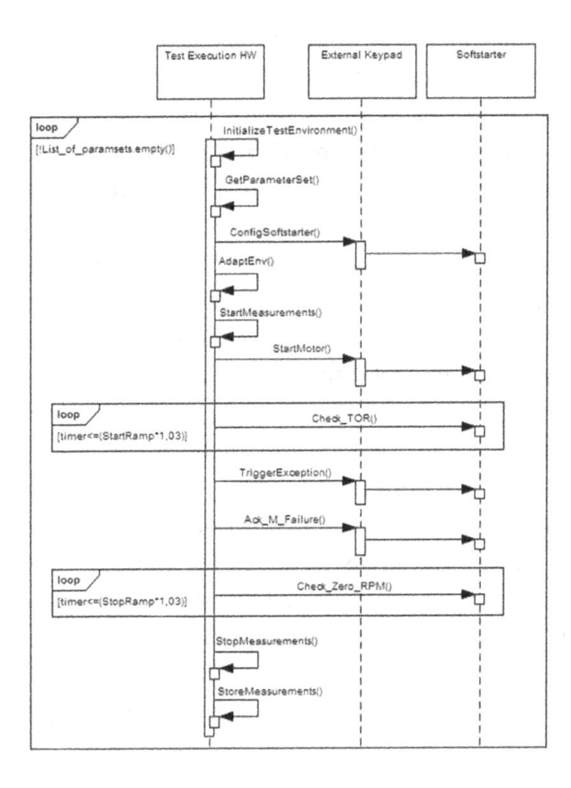

The test analysis evaluates the observable reactions of the system at each test step. Each test case corresponds to a specific transition of the concrete test model and describes the expected system responses. Responses can be discrete (e.g., a relay-switch output for locked rotor) or continuous quantities (e.g., start/stop ramp voltage variation). For continuous outputs, reference shapes were used to ensure efficient test evaluation. In the next step, the test cases were evaluated. A test case fails if at least one of its test steps has detected a failure. A test run is passed if all test cases have not revealed any failure. Finally, the reliability of the test object is estimated based on the test results. During test execution, the number of executions with and without failure is counted for each transition. Based on the usage profile and the failure statistics, the reliability of the test object is estimated after each test run. Reliability in MBST is the probability of a no-failure operation (Miller, Morell, Noonan, Park, Nicol, Murril, & Voas, 1992). In the case of a criticality profile, safety compliance is estimated instead of reliability. The system is released as soon as the reliability reaches a defined value and stays within a certain confidence interval.

EVALUATION OF THE APPROACH

For the evaluation of our test approach in the industrial case study, a measurement program based on the GQM (goal-question-metric) (Basili, Caldiera, & Rombach, 1994) approach was prepared. The GQM approach allows determining case study- and company-specific goals and refining them into measurable metrics. Metrics are used to improve the software development and the testing process (and its resulting software products) while maintaining alignment with the organization's business and technical goals. In the first step, business and improvement goals are analyzed and metrics are defined according to the development and testing process. The result is a GQM plan that collects all goals, questions, and metrics together with details about who, when, and how data are to be collected. This information is then used for designing and defining the data collection procedures, i.e., instrumentation. Afterwards, measurement is performed and raw data are collected, validated, and analyzed according to the GQM plan. Initial feedback is provided to the interested parties. Then, the interested parties draw conclusions and consequences in a post-mortem analysis in accordance with their analysis and interpretations. Finally, the analysis, interpretations, conclusions, and consequences are summarized and/or reported and collected as experience.

Two evaluation rounds were planned and conducted for measuring and assessing both approaches; the established test approach and the new model-based test approach. The main goal of the case study evaluation was the comparison of the model-based testing and non-model-based testing approaches in an industrial environment with respect to the categories development quality, test case quality, and test effort. For effort-related measurement, different metrics have been defined so that the different cost drivers for the two approaches can be taken into account. Measurement plans with responsibilities, schedules, and data collection methods for the categories development quality, test case quality, and test effort were prepared. Based on these plans the required data was collected in both evaluation rounds. Finally, the data was analyzed and interpreted together with the ABB research department and the soft-starter development team. The model-based test approach showed improvements in the categories shown in Table 4.

In the category "Development Quality", the hours to locate and fix a defect, including the effort for detecting failures, locating their causes (faults), and correcting them could be reduced by 20% with the model-based approach and the testing environment.

In the category "Test Case Quality", for the first metric "Test Case Reuse" the reuse of test cases was already present in the non-model-based testing approach. In addition to the existing test cases, formal models were used in the model-based testing approach as a basis for automated test case generation. With the existing test case reuse and the reuse of generated test cases, the overall test case reuse could be improved by 10%. For the metric "Model Reuse", modeling changed to be the central concept for the test phase, which resulted in an 80% improvement for the level of model reuse. In the non-model-based approach, models were used to refine test cases but not to derive test cases in an automated way.

Table 4. Case study improvements

Quality Focus	Metric	Improvement
Development Quality	Hours to locate and fix a defect (old/new) [estimates of #defects and #hours to locate and fix all defects]	20%
Test Case Quality	Test Case Reuse [% Test Cases reused]	10%
	Model Reuse [% States Reused/% Edges Reused/% Annotations Reused]	80%
Test Effort	Test Case Definition [Person Hours]	0%
	Manual Test Case Execution [Person Hours]	50%
	Manual Test Case Analysis [Person Hours]	50%

In the category "Test Effort", for the metric "Test Case Definition" the effort needed for model construction and automated test case generation with the new approach remained constant due to the high effort for learning and technology transfer, compared to the effort for manual test case definition in the non-model-based approach. Thus, the improvement for test case definition was 0%. In further iterations, high effort savings for model construction are estimated by domain experts to be approximately 40% after the third iteration. For the metric "Manual Test Case Execution", the effort for test case execution and evaluation could be reduced by 50% for the model-based approach as a result of the highly automated execution and evaluation environment. For the metric "Manual Test Case Analysis", the effort for analyzing test case results and tracing them back to requirements and design elements could be reduced by 50% with the model-based approach and the test environment.

All the results from the measurements have the same importance for the ABB softstarter development team and the corporate research department. Thus, the calculated average improvement achieved in the softstarter case study was 35%. It is important to note that this improvement will be invested into the higher complexity of future devices. The development time can be kept stable, with more features and customer requirements being addressed within the same time frame.

RELATED WORK

The automated generation of test cases is often achieved by using mathematical models. Model-based testing techniques have been developed that differ in terms of modeling notations and test case generation approaches (Utting & Legeard, 2006). Model-based testing refers to software and system testing where test cases are derived as a whole or in part from a model that describes selected, often structural, functional, or non-functional aspects of the test object. Examples of modeling notations are transition-based notations e.g., finite state machines (Gill, 1962), Pre/Post-models e.g., OCL (Object Management Group, 2000), Z (Davies, & Woodcock, 1996), history-based notations e.g., sequence diagrams (Harel, & Thiagarajan, 2003), operational notations e.g., CSP (Hoare, 1985), Petri Nets (Peterson, 1981), and statistical notations e.g., Markov Chains (Prowell, 2005).

An important task of the softstarter is to reduce the load and torque in the power train of the motor in the starting and stopping phase by controlling the voltage and the current of the electrical device. The control logic can be described well by a state-based model. The automated testing of non-functional quality properties such as reliability and safety is only supported by few model-based test approaches. A mature and systematic approach is model-based statistical testing (MBST) (Prowell, 2005; Kloos, & Eschbach, 2009; Prowell, Trammell, Linger, & Poore, 1999; Zimmermann, Eschbach, Kloos, & Bauer, 2009), which uses a state-based statistical test model to reflect importance, frequency of use, risk, and criticality. Test cases are automatically generated focusing on selected non-functional quality properties. Reliability testing is provided by considering the operational profile of the test object. The operational profile is a quantitative characterization of how a system will be used (Musa 1993). Testing according to the actual usage of the system allows predicting the future system's reliability in the field. Musa defined a stepwise approach for the determination of appropriate operational profiles from customer and user profiles. Approaches for incorporating existing risk analysis models are described in (Zimmermann, Kloos, Eschbach, & Bauer, 2009; Bauer, Stallbaum, Metzger, & Eschbach 2008).

The systematic and efficient testing of configurable systems is a current research topic. Most of the work is related to product line testing. The main aspect in the work of Cohen, Dwyer, and Shi (2006) is the reusability of test cases for different configurations. In McGregor (2001), the focus is on specific variations points between products that are the basis for the composition of test cases. Abstract test scenarios are modeled as use cases representing the product requirements model. The scenarios are parameterized with characteristic values to instantiate the different product variants. A particular model-based testing approach is introduced in Kamsties, Reuys, Pohl, and Reis (2004) and Olimpiew and Gomaa (2005). It describes the representation of use cases with UML extensions by considering the variability of software product lines. This method supports derivation of test cases from models. Another test approach (Reuys, Kamsties, Pohl, & Reis, 2005) provides the derivation of application-specific test scenarios describing the variation points of the product from use cases and activity charts on the domain level.

Other approaches use combinatorial techniques, especially in the field of experiment design (Cohen et al., 2006). The objective is to find an optimal set of configurations that satisfies maximal coverage criteria for each test case. The dependency between input parameters of system functions has been investigated in Kuhn, Wallace, and Gallo (2004). Their empirical study showed a dependency of two to six input parameters for different system functions.

As presented in the section "Softstarter Example", the behavior of the softstarter is tightly bound to its parameters. The behavioral part of the softstarter testing challenge can be covered by state-based statistical test models, whereas the parameters caused a state explosion. Thus, a second approach to handle the huge amount of parameter permutations was needed and found, by using combinatorial techniques.

SUMMARY AND OUTLOOK

In this article, testing of highly configurable embedded systems has been discussed by means of a case study with embedded systems from the automation domain. The basic behavior of such systems is rather simple but dependent on a large number of parameters. The permutation range resulting from the parameters is dramatically higher than the possible permutations of the few use cases of the system. Thus, a new testing process was developed with two major models, a parameter model in combination with a test model, which is the basis for the automated generation of test cases. The testing process is a new combination and realization of state of the art methods. It was applied and evaluated in an industrial case study. The resulting improvement of 35% shows the quality of the approach. While developing the new testing process, some issues arose and lead to the lessons learned. First, the testing process cannot replace expert knowledge nor will it replace existing test cases. Instead, generated test cases complement existing test cases to address coverage requirements efficiently. In addition, the testing process offers a structured way to handle the huge number of possible test cases. Second, the process requires initial efforts for its introduction. A further reduction of the development time seems feasible in case the testing process is used for more than three projects or project iterations. This requires a validation within a time frame beyond five years. Third, the parameter reduction approach of the testing process is also usable for existing and manually developed test cases, although the potential is reduced to address coverage requirements efficiently.

An increased presence of embedded product lines requires the extension of this testing approach towards product lines. The variability of the hardware and the software architecture requires variability in the test model as well, which will be the subject of future research efforts.

For analyzing defects that were not discovered during testing with the approach presented here, a method is discussed in Ruschival, Nenninger, Kantz, and Streitferdt (2009). The contribution describes a systematic approach to modify test cases based on user stories, which led to identify root-causes of device failures. This approach needs to be further elaborated and integrated into a testing tool chain. Clearly, an important topic is continuous research on further automation potentials of the testing tool chain. Usability, efficiency, modularity, and ease of integration are the top issues for future tool chains. Finally, the integration of additional behavioral models currently used for simulation purposes is a promising approach for enhancing the quality of the test model (Zander & Schieferdecker, 2010). The integration of such behavioral models will be the subject of future research. Additionally, the similarity of the concrete test models is an interesting fact that should be analyzed. The goal is to exploit the similarity of the model structure and use this information to reduce the set of test cases that have to be executed for a specific configuration.

REFERENCES

Basili, V. R., Caldiera, G., & Rombach, H. D. (1994). The goal question metric approach. In *Encyclopedia of software engineering*. New York, NY: John Wiley & Sons.

Bauer, T., Böhr, F., Landmann, D., Beletski, T., Eschbach, R., & Poore, J. H. (2007). From requirements to statistical testing of embedded systems. In *Proceedings of the Software Engineering for Automotive Systems Workshops*, Minneapolis, MN.

Bauer, T., Eschbach, R., Groessl, M., Hussain, T., Streitferdt, D., & Kantz, F. (2009). Combining combinatorial and model-based test approaches for highly configurable safety-critical systems. In *Proceedings of the 2nd Workshop on Model-based Testing in Practice at the 5th European Conference on Model-Driven Architecture Foundations and Applications*, Enschede, The Netherlands.

Bauer, T., Stallbaum, H., Metzger, A., & Eschbach, R. (2008). Risikobasierte Ableitung und Priorisierung von Testfällen für den modellbasierten Systemtest. In *Proceedings of the Software Engineering Conference*, Munich, Germany.

Cohen, M. B., Dwyer, M. B., & Shi, J. (2006). Coverage and adequacy in software product line testing. In *Proceedings of the ISSTA Workshop on Role of Software Architecture for Testing and Analysis* (pp. 53-63).

Davies, J., & Woodcock, J. (1996). *Using Z: Specification, refinement and proof.* Upper Saddle River, NJ: Prentice Hall.

Gill, A. (1962). *Introduction to the theory of finite-state machines*. New York, NY: McGraw-Hill.

Harel, D., & Thiagarajan, P. S. (2003). *Message sequence charts*. Retrieved from http://www.comp.nus.edu.sg/~thiagu/public_papers/surveymsc.pdf

Hoare, C. A. R. (1985). *Communicating sequential processes*. Upper Saddle River, NJ: Prentice Hall.

Kamsties, E., Reuys, A., Pohl, K., & Reis, S. (2004). Testing variabilities in use case models. In F. van der Linden (Ed.), *Proceedings of the 5th International Workshop on Software Product-Family Engineering* (LNCS 3014, pp.6-18).

Kantz, F., Ruschival, T., Nenninger, P., & Streitferdt, D. (2009). Testing with large parameter sets for the development of embedded systems in the automation domain. In *Proceedings of the 2nd International Workshop on Component-Based Design of Resource-Constrained Systems at the 33rd Annual IEEE International Computers, Software and Applications Conference*, Seattle, WA (pp. 504-509).

Kloos, J., & Eschbach, R. (2009). Generating system models for a highly configurable train control system using a domain-specific language: A case study. In *Proceedings of the 5th Workshop on Advances in Model Based Testing*, Denver, CO.

Kuhn, D., Wallace, D., & Gallo, A.M., J. (2004). Software fault interactions and implications for software testing. *IEEE Transactions on Software Engineering*, *30*(6), 418–421. doi:10.1109/TSE.2004.24

McGregor, J. D. (2001). *Testing a software product line* (Tech. Rep. CMU/SEI-2001-TR-022). Pittsburgh, USA, Carnegie Mellon University

Miller, K., Morell, L., Noonan, R., Park, S., Nicol, D., & Murril, B. (1992). Estimating the probability of failure when testing reveals no failures. *IEEE Transactions on Software Engineering*, *18*(1), 33–43. doi:10.1109/32.120314

Musa, J. D. (1993). Operational profiles in software-reliability engineering. *IEEE Software*, *10*(2), 14–32. doi:10.1109/52.199724

Object Management Group (OMG). (2000). *Object constraint language specification*: *OMG unified modeling language specification, version* 1.3. Retrieved from http://www.omg.org/spec/UML/1.3

Olimpiew, E. M., & Gomaa, H. (2005). Model-based testing for applications derived from software product lines. In *Proceedings of the 1st International Workshop on Advances in Model-Based Testing* (pp. 1-7).

Peterson, J. L. (1981). *Petri net theory and the modeling of systems*. Upper Saddle River, NJ: Prentice Hall.

Prowell, S. (2005). Using Markov chain usage models to test complex systems. In *Proceedings of the 38th Annual Hawaii International Conference on System Sciences* (p. 318).

Prowell, S., & Poore, J. H. (2003). Foundations of sequence-based software specification. *IEEE Transactions on Software Engineering*, *29*(5), 417–429. doi:10.1109/TSE.2003.1199071

Prowell, S., Trammell, C., Linger, R., & Poore, J. (1999). *Cleanroom software engineering: Technology and process*. Reading, MA: Addison-Wesley.

Reinhold, M. (2003). *Praxistauglichkeit von Vorgehensmodellen: Specification of large IT-systems – integration of requirements engineering and UML based on V-Model'97*. North Rhine-Westphalia, Germany: Shaker Verlag.

Reuys, A., Kamsties, E., Pohl, K., & Reis, S. (2005). Model-based system testing of software product families. In O. Pastor & J. Falcão e Cunha (Eds.), *Proceedings of the 17th International Conference on Advanced Information Systems Engineering* (LNCS 3520, pp. 519-534).

Ruschival, T., Nenninger, P., Kantz, F., & Streitferdt, D. (2009). Test case mutation in hybrid state space for reduction of no-fault-found test results in the industrial automation domain. In *Proceedings of the 2nd International Workshop on Industrial Experience in Embedded Systems Design at the 33rd Annual IEEE International Computers, Software and Applications Conference*, Seattle, WA (pp. 528-533).

Utting, M., & Legeard, B. (2006). *Practical model-based testing: A tools approach*. San Francisco, CA: Morgan-Kaufmann.

Zander, J., & Schieferdecker, I. (2010). Model-based testing of embedded systems exemplified for the automotive domain. In Gomes, L., & Fernandes, J. M. (Eds.), *Behavioral modeling for embedded systems and technologies: Applications for design and implementation* (pp. 377–412). Hershey, PA: IGI Global. doi:10.4018/978-1-60566-750-8.ch015

Zimmermann, F., Eschbach, R., Kloos, J., & Bauer, T. (2009). *Risiko-basiertes statistisches Testen*. [TAV group Meeting of the Gesellschaft für Informatik]. Retrieved from http://pi.informatik.uni-siegen.de/stt/29_4/01_Fachgruppenberichte/TAV28P6Zimmermann.pdf

Zimmermann, F., Kloos, J., Eschbach, R., & Bauer, T. (2009). Risk-based statistical testing: A refinement-based approach to the reliability analysis of safety-critical systems. In *Proceedings of the European Workshop on Dependable Systems*, Toulouse, France.

This work was previously published in the International Journal of Embedded and Real-Time Communication Systems, Volume 2, Issue 2, edited by Seppo Virtanen, pp. 22-41, copyright 2011 by IGI Publishing (an imprint of IGI Global).

Chapter 7
Adoption of Model-Based Testing and Abstract Interpretation by a Railway Signalling Manufacturer

Alessio Ferrari
University of Florence, D.S.I., Italy

Daniele Grasso
General Electric Transportation Systems, Italy

Gianluca Magnani
General Electric Transportation Systems, Italy

Alessandro Fantechi
University of Florence, D.S.I., Italy

Matteo Tempestini
General Electric Transportation Systems, Italy

ABSTRACT

Introduction of formal model-based practices into the development process of a product in a company implicates changes in the verification and validation activities. A testing process that focuses only on code is not comprehensive in a framework where the building blocks of development are models, and industry is currently heading toward more effective strategies to cope with this new reality. This paper reports the experience of a railway signalling manufacturer in changing its unit level verification process from code-based testing to a two-step approach comprising model-based testing and abstract interpretation. Empirical results on different projects, on which the overall development process was progressively tuned, show that the change paid back in terms of verification cost reduction (about 70%), bug detection, and correction capability.

INTRODUCTION

The adoption of modelling technologies into the different phases of development of software products is constantly growing within industry (Mohagheghi & Dehlen, 2010; Miller, Whalen, & Cofer, 2010). Designing model abstractions before getting into hand-crafted code helps highlighting concepts that can hardly be focused otherwise, enabling greater control over the system under development. This is particularly true in the case of embedded safety-critical applications such as

DOI: 10.4018/978-1-4666-2776-5.ch007

aerospace, railway, and automotive ones, that, besides dealing with code having increasing size and therefore an even more crucial role for safety, can often be tested only on the target machine or on *ad-hoc* expensive simulators. For these reasons, the industry of safety-critical control systems has been the first in line to adopt modelling technologies, such as the SysML™/UML® graphical languages (Object Management Group, 2010a, 2010b, 2010c) for high-level design, and modelling and simulation platforms like Simulink®/Stateflow®1 or the SCADE Suite®[2] for lower-level design, to support the development of their applications.

General Electric Transportation Systems (GETS)[3] develops embedded platforms for railway signalling systems and, inside a long-term effort for introducing formal methods to enforce product safety, employed modelling first for the development of prototypes (Bacherini, Fantechi, Tempestini, & Zingoni, 2006) and afterwards for requirements formalization and automatic code generation (Ferrari, Fantechi, Bacherini, & Zingoni, 2009). Within the new development context also the verification and validation activities have experienced an evolution toward a more formal approach.

In particular, the code-based unit testing process guided by structural coverage objectives, which was previously used by the company to detect errors in the software before integration, has been completely restructured to address the new model-based paradigm. The process refactoring has been driven by three main reasons:

- The traditional approach based on exercising the code behaviour resulted in being too costly to be applied to a code that saw a size increment of four times for the same project within two years. This fast growth was partly due to the increase of the actual projects size and partly to the code generators that, as known, produce more redundant code than the one that could be produced by manual editing.
- When an automatic tool is used to translate from a model to software, it has to be ensured that the latter is actually compliant with the intended behaviour expressed by the model.
- Unit testing alone, whether model-based or code-based, cannot cover all the possible behaviours of the code in terms of control flow and data flow. Most notably, it lacks in detecting all those runtime errors, such as division by zero and buffer overflow, that might occur only with particular data sets.

The restructured unit level verification process is based on two phases, namely model-based testing and static analysis by means of abstract interpretation. The first phase is used to exercise the functional behaviour of models and code, and, at the same time, to ensure that the synthesized code conforms to the model behaviour. The second phase is used to ensure that the code is free from runtime errors. Unit level verification costs were in the end reduced of about 70%, while notably decreasing the man/hours required for bug detection and correction.

The rest of the paper is structured as follows. In the first section we review the literature on industrial applications of abstract interpretation and model-based testing. The second section describes the unit testing process adopted, introducing the normative context of the railway industry and giving details on the verification phases. The third section analyses the results obtained applying the approach on two case studies, and the last one draws conclusions on the presented experience.

The article is an extended version of a conference contribution (Grasso, Fantechi, Ferrari, & Becheri, 2010) augmented with additional material from a recent case study (Ferrari, Magnani, Grasso, Fantechi, & Tempestini, 2010). Compared with the original work, the current one completes it with the following contents: (1) description of the evolution of the model-based testing phase, where a new tool, called 2M-TVF, has been introduced to counter the objections about the non-trusted code generator; (2) full description, with examples, of the abstract interpretation phase, previously presented only in terms of process tasks; (3) empirical results merged with those of the recent case study presented in (Ferrari, Magnani, Grasso, Fantechi, & Tempestini, 2010); (4) a literature review on related industrial experiences.

RELATED WORK

In this section the technologies of model-based testing and abstract interpretation are introduced, together with references to notable industrial applications.

Model-Based Testing

In modern industrial software development, modelling plays a role in many steps of the life cycle, from requirements elicitation and architecture definition, to software implementation and verification. Models are used as prototypes to interact with the customer and help formalizing requirements. Models are used in the architecture development phase to define graphical abstract specifications for a system or for system components. During the code production phase, they can be adopted to automatically synthesize code or to guide software implementation. Finally, models can be employed during the verification phase, to automatically generate tests or anyway to guide test definition and test results analysis. Model-based Testing (MBT) is a general term that deals with the latter usage of software models, namely employing models to support any task related to the testing activities of an application (El-Far & Whittaker, 2002; Broy, Jonsson, Katoen, Leucker, & Pretschner, 2005; Hierons, Bowen, & Harman, 2008). In Hierons et al. (2009) a classification of the different MBT approaches is given according to the modelling language adopted, while an extensive introduction to methods and tools can be found in Utting and Legeard (2007).

We adopt the classification of Pretschner and Phillips (2005), where four main strategies are identified:

- The most common one, often used in the literature for defining MBT itself (Dalal et al., 1999), starts from the code to extract an abstract model of it. The model is then used to define abstract test cases (i.e., test cases at model level made of sequences of input and expected output) that are then transformed into actual test cases for the implementation under test.
- With the second strategy, the model to define tests is created manually, starting from the requirements and the software specification that have been used to produce the code.
- The third approach uses a common model to generate code and to produce tests. As noticed in Pretschner et al. (2005), if the model used for automatic code synthesis is rather close to the actual software implementation, its usage for testing activities adds little improvements in terms of effort reduction during test definition. Nevertheless, repeating the execution of tests at model level and at code level helps increasing confidence on the translator, which is a central issue for safety-critical software, where every translation step, from model to source code to binary code, has to be ensured to be fully consistent. The methodology presented in this paper is inspired to this third strategy and in the subsequent sections we will delve the aspects pointed out here.
- A last, more complex, approach uses two models: one, called design model, is used for code generation, and the other one, called test model and defined at a more abstract level, for test case generation. As presented in Aydal, Paige, Utting, and Woodcock (2009), where the Alloy lan-

guage is used for the test model and the Z language is used for the design one, this approach shows promising results for detecting requirements flaws. However, the effort required for building two models complicates the adoption of this strategy by industry, for which time-to-market is also a pressing issue.

The reluctance of the industry to cope with the cost of the paradigm shift is actually extended to MBT in general: only in the latest years we have experienced a greater interest of companies towards this technology. Early industrial applications of MBT have been surveyed in Prenninger, El-Ramly, and Horstmann (2005). The authors state that the method is mature enough for industry, but the collected papers do not give any evidence concerning cost reduction in comparison with traditional testing.

Among the methodologies adopted in the first industrial applications of MBT, the approach presented in Pretschner, Slotosch, Aiglstorfer, and Kriebel (2004), dealing with the smart cards domain, stands out. The CASE tool AutoFocus is used for behavioural modelling, and tests are generated through symbolic execution based on Constraint Logic Programming (CLP). The same approach is applied in Pretschner et al. (2005) on an automotive network controller, to enlighten the benefits of MBT in detecting requirements errors.

The other sectors where MBT has been employed more recently, and valuable industrial research papers have been published, are the telecommunications, the automation and the aerospace ones.

In the telecommunications domain, the work presented in Hessel & Pettersson (2006) shows an application of MBT to a Wireless Application Protocol Gateway developed by Ericcson, where timed automata are used as specification language and a tool based on the Uppaal model checker generates the abstract test suite. Previous works (Bozga et al., 2000; Larsen, Mikucionis, Nielsen, & Skou, 2005) had already shown the interest of the telecommunication industry to the MBT technology.

The ABB Group, from the automation domain, constrained by process standards such as the IEC 61508, has recently started to deal with the integration within its development process of an automated testing approach that uses UML and TTCN-3 (Streitferdt et al., 2008). As in our paper, the problem is analysed from a process point of view, which is a common issue in any standard regulated domain.

Concerning flight critical software, (Pasareanu et al., 2009) presents a novel strategy developed at the NASA Ames Research Center, that uses Simulink as specification language, and extract tests from a Java symbolic representation of the model, to test the software that is auto-generated from the Simulink specification with Real-Time Workshop® (RTW) Embedded Coder™[4]. This last state-of-the-art experience where the design tool-chain is the same as the one used in this work, gives an alternative way of combining code generation and model-based testing, another common issue in the development of safety-critical systems.

Abstract Interpretation

In industrial software verification and validation processes, dynamic analysis techniques such as testing, that are mainly focused on checking functional and control flow properties of the code, are normally complemented with static analysis, that aims at automatically verifying properties of the code without actually executing it. The properties checked by static analyzers range from coding standard adherence and programming style checking, to absence of runtime errors. Examples of techniques traditionally adopted for static analysis are data flow analysis (Fosdick & Osterweil, 1976), program slicing (Weiser, 1981), constraint solving (Aiken, 1999), and, with an increasing spread in the latest years, abstract interpretation. Abstract interpretation (Cousot & Cousot, 1977)

(Cousot & Cousot, 1979, 1992) is based on the theoretical framework developed by Patrick and Radhia Cusot in the seventies. However, due to the absence of effective analysis techniques and to the lack of sufficient computer power, only after twenty years software tools have been developed to support it, so that applications of the technology at industrial level could take place. Companies of the automotive and aerospace sector have applied abstract interpretation to evaluate worst-case execution time (WCET) of programs (Thesing et al., 2003; Barkah, Ermedahl, Gustafsson, Lisper, & Sandberg, 2008) and for stack overflow prevention (Ferdinand, Heckmann, & Franzel, 2007) (Regehr, Reid, & Webb, 2005), adopting both ad-hoc techniques and commercial software such as the WCET tool aiT and the StackAnalyzer package, distributed by AbsInt[5]. Our focus is the application of the technology in the analysis of source code for runtime error detection, which means detecting variables overflow/underflow, division by zero, dereferencing of non-initialized pointers, out-of-bound array access and all those errors that, might them occur, would bring to an undefined behaviour of the program.

In this context, abstract interpretation is a methodology that aims at ensuring the correctness of a program in terms of runtime errors. Since the correctness of the source is not decidable at the program level, the tools implementing abstract interpretation work on a conservative and sound approximation of the variable values in terms of ranges of values, and consider the state space of the program at this level of abstraction. The problem boils down to solve a system of equations that represent an over-approximate version of the program state space. Finding errors at this higher level of abstraction does not imply that the bug also holds in the real program. The presence of false positives after the analysis is actually the drawback of abstract interpretation that hampers the possibility of fully automating the process. Uncertain failure states (i.e., statements for which the tool cannot decide whether there will be an error or not) have normally to be checked manually and several approaches have been put into practice to automatically reduce these false alarms. Besides academic studies on statistical approaches for false positive ranking (Kremenek & Engler, 2003; Jung, Kim, Shin, & Yi, 2005) and strategies based on refinements of the abstract interpretation analysis (Rival, 2005; Gulavani & Rajamani, 2006), it is an interest of safety-critical systems manufacturers to address this problem in an effective manner.

In Brat and Klemm (2003) the first experimentation of NASA with Polyspace®[6] (Deutsch, 2004) for the analysis of the Mars Exploration Rover flight software is presented. The presence of a 20% of warnings to be manually checked, that requires a time consuming activity on programs of several hundreds of thousands of lines of code (from now on, KLOC), and the slow performances of the tool, led the organization to develop a prototype program called C Global Surveyor, particularly targeted for buffer overflow detection (Venet & Brat, 2004). The precision achieved did not outperform the one of Polyspace, this suggesting that the bound is possibly inherently related to the abstract interpretation framework itself. However, the possibility of tailoring the tool to the type of software analysed, an option that was not available with proprietary software like Polyspace, radically decreased the analysis time.

In the avionics sector, successful experiments for the suppression of warnings (Delmas & Souyris, 2007) have been performed using the tool Astrée (Blanchet et al., 2003), currently distributed by AbsInt. Thanks to a tight collaboration between application and tool developers they have achieved a reduction to zero of the number of false alarms issued by the tool. An alternative approach, that consists in complementing the Polyspace analysis with bounded model checking by means of CBMC, has been experimented in the automotive domain (Post, Sinz, Kaiser, & Gorges, 2008). The authors state that the strategy reduces of about 70% the warnings issued by Polsyspace (no distinction between true and false alarm is given). The main weakness of this study is the fact that they bound the execution time of Polyspace to seven hours only, since they focus on the cost of the bounded model checking step, but this allows them to give only partial conclusions on the overall approach.

Though experimented on open-source software only, and oriented solely to buffer overflow detection, it is worth mentioning also the recent study presented in Kim, Lee, Han, and Choea (2010), that combines abstract interpretation with symbolic execution by means of satisfiability modulo theory solvers, and is able to filter out 68% of the false alarms on average.

The strategies applied in Delmas and Souyris (2007), Post et al. (2008), and Kim et al. (2010) are based on multiple-step refinement approaches that are conceptually close to the one presented in this paper: after the first analysis, constraints are deduced and further analyses are performed. Compared to the other methodologies reported in the literature, the main strengths of our strategy are: (1) it is currently implemented in the development process of the company, and therefore it is an assessed approach; (2) we use one single commercial general-purpose tool without customizations, and this might be useful for the application of the approach to other domains.

THE ADOPTED UNIT TESTING PROCESS

GETS adopted a two-phase process for the software unit testing: the first phase consists of model-based testing, and aims to verify whether the software is compliant with the given requirements. The second phase is based on abstract interpretation and has the objective to detect runtime errors that may be caused by data flow deficiencies of the implementation. In order to discuss the two phases, we need first to introduce the normative context within which GETS operates.

The CENELEC Norms

In the context of European railway signalling and control, CENELEC created standards as references for the safety certification of products. There, safety is defined as the absence of unacceptable risk levels (Comité Européen de Normalisation en ÉLectronique et en ÉleCtrotechnique) (CENELEC, 1999).

CENELEC EN 50128 (CENELEC, 2001) is the norm that specifies the procedures and the technical requirements for the development of programmable electronic devices to be used in railway control and signalling protection. This norm is part of a family, and it refers only to the software components and to their interaction with the whole system (Figure 1). The basic concept of the norm is the SSIL (Software Safety Integrity Level): the higher is the integrity level of a system and the graver are the consequences of a failure. Integrity levels range from 0 to 4, where 0 is the lower level, which refers to software with no effects on the safety of a system, and 4 refers to the most critical software.

Quantitative safety verification cannot be conducted on software, since it is not possible to reliably quantify failure rates and the contribution software has given to an occurred accident. For these reasons software is verified using qualitative techniques: the EN 50128 norm defines an

Figure 1. Scopes of CENELEC norms

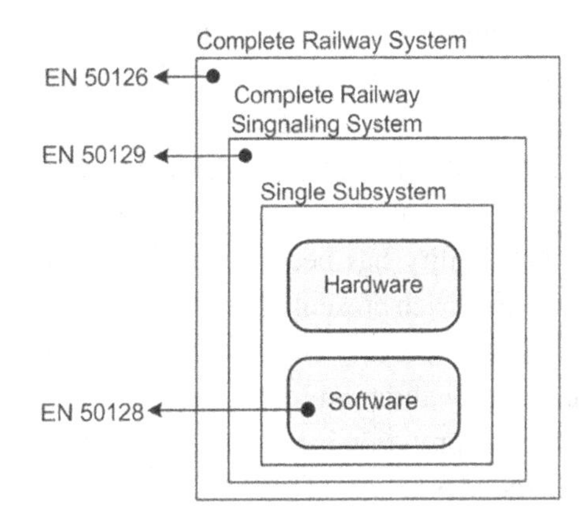

elaborated life cycle for the definition of specifications, the development and the verification of the software. The norm encourages the usage of models and formal methods in every phase of the software development cycle, starting from the design to the verification. The rationale is that models are more related to abstract concepts than the technologies used for their implementation into code, and are therefore closer to the domain of the problem. On the other hand, sufficiently detailed models can be used for automatic generation of code: this possibility has been adopted by GETS, but automatic code generation, according to EN 50128, requires that the code generator itself is somehow trusted. The norm lists two criteria that can be used to achieve trust in model-to-code translator. The first criterion requires a *certification* for the tool: in this way the translator will be qualified for the usage in a safety-critical development process. The second criterion (*proven in use* translator) is applicable to those translators that have been used in many previous applications and that have been monitored during their usage to assess their trustfulness.

Since the chosen code generator, namely RTW Embedded Coder, is not certified according to EN 50128, the only possibility was to fulfill the requirements of the proven-in-use criterion. This entails a workflow of several tasks of bug recordings and continuous reviews. As a core part of the process to achieve the proven-in-use property for the translator, we found useful to introduce a technique named *translation validation* (Conrad, 2009). This consists in verifying the functional equivalence between the models and the generated software by executing the same tests on the model and the code, and afterwards performing further structural analysis to ensure that no additional functionality has been introduced. The MBT testing approach presented as the first phase of our verification process is an implementation of the translation validation technique.

Abstract interpretation is not currently part of the recommended practices of the EN 50128 norm, since this has been published before abstract interpretation became a mature tool supported technique. However, due to the evident benefits that it can bring in terms of runtime errors detection, companies started practising it as a completion of the verification process to enforce the safety of its products. As a result of this growing interest, the Polyspace tool has been recently certified for usage in the context of the EN 50128 norm. The abstract interpretation approach presented as the second phase of our verification process shows how the company has employed Polyspace in its application domain.

Phase 1: Model-Based Testing

The adopted approach for MBT comprises two steps. The first one is a form of model/code back-to-back (B2B) testing (Vouk, 1990), where both the model and the related generated code are tested using the same stimuli as inputs, and the numerical results obtained as output are checked for equivalence. The second one consists of an additional evaluation to grant the absence of unwanted and unexpected behaviours introduced by the model-to-code translation process. This evaluation is basically the comparison of the measures of structural coverage reached on both the code and the model. The two phases could be seen respectively as a duplicated black-box testing (output comparison) and a duplicated white box testing (coverage comparison).

The B2B testing step in its first application is depicted in Figure 2 (a).

First, a Stateflow model is created starting from the unit requirements. This model represents the behavioural view of the system. In order to verify if this behaviour is compliant with the specification, a test suite for the model is manually derived from the specification, according to the full requirement coverage criterion (i.e., at least one test for each functional requirement). The outputs given by the simulation with the given test data are visually checked to assess those behaviours which do not comply to the specifications. The choice to manually define and verify

Figure 2. Approaches to the automation of MBT

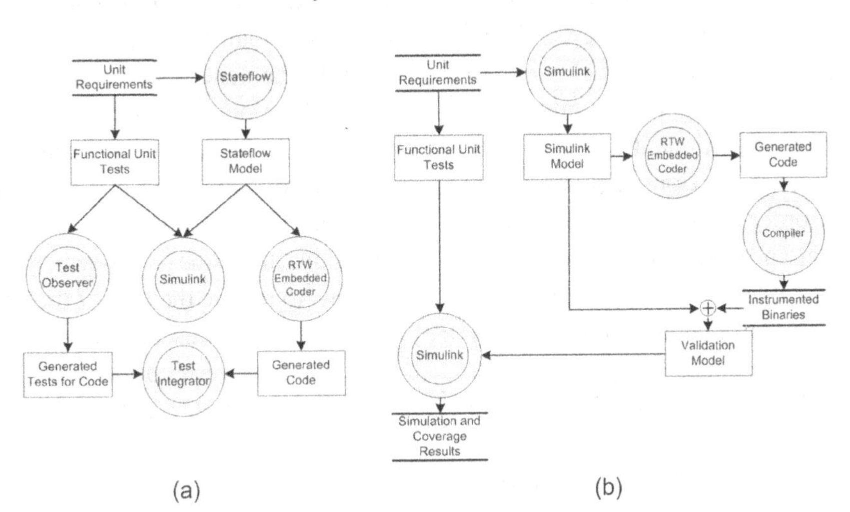

tests instead of using automatic test generation is due to the fact that the latter technique requires models with a higher level of abstraction than the ones GETS uses to generate the code.

A tool called *Test Observer* was developed to automatically translate the functional unit tests written for the Stateflow model into an appropriate form to be used as tests for the generated code. *Test Observer* records both input and output sequences of the model during the simulation in the form of Simulink time series, that are Simulink data objects composed by pairs (*time*, *value*), and then translates the time-series into given input/expected outputs matrices for the generated module. C code is then generated starting from the model, using RTW Embedded Coder.

For each test case another tool, called *Test Integrator*, produces an executable file that embeds the given input/expected output matrices, together with a set of functions to check if the output of the generated code matches the expected output. If the code executes without errors, then it can be stated that, for the given functional unit test, the generated code shows the same behaviour of the model (Grasso, Fantechi, Ferrari, Becheri, & Bacherini, 2010).

Recently the B2B testing task has been improved and integrated in a new framework, called 2M-TVF (*Matlab Model to Validation Framework*), whose workflow is depicted in Figure 2 (b). The 2M-TVF framework offers functionalities of validation of the code generation process and of the generated code itself. It takes as inputs the model and the manually developed tests, and creates a validation model that contains both the Simulink models to be tested and the code produced starting from those models. The code to be tested, included in the validation model, is first instrumented to permit the evaluation of the coverage after the execution of the tests, and then it is compiled to obtain the executable file. The framework does not aim to check for runtime errors, which are handed over to the abstract interpretation phase, although dynamic testing can discover them if present. Furthermore, the input model is never modified during the validation, to maintain it as it was used to generate the code. The major advantage of the adoption of 2M-TVF is that the whole verification process is conducted internally to the Simulink environment, and it is completely automated.

The validation process conducted by the framework consists of two phases: the first one is analogous to the B2B phase in its first form. If this phase terminates without errors, it means that the model and the generated code will show the same behaviour in response to the same stimuli. The second phase consists of the evaluation, both on the model and on the code, of the coverage reached after the execution of the tests. Every coverage difference detected must be assessed, since the code generator might have introduced unexpected spurious functionalities during the model-to-code translation. Comparable metrics must be used in order to evaluate the similarity between model and code coverage (Baresel, Conrad, Sadeghipour, & Wegener, 2003). In our case, it was chosen to use decision coverage for both the model and the code. The choice of this metric was due to the fact that the EN 50128 norm requires at least statement coverage.

After its execution, the 2M-TVF framework shows a report of the conducted activity, which contains the results of output comparison and achieved coverage. This report is then subject to a manual assessment of every detected difference.

Phase 2: Abstract Interpretation

Concerning the abstract interpretation phase of the verification process, GETS has adopted Polyspace, a commercial tool provided by The MathWorks. From an industrial perspective, having the same producers for several tools employed in the development process gives more confidence on their compatibility, and simplifies the interface with the tool providers.

Polyspace analyses the C code and detects the statements that could produce errors during the execution of the code. The tool presents its results through chromatic marks on the analysed code:

- **Green:** If the statement can never lead to a runtime error.
- **Orange:** If the statement can produce an error under certain conditions.
- **Red:** If the statement leads to a runtime error in every run.
- **Grey:** If the statement is not reachable.

In order to perform static analysis on the code, the tools based on abstract interpretation techniques, such as Polyspace, build an abstract domain that represents an over-approximation of the real domain. The abstraction process might bring to the generation of false positives during the verification: this behaviour is caused by errors raised in those runs which are allowed only in the extended domain, but not in the original one. For this reason, it is essential, for the adoption of this technique, to define a well-structured process that permits to reduce the cost of the analysis of false positives that represents the price to pay to obtain the exhaustive verification of the code behaviour.

In order to address the problem and, at the same time, to obtain a substantial improvement of the confidence on the correctness of the code, GETS decided to adopt a two step process (Figure 3). The first step is performed with a very large over-approximation set. The second one capitalizes the information obtained by the analysis of the previous one, and executes the verification with the use of a finer approximation set.

The purpose of the first step is mainly to detect systematic runtime errors (red), that is, errors which arise in all the runs considered in the verification, and unreachable statements (grey). Examples of systematic runtime errors are infinite loops, out of bound array accesses and usage of not initialized variables.

Although unreachable code might seem to be a low severity problem, in our experience grey marks are often indications of erroneous modelling of the specifications: a code block that is never executed might be the translation of

Figure 3. Polyspace process

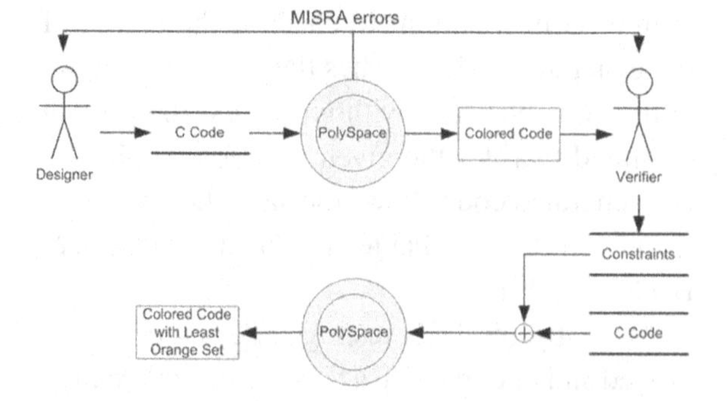

an unreachable state in a Stateflow chart. Only in some limited cases, grey marks are related to additional *defensive-programming* instructions introduced by the translator to maintain control even in presence of completely unexpected input, which may be due to hardware or software faults. For example, the *default* statement in a *switch/case* block (which is the natural translation of a state-machine), is likely to be never executed, and therefore will always be marked with grey. Obviously, these grey marks do not harm the safety of the code because they represent collectors used to handle unexpected behaviours.

The first step is performed using all the possible over-approximations settings provided by Polyspace:

- **Full-Interleaving:** The tool automatically generates the function calls for the public procedures of the module under test if they are not invoked by other functions defined in the same module. All the possible interleaving of the automatic function calls are analysed in the verification.
- **Static Variables Initialization:** The static variables defined in the module in every run are initialized with all the values of their type range (in the following we will refer to this kind of approximation as the *full-range* initialization).
- **Global Variables Initialization:** The global variables defined in the module are managed in the same way of static variables.
- **Generation of Function Calls:** The formal parameters of the function for which the tool generates the call are initialized at full-range.

Since these approximations are used in the first step in order to be sure that the analysed runs include all the actual runs, the results obtained are not selective enough. The large set of spurious runs that are analysed in this step leads to an outstanding number of orange checks.

For example, in Figure 4 the statement highlighted by the arrow could raise a runtime error, in particular the function could access a memory location which is outside the bound of the array named *buffer*.

In the example we consider that the function is static and that in the analysed module there is a function call to the *get_value* procedure. In this condition Polyspace does not generate an automatic call to the function, but makes use of the existing call to verify the function code. Otherwise, the code reads a location of the array *buffer* indexed by the global variable *index*. In the first step, Polyspace automatically initializes all the global variables with full-range values. For this reason, when the tool analyses the statement highlighted in Figure 4, it finds that for some values of the variable *index* there is an out of bound access to the array. The result suggests that on the values that the variable *index* can assume, narrower bounds have to be introduced in order to reduce orange marks.

In the example, Polyspace signals a possible erroneous behaviour on the array bracket, but the actual cause of the orange mark is the full range initialization of the variable *index*. It is on this variable that one has to work in order to avoid the warning. This situation is similar for any orange mark coming from the first step: in order to narrow the approximation for the subsequent step of Polyspace, each orange mark has to be

Figure 4. Example of unsafe statement (array out of bound exception)

```
[...]
/* Global Variables Declaration */
int buffer[100];
int index;
[...]
/* Function Definition */
int get_value(int *buffer)
{
   return buffer[index];
}
[...]          ↑
```

related to the cause that produced it. The generic classes of causes that generate the orange marks are well known, and can be referred to the over-approximations settings that have been listed before. Therefore, an analyst with a minimum proficiency with the tool can easily evaluate the orange marks and quickly classifies their causes.

In the case of the example, the analyst recognizes the orange mark on a bracket as referring to the *global variable initialization* setting, so can pinpoint the variable that has been initialized full range. Another case might be the one in which one module has two interface functions, the first to initialize static variables, and the second to actually perform the functionalities required to the module (this is actually the normal structure of the generated code). In the actual usage of the program, the initialization function will always be called before the other one. However, the tool will issue orange marks on all the static variables used by the execution function: due to the *full-interleaving* over-approximation, the tool assumes that the second function might be called before the initialization one, leaving the static variables without an initial value. Also in this case, the analyst recognizes a bunch of oranges on static variables, and can associate them to the full-interleaving class. Then, he gives constraints to the tool concerning the order of execution of the functions.

As exemplified, the identified classes are used to define input constraints to be given to the tool to restrict the analysed abstract domain of the program. Sometimes, editing the configuration file that defines the constraints might require advices from the developers, since the analyst is often not aware of the actual domains of the variables, or of the program context in which a certain function is used. However, we experienced that the analyst is much more independent if (s)he has to deal with the automatically generated code, since the repetitive structure of the software simplifies the review task.

The second Polyspace step, performed with the restrictive settings, allows a finer approximation of the real domain of the program and then a reduction of the number of false positives. At the end of this step, the remaining orange marks are due to the complex interactions between variables that cannot be constrained by simply introducing finer approximation bounds.

As an example, consider the code segment depicted in Figure 5 that describes a typical software procedure present in the railway signalling context: it deals with a train receiving messages from the car-borne equipment at every given distance, in proximity of a so called *information point*. Every time the train passes by the information point and receives a message, the code assigns the current value of the space, maintained in the variable *current_space,* to the variable *last_msg_space.* Once the train gets by the information point, it uses the procedure in Figure 5 to compute the space covered from the last message received.

Polyspace produces an orange mark that signals the risk that the described statement could raise an underflow. This orange mark is reported also in the second step of the Polyspace verification, because in this case the constraint on the possible values assumed by the variables does not handle the particular bound that makes impossible the underflow.

According to our experience, the overall time employed for the configuration and set-up activities is 20% more than the time Polyspace takes to actually execute the two steps. The most time consuming task is the first review of the orange marks, which takes about the 48% of the overall

Figure 5. Example of unsafe statement (underflow exception)

```
[...]
/* Function Definition */

uint32 compute_distance(uint32 current_space, uin32 last_msg_space)
{
   return (current_space - last_msg_space);
}
[...]
```

time required for the whole process. Due to the low number of residual oranges after the second step (normally about 2.6% of the total), the cost of the second review is basically negligible. Nevertheless, one has to consider that the absolute overhead of the orange review is acceptable: about 5 minutes for each orange, in average. The generated code is characterized by a limited number of different classes of motivations for the orange marks, and this makes most of the review a rather systematic activity.

CASE STUDIES

In this section we describe the results achieved on two industrial case studies where the verification approach was applied. The first case study concerns the Baseline3 (BL3) project, an Automatic Train Protection (ATP) system developed by GETS in 2008. The second case study is about the development of the SSC Metrô Rio ATP system, developed in 2009.

GETS first employed the code generation technology in the BL3 project, after the positive results of specific feasibility studies performed in the previous years in tight collaboration with the University of Florence. In the BL3 project code generation was limited to the unit modules (i.e. single Stateflow charts), which were subsequently integrated together by means of hand-crafted code. The behaviour of the system was modelled by means of 21 Stateflow charts, one for each specific function. The code generated from these models was about 150 KLOC in total. The highest cyclomatic complexity reached was about 60, and some functions showed more than 700 paths[7]: A set of modelling guidelines was introduced to disambiguate Stateflow semantics and to achieve the compliance of the code to the CENELEC standards and to the internal company code guidelines (Ferrari, Fantechi, Tempestini, & Zingoni, 2009). Automatic code generation led to the necessity of new verification techniques: the increased complexity of the control flow made impossible the use of former methods based on structural testing aimed at obtaining full path coverage. The verification phase was thus modified introducing the methodology composed by the model-based testing step and the abstract interpretation step.

For the MBT phase the first version of the B2B testing approach in its first version was adopted. Test cases were provided for each Stateflow model, according to the requirement coverage criterion. The test suite was composed by 327 manually defined test-cases, which covered the 100% of the modelled functional requirements. By executing the provided test-suite, we were able to detect 42 errors at model level, mostly related to the misinterpretation of the requirements expressed in natural language. *Test Observer* and *Test Integrator* were used in the verification phase: the execution of the test suite on generated code and the subsequent comparison of both model and code outputs did not highlight any discrepancy between model behaviour and related code behaviour. Moreover, the evaluation of the model coverage according to the decision coverage metrics resulted in being around 97% in average for each model; in most cases the code coverage coincided with the model coverage.

Concerning the second verification phase, the use of Polyspace, introduced for the first time in the BL3 project, permitted the substitution of the dynamic analysis phase, with increased confidence on the verification results, thanks to its exhaustiveness and, at the same time, reduced the cost of the verification itself. The first step of the Polyspace verification led to the results depicted in Figure 6 (a): a high number of orange marks were found, around 15% of the total checks. The structure of the automatically generated code, rather complex but sufficiently repetitive, permitted to perform the analysis of the causes of the high number of orange marks in a relatively short time: this analysis led to a series of constraints to be used during the execution of the second Polyspace step.

Thanks to these constraints the second step of abstract interpretation analysis was able to reduce the orange marks (Figure 6 (b)), with great advantages in terms of the effort spent in the investigation of the causes of the remaining orange marks.

The new verification approach permitted a reduction of the overall verification cost of 70%, as reported in Table 1: the table compares the verification cost of the BL3 project to the effort spent for traditional structural testing on code (according to 100% boundary-interior path coverage), which was applied in a previous project of comparable size in terms of modules.

A more recent project developed by GETS is the ATP for the metro of Rio de Janeiro. For the development phase of this project, model-based development was used also at the modules integration level: this led to an increased use of Simulink as modelling environment to connect single Stateflow charts (i.e., system modules), structuring the system through a multiple-level hierarchical model, in which the different levels are intended for different development stages, from a more abstract to a more detailed view. Furthermore, a new set of modelling guidelines was introduced extending the MAAB guidelines (Matlab Automotive Advisory Board) (MAAB, 2007) adopted in the BL3 project, in order to obtain a more accurate restriction of the Stateflow language to a semantically unambiguous subset.

The SSC Metrô Rio ATP is composed by 13 Stateflow models for a total amount of approximately 120 KLOC. The test suite, provided according to the functional requirements coverage, consisted of 238 manually defined test-cases that covered 100% of functional requirements on the models; other test cases were provided to reach 100% of decision coverage on the models[8].

The validation framework 2M-TVF was used to perform the comparison of the behaviour of models and related code: the models, together with the test suites, served as inputs for the framework that automatically performed the validation of the code for each module and gave the measures of coverage on both model and code. A report was produced after the execution of the MBT phase, containing all the useful information on the performed verification activity.

The advantages of the new unit testing approach can be expressed in terms of bugs found and time saved during the verification phase, also thanks to the automation of the verification activities.

The modifications introduced in the development of SSC Metrô Rio ATP did not have a negative impact on the abstract analysis phase: the results obtained by the execution of the first Polyspace step shows a similar behaviour with respect to the ones obtained in the context of the BL3 project, as depicted in Figure 7; also the results of the second step are coherent with the BL3 ones. Furthermore, the adoption of the new guidelines and the multi-level architecture led to

Figure 6. Results of the Polyspace verification phase on BL3 project

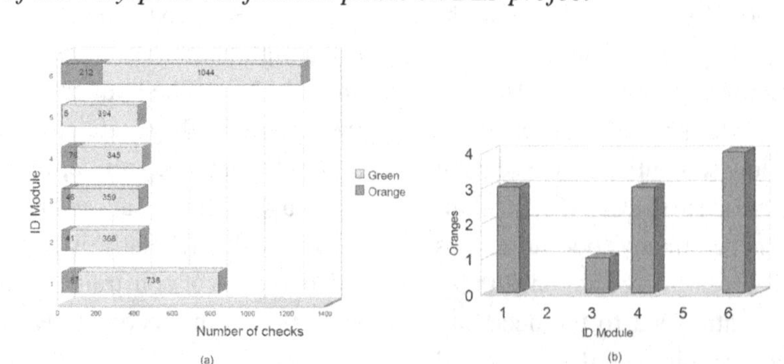

Table 1. Comparison of verification cost of BL3 against a comparable project

Verification Process	Modules	Paths	Hours
Structural Testing	19	2274	**728**
MBT + Abstract Interpretation	21	>8000	**227** (162 + 75)

a better structured generated code, with a great improvement for the human verification activities after the Polyspace analysis. Indeed, in the SSC Metrô Rio ATP project we have experienced a sensible reduction of the analysis time for the justification of orange marks: from 3 hours and a half in average for each module employed in the BL3 project, to 3 hours and ten minutes.

Table 2 compares the results of the verification activities performed on SSC Metrô Rio with the results of a previous ATP project named SSC BL1Plus, where model-based development was also employed. In SSC BL1Plus, the extended modelling guidelines, the 2M-TVF validation framework and the abstract interpretation technique (Ferrari, Grasso, Magnani, Fantechi, & Tempestini, 2010) were not yet introduced. Since the two projects are comparable in terms of decision coverage (see DC column) achieved with the tests, and therefore the code behaviour has been consistently exercised, it can be hypothesized that the considerable reduction of detected bugs (from 10 to 3 bugs per module) is due to the improvements of the development process. In the last column of Table 2 we have considered the overall time required to pinpoint the bugs and provide corrections. The well defined architecture derived from the multi-layer approach was allowed to reduce this time from 55 minutes to half an hour per bug in average.

Figure 7. Results of the first step of Polyspace verification phase on SSC Metrô Rio project

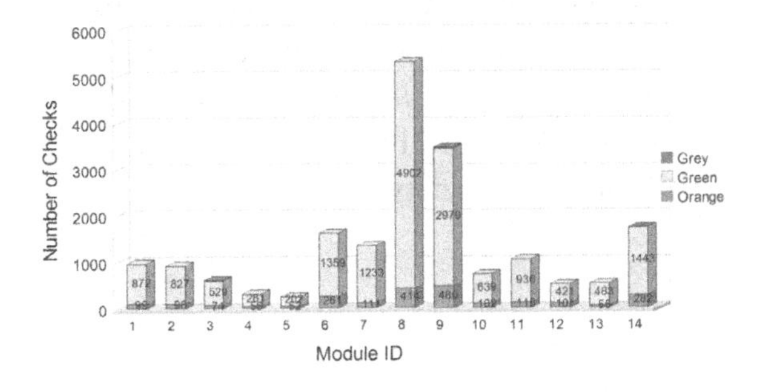

Though the automation of the MBT phase through the 2M-TVF framework has improved the validation task in comparison with the BL3 project, the bottleneck of the activity is still the manual definition of tests for the models, which takes about the 60-70% of the cost of the verification process. In order to address this issue, the company is currently evaluating strategies for test generation that can be considered compatible with the current process.

CONCLUSION

This paper has presented the unit testing process adopted by GETS, a railway signalling manufacturer, and the path the company followed to consolidate the described process.

Thanks to research activities, also in collaboration with University of Studies of Florence, GETS was able to adopt a whole set of advanced techniques in order to improve the reliability and the safety of the products. Model-based development is routinely used within the company and the introduction of model-based testing and abstract interpretation led to a further improvement in terms of development time and verification accuracy. Since GETS operates in a safety context regulated by the CENELEC norms, a set of techniques have

Table 2. Bug detection and correction costs for SSC Metrô Rio and a comparable project

Project	Modules	LOC	DC	Bugs	Man/H
SSC Metrô Rio	13	120K	100,00%	33	**16**
SSC BL1lus	12	40K	93,00%	114	**105**

been adopted to perform validation activities, such as the validation of the automatically generated code by means of a new developed framework. The use of abstract interpretation permits the full exploration of the state space of the program that is a prohibitive goal for traditional testing, thanks to the fact that the code is not executed but formally analysed; guidelines for the use of Polyspace have already been defined by the company.

The effectiveness of the new unit testing process has been shown by the results obtained in different projects run by GETS during the last years, namely the ATP systems BL3 and SSC Metrô Rio. In the first project the described process was used for the first time: the results of this application demonstrated that the adoption of the new approach could conduct to a notable decrease of verification and testing costs, furthermore increasing the confidence on the correctness of the code. The evolution in the design approach, included the usage of more restrictive modelling guidelines, brought to a reduction of code defects that has implied considerable time savings for bugs correction. The positive results of the two reported experiences make us convinced that the approach is mature to be introduced as company-standard for the development of safety-critical systems.

REFERENCES

Aiken, A. (1999). Introduction to set constraint-based program analysis. *Science of Computer Programming*, *35*(2-3), 79–111. doi:10.1016/S0167-6423(99)00007-6

Aydal, E. G., Paige, R. F., Utting, M., & Woodcock, J. (2009). Putting formal specifications under the magnifying glass: Model-based testing for validation. In *Proceedings of the 2nd International Conference on Software Testing Verification and Validation*, Denver, CO (pp. 131-140). Washington, DC: IEEE Computer Society.

Bacherini, S., Fantechi, A., Tempestini, M., & Zingoni, N. (2006). A story about formal methods adoption by a railway signaling manufacturer. In J. Misra, T. Nipkow, & E. Sekerinski (Eds.), *Proceedings of the 14th International Symposium on Formal Methods*, Hamilton, ON, Canada (LNCS 4085, pp. 179-189).

Baresel, A., Conrad, M., Sadeghipour, S., & Wegener, J. (2003). The interplay between model coverage and code coverage. In *Proceedings of the 11th European International Conference on Software Testing, Analysis and Review*, Amsterdam, Netherlands.

Barkah, D., Ermedahl, A., Gustafsson, J., Lisper, B., & Sandberg, C. (2008). Evaluation of automatic flow analysis for WCET calculation on industrial real-time system code. In *Proceedings of the 20th Euromicro Conference on Real-Time Systems*, Prague, Czech Republic (pp. 331-340). Washington, DC: IEEE Computer Society.

Blanchet, B., Cousot, P., Cousot, R., Feret, R., Mauborgne, R., Miné, A., et al. (2003). A static analyzer for large safety-critical software. In *Proceedings of the ACM SIGPLAN Conference on Programming Language Design and Implementation*, San Diego, CA (pp. 196-207). New York, NY: ACM Press.

Bozga, M., Fernandez, J. C., Ghirvu, L., Jard, C., Jéron, T., & Kerbrat, A. (2000). Verification and test generation for the Sscop protocol. *Science of Computer Programming*, *36*(1), 27–52. doi:10.1016/S0167-6423(99)00017-9

Brat, G., & Klemm, R. (2003). Static analysis of the mars exploration rover flight software. In *Proceedings of the 1st International Space Mission Challenges for Information Technology*, Pasadena, CA (pp. 321-326). Washington, DC: IEEE Computer Society.

Broy, M., & Jonsson, B. Katoen, Philipps, J., Leucker, M., & Pretschner, A. (eds.). (2005). *Model-based testing of reactive systems* (LNCS 3472, pp. 281-291). Berlin, Germany: Springer-Verlag.

Comité Européen de Normalisation en Électronique et en Électrotechnique. (1999). *EN 50126: Railway applications - the specification and demonstration of dependability, reliability, availability, maintainability and safety (RAMS)* (Tech. Rep. No. TX9X). Brussels, Belgium: CENELEC.

Comité Européen de Normalisation en Électronique et en Électrotechnique. (2001). *EN 50128: Railway applications - communications, signalling and processing systems - software for railway control and protection systems*. Brussels, Belgium: CENELEC.

Conrad, M. (2009). Testing-based translation validation of generated code in the context of IEC 61508. *Formal Methods in System Design*, *35*(3), 389–401. doi:10.1007/s10703-009-0082-0

Cousot, P., & Cousot, R. (1977). Abstract interpretation: A unified lattice model for static analysis of programs by construction or approximation of fixpoints. In *Proceedings of the 4th ACM SIGACT-SIGPLAN Symposium on Principles of Programming Languages*, Los Angeles, CA (pp.238-353). New York, NY: ACM Press.

Cousot, P., & Cousot, R. (1979). Systematic design of program analysis frameworks. In *Proceedings of the 6th Annual ACM SIGPLAN-SIGACT Symposium on Principles of Programming Languages*, San Antonio, TX (pp. 269-282). New York, NY: ACM Press.

Cousot, P., & Cousot, R. (1992). Abstract interpretation frameworks. *Journal of Logic and Computation*, *2*(4), 511–547. doi:10.1093/logcom/2.4.511

Dalal, S. R., Jain, A., Karunanithi, N., Leaton, J. M., Lott, C. M., Patton, G. C., et al. (1999). Model-based testing in practice. In *Proceedings of the 21st International Conference on Software Engineering*, Los Angeles, CA (pp. 285-294). New York, NY: ACM Press.

Delmas, D., & Souyris, J. (2007). Astrée: From research to industry. In H. R. Nielson & G. Filé (Eds.), *Proceedings of the 14th International Static Analysis Symposium*, Lyngby, Denmark (LNCS 4634, pp. 437-451).

Deutsch, A. (2004). *Static verification of dynamic properties*. Retrieved from http://www.sigada.org/conf/sigada2003/SIGAda2003-CDROM

El-Far, I. K., & Whittaker, J. A. (2002). Model-based software testing. In Marciniak, J. J. (Ed.), *Encyclopedia of software engineering* (*Vol. 1*, pp. 825–837). New York, NY: John Wiley & Sons.

Ferdinand, C., Heckmann, R., & Franzel, B. (2007). Static memory and timing analysis of embedded systems code. In *Proceedings of the 3rd European Symposium on Verification and Validation of Software Systems*, Eindhoven, The Netherlands (pp. 153-163).

Ferrari, A., Fantechi, A., Tempestini, M., & Zingoni, N. (2009). Modeling guidelines for code generation in the railway signaling context. In *Proceedings of the 1st NASA Formal Methods Symposium*, Moffet Field, CA (pp. 166-170).

Ferrari, A., Grasso, D., Magnani, G., Fantechi, A., & Tempestini, M. (2010, September). The Metrô Rio ATP case study. In S. Kowalewski & M. Roveri (Eds.), *Proceedings of the 15th International Workshop on Formal Methods for Industrial Critical Systems*, Antwerp, Belgium (LNCS 6371, pp. 1-16).

Fosdick, L. D., & Osterweil, L. J. (1976). Data flow analysis in software reliability. *ACM Computing Surveys*, *8*(3), 305–330. doi:10.1145/356674.356676

Grasso, D., Fantechi, A., Ferrari, A., Becheri, C., & Bacherini, S. (2010). Model based testing and abstract interpretation in the railway signaling context. In *Proceedings of the Third International Conference on Software Testing, Verification and Validation, Paris, France* (pp. 103-106). Washington, DC: IEEE Computer Society.

Gulavani, B. S., & Rajamani, S. K. (2006). Counterexample driven refinement for abstract interpretation. In H. Hermanns & J. Palsberg (Eds.), *Proceedings of the 12th International Conference on Tools and Algorithms for the Construction and Analysis of Systems*, Wien, Austria (LNCS 3920, pp. 474-488).

Harel, D., Lachover, H., Naamad, A., Pnueli, A., Politi, M., Sherman, R., et al. (1988). Statemate: A working environment for the development of complex reactive systems. In *Proceedings of the 10th International Conference on Software Engineering*, Raffles City, Singapore (pp. 396-496). Washington, DC: IEEE Computer Society.

Hessel, A., & Pettersson, P. (2006). Model-based testing of a WAP gateway: An industrial study. In L. Brim & M. Leucker (Eds.), *Proceedings of the 11th International Workshop on Formal Methods for Industrial Critical Systems*, Bonn, Germany (LNCS 4346, pp. 116-131).

Hierons, R. M., Bogdanov, K., Bowen, J. P., Cleaveland, R., Derrick, J., & Dick, J. (2009). Using formal specifications to support testing. *ACM Computing Surveys*, *41*(2), 1–9. doi:10.1145/1459352.1459354

Hierons, R. M., Bowen, J. P., & Harman, M. (Eds.). (2008). *Formal methods and testing* (LNCS 4949, pp. 1-38). Berlin, Germany: Springer-Verlag.

Howden, W. E. (1975). Methodology for the generation of program test data. *IEEE Transactions on Computers*, *24*(5), 554–560. doi:10.1109/T-C.1975.224259

Jung, Y., Kim, J., Shin, J., & Yi, K. (2005). Taming false alarms from a domain-unaware c analyzer by a Bayesian statistical post analysis. In C. Hankin & I. Siveroni (Eds.), *Proceedings of the 12th International Static Analysis Symposium*, London, UK (LNCS 3672, pp. 203-217).

Kim, Y., Lee, J., Han, H., & Choea, K. M. (2010). Filtering false alarms of buffer overflow analysis using SMT solvers. *Information and Software Technology*, *52*(2), 210–219. doi:10.1016/j.infsof.2009.10.004

Kremenek, T., & Engler, D. R. (2003). Z-Ranking: Using statistical analysis to counter the impact of static analysis approximations. In R. Cousot (Ed.), *Proceedings of the 10th International Static Analysis Symposium*, San Diego, CA (LNCS 2694, pp. 295-315).

Larsen, K. G., Mikucionis, M., Nielsen, B., & Skou, A. (2005). Testing real-time embedded software using UPPAAL-TRON - an industrial case study. In *Proceedings of the 5th ACM International Conference on Embedded Software*, Jersey City, NJ (pp. 299-306). New York, NY: ACM Press.

Matlab Automotive Advisory Board. (2007). *Control algorithm modeling guidelines using Matlab, Simulink and Stateflow, version 2.0.* Retrieved from http://www.mathworks.com/automotive/standards/maab.html;jsessionid=WTtvNL6XcCkNpFBMJQsPdQ6tnhcNsPQXJx9Jxffh1rLDnTBKKh7w!892165066

Miller, S. P., Whalen, M. W., & Cofer, D. D. (2010). Software model checking takes off. *Communications of the ACM*, *53*(2), 58–64. doi:10.1145/1646353.1646372

Mohagheghi, P., & Dehlen, V. (2010). Where is the proof? A review of experiences from applying MDE in industry. In I. Schieferdecker & A. Hartman (Eds.), *Proceedings of the International Conference on Model Based Testing of Reactive Systems* (LNCS 5095, pp. 432-443).

Object Management Group. (2010a). *OMG systems modeling language (OMG SysML™), Version 1.2*. Retrieved from http://www.omg.org/spec/SysML/1.2

Object Management Group. (2010b). *OMG unified modeling language™(OMG UML), Infrastructure, Version 2.3*. Retrieved from http://www.omg.org/spec/UML/2.3/Infrastructure

Object Management Group. (2010c). *OMG unified modeling language™(OMG UML), Superstructure, Version 2.3*. Retrieved from http://www.omg.org/spec/UML/2.3/Superstructure

Pasareanu, C. S., Schumann, J., Mehlitz, P., Lowry, M., Karsai, G., Nine, H., et al. (2009). Model based analysis and test generation for flight software. In *Proceedings of the 3rd IEEE International Conference on Space Mission Challenges for Information Technology*, Pasadena, CA (pp. 83-90). Washington, DC: IEEE Computer Society.

Post, H., & Sinz, C. Kaiser, A., & Gorges, T. (2008). Reducing false positives by combining abstract interpretation and bounded model checking. In *Proceedings of the 23rd IEEE/ACM International Conference on Automated Software Engineering*, L'Aquila, Italy (pp. 188-197). Washington, DC: IEEE Computer Society.

Prenninger, W., El-Ramly, M., & Horstmann, M. (2005). Case studies. In M. Broy, B. Jonsson, J. P. Katoen, M. Leucker, & A. Pretschner (Eds.), *Proceedings of the International Conference on Model Based Testing of Reactive Systems* (LNCS 3472, pp. 439-461).

Pretschner, A., & Phillips, J. (2005). Methodological issues in model-based testing. In M. Broy, B. Jonsson, J. P. Katoen, M. Leucker, & A. Pretschner (Eds.), *Proceedings of the International Conference on Model Based Testing of Reactive Systems* (LNCS 3472, pp. 281-291).

Pretschner, A., Prenninger, W., Wagner, S., Kühnel, C., Baumgartner, M., Sostawa, B., et al. (2005). One evaluation of model-based testing and its automation. In *Proceedings of the 27th International Conference on Software Engineering*, St. Louis, MO (pp. 392-401). New York, NY: ACM Press.

Pretschner, A., Slotosch, O., Aiglstorfer, A., & Kriebel, S. (2004). Model-based testing for real. *Software Tools for Technology Transfer*, *5*(2-3), 140–157. doi:10.1007/s10009-003-0128-3

Regehr, J., Reid, A., & Webb, K. (2005). Eliminating stack overflow by abstract interpretation. *ACM Transactions on Embedded Computing Systems*, *4*(4), 751–778. doi:10.1145/1113830.1113833

Rival, X. (2005). Understanding the origin of alarms in Astrée. In C. Hankin & I. Siveroni (Eds), *Proceedings of the 12th International Static Analysis Symposium*, London, UK (LNCS 3672, pp. 303-319).

Streitferdt, D., Nenninger, P., Bilich, C., Kantz, F., Bauer, T., & Eschbach, R. (2008). Model-based testing in the automation domain, safety enabled. In *Proceedings of the 1st Workshop on Model-based Testing in Practice*, Berlin, Germany (pp. 83-89).

Thesing, S., Souyris, J., Heckmann, R., Randimbivololona, F., Langenbach, M., Wilhelm, R., et al. (2003). An abstract interpretation-based timing validation of hard real-time avionics software systems. In *Proceedings of the International Conference on Dependable Systems and Networks*, San Francisco, CA (pp. 625-632).Washington, DC: IEEE Computer Society.

Utting, M., & Legeard, B. (2007). *Practical model-based testing: A tools approach.* San Francisco, CA: Morgan Kaufmann.

Venet, A., & Brat, G. (2004). Precise and efficient static array bound checking for large embedded C programs. In *Proceedings of the ACM SIGPLAN Conference on Programming Language Design and Implementation*, Washington, DC (pp. 231-242). New York, NY: ACM Press.

Vouk, M. A. (1990). Back-to-back testing. *Information and Software Technology, 32*(1), 34–45. doi:10.1016/0950-5849(90)90044-R

Weiser, M. (1981). Program slicing. In *Proceedings of the 5th International Conference on Software Engineering*, San Diego, CA (pp. 439-449). Washington, DC: IEEE Computer Society.

ENDNOTES

1. Latest available version: Simulink 7.6, Stateflow 7.6, released on September 3rd, 2010. Simulink and Stateflow are registered trademarks of The MathWorks®, Inc. (http://www.mathworks.com/).
2. Latest available version: SCADE Suite 6.1, released on January 8th, 2009. SCADE Suite is a registered trademarks of Esterel Technologies, Inc. (http://www.esterel-technologies.com/).
3. http://getransportation.com/
4. Latest available version: Real-Time Workshop Embedded Coder 5.6, released on September 3rd, 2010. Real-Time Workshop is a registered trademarks of The MathWorks, Inc. (http://www.mathworks.com/).
5. http://www.absint.de/
6. Latest available version: Polyspace 8.0, released on September 3rd, 2010. Polyspace is a registered trademarks of The MathWorks, Inc. (http://www.mathworks.com/).
7. In the context of the paper, we consider that the number of paths is always evaluated according to the boundary-interior method (Howden, 1975).
8. Coverage measurements were obtained by the means of Model Coverage Tool in Simulink Verification and Validation™ for the models (latest available version: Simulink Verification and Validation 3.0, released on September 3rd, 2010) and Gcov Tool in GNU Compiler Collection for the code (latest available version: GNU Compiler Collection 4.4.5, released on October 1st, 2010).

This work was previously published in the International Journal of Embedded and Real-Time Communication Systems, Volume 2, Issue 2, edited by Seppo Virtanen, pp. 42-61, copyright 2011 by IGI Publishing (an imprint of IGI Global).

Section 3
Technologies and Design Methods for Network-on-Chip Communication

Chapter 8
Analysis of Monitoring Structures for Network-on-Chip:
A Distributed Approach

Ville Rantala
University of Turku, Finland

Teijo Lehtonen
University of Turku, Finland

Pasi Liljeberg
University of Turku, Finland

Juha Plosila
University of Turku, Finland

ABSTRACT

Monitoring services are essential for advanced, reliable NoC systems. They should support traffic management, system reconfiguration and fault detection to enable optimal performance and reliability of the system. The paper presents a thorough description of NoC monitoring structures and studies earlier works. A distributed monitoring structure is proposed and compared against the structures presented in previous works. The proposed distributed network monitoring system does not require centralized control, is fully scalable and does not cause significant traffic overhead to the network. The distributed structure is in line with the scalability and flexibility of the NoC paradigm. The paper studies the monitoring structure features and analyzes traffic overhead, monitoring data diffusion, cost and performance. The advantages of distributed monitoring are found evident and the limitations of the structure are discussed.

INTRODUCTION

Network-on-Chip (*NoC*) is a promising interconnection paradigm for future high-performance integrated circuits (Dally & Towles, 2001; Benini & De Micheli, 2002; Nurmi, Tenhunen, Isoaho, & Jantsch, 2004). To enable the full potential of the NoC there is a need for monitoring services to diagnose the system functionality, to optimize the performance and to do run-time system reconfiguration. The reconfiguration is required to keep the system working with reasonable performance regardless of faults and unbalanced load in the system (Tagel, Ellervee, & Jervan, 2009).

DOI: 10.4018/978-1-4666-2776-5.ch008

Monitoring services can be divided to system diagnostics and traffic management. System diagnostics is used to improve the performance and reliability of the computational parts of the system. It includes fault detection, performance monitoring as well as computation load management. The other part of the monitoring services, the traffic management, focuses to the communication infrastructure which enables the interaction between computational components. Traffic management contains features to maximize communication infrastructure performance and reliability while optimizing its power consumption. It is used to balance the utilization of communication resources and to avoid congestion in the network as well as to reconfigure the routing in the network in case of faults or congestion related problems. The core of traffic management is network monitoring which observes network components and delivers the observed information to be used in traffic management.

This paper focuses on monitoring structures which are used to implement monitoring services in NoC. The principles of NoC monitoring structures are discussed and previously presented implementations are studied. A novel distributed monitoring approach is also presented. The paper is organized as follows. At first the general principles of NoC monitoring are addressed. Then NoC monitoring system structures are described and related work is discussed. Our monitoring approach is presented, analyzed and discussed in the last half of the paper, and finally conclusions are drawn.

MONITORING IN NOC

An NoC monitoring system has the same main requirements as the NoC itself; it should be flexible, scalable and it should also be capable of real-time operation (Ciordas, Basten, Radulescu, Goossens, & Meerbergen, 2005). Scalability and flexibility ensure that a monitoring system can be used in different sized NoCs without a time consuming redesign process. A monitoring system itself should also be fault tolerant.

A network monitoring structure consists of *monitoring units* (*monitors*) and *probes* as well as *communication resources*, or *wires*, to connect these components to each other and to the NoC, as illustrated in Figure 1. The probes are connected to network components, e.g., *routers*, *network interfaces* or *links*, whose functionality they observe and deliver the observed data to the monitoring units. If there are dedicated resources for communication between the probes and the monitors, the probes can be controlled from the monitors. Otherwise the probes have to be more autonomous to be able to communicate over the shared packet switched network. Monitors collect the monitoring data, exchange the data with other monitors and provide information for network components which can use it to reconfigure their operation. The system diagnostics can share the resources of the traffic management but may also require dedicated components, e.g., probes which are attached to the computational resources and centralized control devices which are not necessarily required in the traffic management.

Traffic Management

To enable the maximum throughput the communication resource utilization has to be balanced so that all the resources are utilized and the congestion is avoided. Adaptive routing and route reconfiguration are keys to balance traffic and to avoid congestion (Dally & Towles, 2004; Duato, Yalamanchili, & Ni, 2003; Rantala, Lehtonen, & Plosila, 2006). Implementation of these requires network monitoring; there should be a monitoring system which collects the information needed to optimize the routing. Basically it is not necessary to have complete knowledge of the network status in each router; a router should have information on the network state at the region around it so that it can direct the traffic to the least congested directions.

Fault Detection

Modern integrated circuits are sensitive to transient faults. Because of a complex manufacturing process it is also possible that the circuits can contain permanent manufacturing faults which can occur as run-time errors, e.g., due to electromigration. (Rabaey, 1996; Tianxu & Xuchao, 2003; Wong, Frank, Solomon, Wann, & Welser, 2005) The objective is to design a system that tolerates faults or recovers from errors caused by faults. The circuits should also be capable to reliable operation even though there are a few permanent faults.

The detection of faults requires a monitoring system to locate them and to provide the information for other components. For instance in the case of a manufacturing fault, the system could disable the faulty component and migrate its tasks to some other component. This kind of operation requires redundancy in the system to enable task migration. The fault detection is also a part of the traffic management. When there are faults in the communication infrastructure, the monitoring system can detect them and inform the system to reconfigure the routing through faultless communication resources.

Resource Allocation

Implementation of the monitoring services affects the system to be monitored. The resource allocation should be taken into account when balancing between manufacturing costs, area overhead, intrusiveness and system performance.

A monitoring system with dedicated resources is separated from the actual system and built separately of the data NoC. The monitoring components and the communication resources, used to move the monitoring data, are only in the use of the monitoring system. A monitoring system with dedicated resources is straightforward to design and it has a minimal impact to system performance. However, this approach results in area overhead and increase the complexity of the system, which typically also means increase in power consumption.

If the system does not include dedicated resources for monitoring a monitoring system can be integrated into the communication system. The monitoring data is transferred in the same communication infrastructure than the actual data of the system. The resource sharing requires virtual channels in the routers to guarantee monitoring system functionality under congested network conditions. A monitoring probe can be connected to the network through a dedicated network interface or it can share a network interface of a processing element (Ciordas, Goossens, Radulescu, & Basten, 2006).

Intrusiveness of a monitoring system defines how much the monitoring process disturbs the functionality of the system which is monitored. The objective is to design as non-intrusive monitoring system as possible. The probes should operate without disturbing the devices which they are probing and the traffic overhead on shared communication resources should be kept low. A monitoring system using shared resources does not require additional communication resources, which limits the area overhead. However, resource sharing can affect communication performance and that way interfere with the actual operation of the system. Shared-resource monitoring systems

Figure 1. Main components of network monitoring systems. Network components to be monitored are e.g. routers, links or network interfaces.

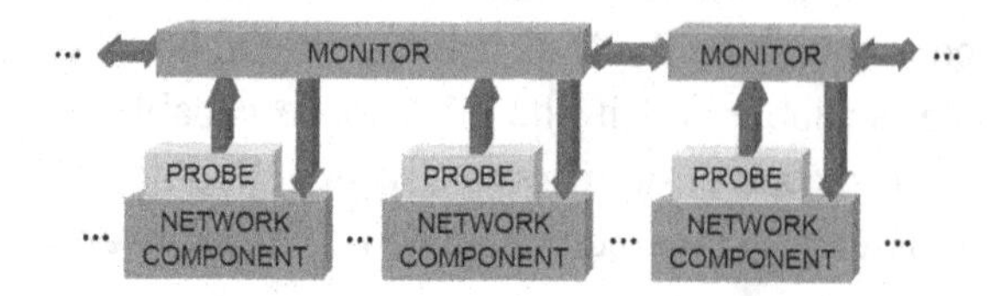

are fascinating because of their minimal added complexity. Nevertheless, due to intrusiveness aspects, shared-resource monitoring systems have to be designed carefully.

Fault Tolerance of Monitoring Systems

The monitoring systems can be used to increase the fault tolerance of NoCs. However, the monitoring systems themselves are also vulnerable to faults. If a monitoring system shares the resources with an NoC, the fault tolerance methods utilized to the core data can be applied also to the monitoring data. Otherwise, if a monitoring system has dedicated communication resources there should be separate fault tolerance mechanisms for these. These methods could be, e.g., triple modular redundancy, or error control codes (Johnson, 1989; Bertozzi, Benini, & De Micheli, 2005; Murali, Theocharides, Vijaykrishnan, Irwin, Benini, & De Micheli, 2005; Lehtonen, 2009).

STRUCTURES OF NETWORK MONITORING SYSTEMS

Depending on the purpose, requirements and operational principles the monitoring system can have centralized, clustered, distributed or localized structure. The structure of the monitoring system defines how the monitoring data is delivered to the other components in the system and which of the components are able to access this information.

Centralized Monitoring

In a system with centralized monitoring there is a central monitoring unit and a number of probes which are attached to the network components. Probes deliver the data to the central monitor which collects the monitoring data and delivers information all around the system.

Centralized monitoring has advantages on performance monitoring and on collecting statistical information from the system. Resource allocation for computational tasks may also require centralized control. However, a centralized monitoring system is inflexible and scales poorly. When the size of a system increases, the average distance between the probes and the central monitoring component increases as well. Long distances between the probes and monitors increase the probability of network faults and cause more traffic to the network especially if the monitoring system uses shared communication resources. In a highly loaded network this additional traffic can increase the probability of congestion. In centrally monitored systems all the monitoring is carried out in a single component which makes it very susceptible to faults.

Clustered and Distributed Monitoring

In clustered and distributed NoC monitoring there is no centralized monitoring unit where the global monitoring data is collected. In the clustered approach the system is divided into subsystems, clusters, each of which has a cluster monitoring unit where the monitoring data from the probes in the cluster is collected. The cluster monitors can exchange information with each other.

In a distributed monitoring structure there is a probe and a monitor unit at each router in the network. A distributed monitoring system is presented in Figure 2 and clustered monitoring system with cluster size of four is presented in Figure 3. The smaller the cluster is, the less effect a faulty monitor has on the system operation. When the monitoring is highly distributed, the impact of faulty monitoring units can be kept low.

Localized Monitoring

Localized monitoring is the simplest monitoring scheme to be implemented in an NoC. Typically there are no actual monitor components but the

Figure 2. Distributed monitoring in NoC with 4 routers. Probes and monitors are integrated into the routers.

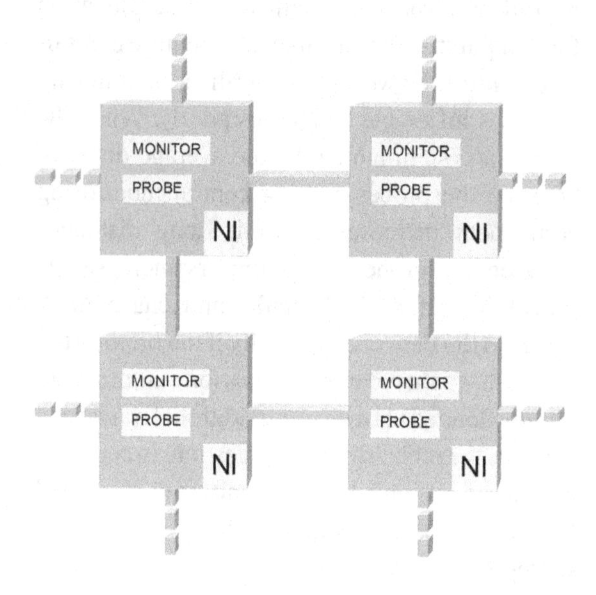

probed values are used locally without further processing or they are delivered to the neighboring routers. The localized monitoring is typically used in adaptive routing where the probes observe output channel reservation or the router utilization levels. The probed data is used locally or delivered to neighbors where routing algorithm analyzes these probed values and uses the information to reconfigure routing.

RELATED WORK

Centralized, clustered and localized NoC monitoring systems have been presented in several papers. An operating-system controlled NoC has been presented in (Nollet, Marescaux, & Verkest, 2004). It has separate *data NoC* and *control NoC* from which the latter is used for centralized monitoring and control. An operating system controls the network management through the dedicated control NoC by collecting data from the processing elements (PE). Another centralized NoC monitoring implementation is presented in (Mouhoub & Hammami, 2006) where a dedicated embedded processor is used to observe FIFO occupancies and packet transfer latencies. The monitoring processor is connected to routers with dedicated links. The collected data is used for partial dynamic reconfiguration purposes. In (Ciordas, Goossens, Basten, Radulescu, & Boon, 2006) a transaction monitoring system for Æthereal NoC (Goossens, Dielissen, & Radulescu, 2005) has been presented. This system, which monitors transactions in NoC components, can be configured to shared or dedicated communication resources. The probes monitor bit-level data which is interpreted as transactions and forwarded to a centralized monitor (*Monitoring Service Access point*). A congestion control system which monitors links and uses shared communication resources is presented in (van den Brand, Ciordas, Goossens, & Basten, 2007). It measures the amount of congestion by monitoring link utilization and adjusts data sources to control congestion. All the above implementations are meant for traffic management while a monitoring system for run-time optimization and resource allocation was presented in (Fiorin, Palermo, & Silvano, 2009). The probes are implemented within NIs and they observe throughput, latency and event occurrences in PEs.

Figure 3. Clustered monitoring (cluster size: 4)

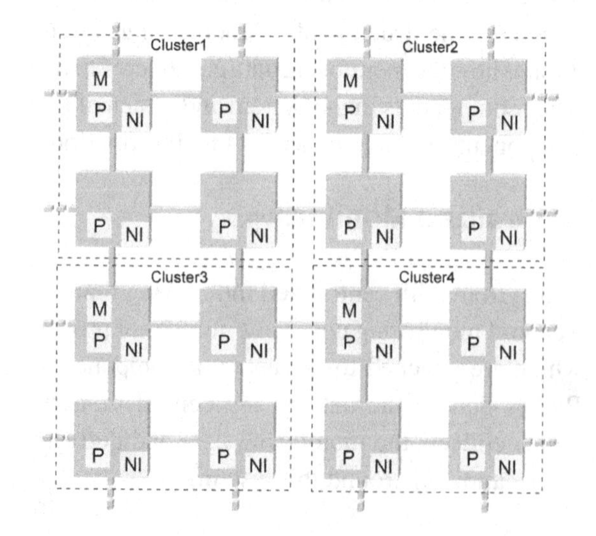

ROAdNoC is a runtime observability infrastructure which uses shared communication resources and basically has a clustered structure but can also utilize centralized control if necessary. It observes routers and NIs and collects error, run-time and functional information. (Al Faruque, Ebi, & Henkel, 2008) A notably similar structure is presented in (Ciordas, Basten, Radulescu, Goossens, & Meerbergen, 2005). Both of these structures utilize event-based communication and the latter can be used to monitor all the NoC components. On event-based communication the probes generate events from the monitoring data which allows the monitoring data abstraction and leads to communication overhead reduction. A clustered monitoring structure is presented also in (Marescaux, Rångevall, Nollet, Bartic, & Corporaal, 2005). It collects information from the routers, identifies which senders cause the congestion, and does traffic shaping by restraining these sender cores. The control data has dedicated communication resources.

Localized monitoring systems have been presented in several papers as a part of adaptive routing mechanism; for example for Nostrum NoC (Nilsson, Millberg, Öberg & Jantsch, 2003). In the basic version of the Nostrum NoC there is a probe on each router. Those provide traffic level information for the neighboring routers. The difference from distributed structures is that there are no separate monitors which collect the information and exchange it with each other.

DYNAMICALLY CLUSTERED DISTRIBUTED MONITORING STRUCTURE

Scalability is an essential feature of NoCs (Benini & De Micheli, 2002). The distributed approach makes it possible to scale the monitoring system similarly as the actual NoC can be scaled. Centralized monitoring solutions scale poorly and are built separately with the NoC. While a centralized structure can be feasible on a small system the limited scalability is against main principles of the NoC paradigm: an NoC should be fully scalable in any size and its performance should not be weakened when the size of the system is increased. Our hypothesis is that the use of distributed monitoring systems has a positive impact to traffic overhead caused by network monitoring, which decreases the demand of dedicated resources. The distributed monitoring structure has potential to be more fault tolerant than the centralized and clustered systems because the faultiness of a single component has an immediate influence only to just a few other components. Faultiness of a centralized component may endanger the functionality of the whole system.

The distributed monitoring system has several potential advantages when compared with the centralized and clustered monitoring. In centralized and clustered monitoring the monitoring data has to be moved to the monitoring unit for processing and then back to the routers before the data can be exploited. In a distributed system, the monitor receives information from its local probe and its neighbors and performs the required processing. It is also possible to implement centralized control at the operating system level while the traffic management uses distributed monitoring.

Monitoring Structure

We propose a NoC monitoring structure which is focused on traffic management, including routing reconfiguration and traffic load balancing, but can also be utilized for system reconfiguration and problem compensation in the case of faults or congestion, for instance. Fortunately, the origin of a problem is not relevant from the traffic management point of view.

Scalable NoC monitoring is based on an idea that there is no need for comprehensive status information of the whole network in a single centralized monitoring unit. To guarantee high and predictable performance the congestion should be avoided beforehand. We propose that efficient traffic management can be achieved without centralized network monitoring. When the traffic overhead of the traffic management is minimized the remaining bandwidth can be used for normal data traffic as well as essential centrally controlled features which cannot be realized with a distributed structure. Most of the traffic management tasks can be executed without centralized control, thus, there is no reason to waste resources to these purposes.

In the proposed approach every router has its own status which is based on the utilization level of the router and its neighbors. Every router has up-to-date information about the statuses of 12 closest neighbor routers around it. This monitoring protocol can be implemented in centralized, clustered or distributed monitoring systems. The distributed monitoring structure is also called the *dynamically clustered* structure. There is a dynamic cluster around every router where the status of the router has been diffused. The router has the knowledge of the statuses of the routers which belong to the dynamic cluster and the router is always placed in the middle of its own dynamic cluster which balances the amount of neighbor status data. Figure 4 illustrates an NoC with 100 routers. Every router has a dynamic cluster but only three of them are illustrated to keep the presentation clear. A router has the status information of the components in its own dynamic cluster. This information drifts from the network components to a monitor along simple neighbor to neighbor information exchange. The clustered structures, presented earlier in the section *Clustered and Distributed Monitoring* are statically clustered, which means that the borders of the clusters are fixed.

In statically clustered systems, the cluster monitor defines an overall status of its local cluster. This overall status is delivered to the neighboring clusters where it is forwarded to the routers. In a dynamically clustered structure every router has an equal amount of neighbor status information. This differs from statically clustered structures where the amount of neighbor data depends on the location of a router in a cluster (edges and corners). In a centralized implementation, the accuracy of monitoring statuses is on the same level with the distributed implementations. The difference is that in the centralized systems the status update requires a large amount of data to be transferred between routers and the central monitoring unit (see the section *Centralized Monitoring*).

Monitoring Systems in System Reconfiguration

The presented monitoring systems can be also used on small system reconfiguration tasks which do not require centralized control. Reconfiguration messages can be broadcasted over the monitoring system all around the network. The traffic diffusion analysis, which is strongly related with broadcasted reconfiguration messages, is presented later.

Figure 4. Dynamic clusters of three routers, marked with numbers 1, 2 and 3 in distributed monitoring. Each of the 100 routers in this network has a similar dynamic cluster around it.

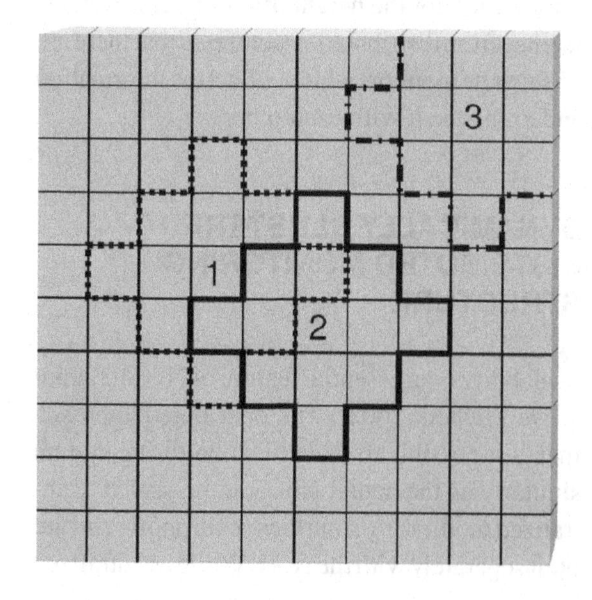

ANALYSIS

We analyzed and compared the different monitoring structures at a conceptual level. The analysis is based on calculations made by hand and with MATLAB. The conceptual level analysis is sufficient in terms of monitoring structure comparison. In our analysis we use the widely known mesh topology and assume that in every case there is an equal amount of network status data on every router and the data update frequency is constant (Dally & Towles, 2004; De Micheli & Benini, 2006). A monitor is always attached to a router. Therefore, the transfer of the status data of this router does not load the network. We also assume that

$$N_{width} = N_{height}$$

which means that an NoC has an equal number of rows and columns. The NoC width and height represent the number of routers next to each other in horizontal and vertical dimensions, respectively. The corresponding assumption for clusters is

$$C_{width} = C_{height}$$

which defines that clusters are squares. The number of routers in an NoC is depicted using term N_{size} and the number of routers in a cluster using term C_{size}. The last assumption is that

$$N_{size} = kC_{size}, k \in \mathbb{N}$$

which defines that an NoC consists of one or many *complete* clusters. The assumptions define typical features of mesh Networks-on-Chip. They are carefully defined to enable structure comparison and analysis without remarkable inaccuracy.

The metrics to be analyzed are traffic overhead caused by the monitoring system, diffusion of the data in the network, maximum path lengths on monitoring data transfers and at a coarse level the cost of centralized, clustered and distributed NoC monitoring systems. The traffic overhead analysis is essential when estimating the intrusiveness of a monitoring system. The rate of the information diffusion in the network has an influence to the reaction speed and it is closely related with the maximum path length analysis. The maximum path length analysis is to some extent also related with monitoring system fault tolerance. The probability of fault related problems increases when the information transfer paths get longer. Additional challenges can be inflicted by the utilized fault tolerance mechanisms which may slow the system operation down at some point e.g. due to retransmissions.

Traffic Overhead

The size and structure of the monitoring system, the monitoring cluster size and the number of clusters have an effect on the traffic overhead caused by the monitoring service. In statically clustered structures there is communication (a) from probes to a monitor, (b) between monitors and (c) from a monitor to routers. Because the distributed system can be seen as a clustered system with cluster size of 1, there is no communication inside a cluster, only between them (b). In centralized structures there is only one large cluster which means that the communication is only between probes and monitors (a) as well as between monitors and routers (c)

It is assumed that the amount of data in a traffic status update between two network components is so small that a single packet can hold all the information which is transferred. One transaction in this analysis means moving a packet from a router to its neighboring router, from a probe to a monitor or from a monitor to a router. The number of transactions required in a complete network status update in a centralized monitoring structure is represented with equation

$$T_{cent} = 2H_{avg}\left(N_{size} - 1\right)$$

where H_{avg} is the average hop count between the monitor and any router and N_{size} is the size of the NoC i.e. the number of routers in the NoC. The multiplier 2 depicts bidirectional traffic between the routers and the monitor. The average hop count can be calculated if a deterministic routing algorithm is used but may vary in systems using adaptive routing with non-minimal routes.

In a clustered monitoring structure the number of required transactions in a complete status update is represented with equation

$$T_{clust} = 2\left(H_{avg}\left(C_{size} - 1\right)\frac{N_{size}}{C_{size}} + C_{borders}\sqrt{C_{size}}\right)$$

where the multiplier of 2 illustrates bidirectional traffic in and between clusters. H_{avg} is the average hop count between any router and the monitor of the same cluster. N_{size} and C_{size} are sizes of an NoC and a cluster, respectively. Hence, the ratio of these is the number of clusters. The first term of the sum represents the traffic inside clusters and the second term between monitors of neighboring clusters. The square root depicts the distance between the monitors. $C_{borders}$ is the number of borderlines between clusters and is calculated using equation

$$C_{borders} = 2\left(\left(\sqrt{\frac{N_{size}}{C_{size}}} - 1\right)\sqrt{\frac{N_{size}}{C_{size}}}\right)$$

where the second square root represents the number of rows in the network which is multiplied with the term representing the number of cluster borderlines in a row. The multiplier of 2 completes the equation to include also the borderlines in the other dimension, i.e. horizontal and vertical borderlines.

In distributed monitoring structure there is traffic only between neighboring routers and the number of transactions is calculated using equation

$$T_{dist} = 4N_{size} - 4\sqrt{N_{size}}$$

The first term of the difference represents the traffic which is sent from each router to four directions. The second term corrects the amount of traffic in the edges of the network where there is traffic only to two or three directions. Note that $T_{dist} = T_{clust}$ when $C_{size} = 1$ and $T_{cent} = T_{clust}$ when $C_{size} = N_{size}$.

The amount of required communication in NoCs of different sizes and with different cluster sizes is illustrated in Figure 5. Figure 5 shows that larger cluster sizes cause more traffic and the centralized system clearly causes the largest traffic overhead. The cluster size of two is also analyzed to complete the analysis. One can notice that it causes the lowest overhead while the overhead caused by the distributed structure and the clustered one with the cluster size of four is just slightly higher. However, in the below section *Diffusion of the Network Information* we conclude that clusters with two routers are not well suited to symmetrically structured networks because they spread the traffic information unevenly.

Figure 5. Number of transactions required in complete network status update

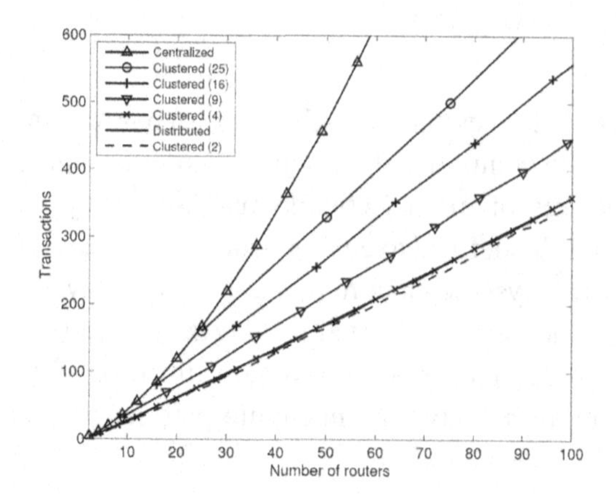

The lengths of paths between routers, probes and monitors affect the delay of a complete update process. The comparison of the longest packet traverse paths was made by calculating the longest productive paths between the possible positions of routers and monitors. This comparison is presented in Figure 6. The figure shows the longest, or worst case, distances between routers and monitors or two neighboring monitors (which one is longer) in different structures. Lengths of the information transfer paths have to be taken into account because the monitoring data transfer delay has a significant impact on the monitoring system performance.

The influence of congestion and other problems cannot be seen in this analysis. If the paths of the monitoring packets are significantly rerouted due to problems in the network, the traverse lengths and hop counts can increase. The severity of a problem to a short path is relatively significant and may disable the path completely. However, the probability of problems is higher for the longer paths and that way the longer the path, the more vulnerable to faults it is.

Diffusion of the Network Information

Network information diffusion in distributed and clustered network monitoring is illustrated in Figure 7. Diffusion of information in the monitoring system is divided into phases so that a phase consists of traffic information collection inside a monitoring cluster, information exchange between neighboring clusters and information delivery to the routers in the local cluster. The local cluster is reached during the first phase, neighbor clusters during the second phase and during the third phase the information spreads to the neighbors of the neighboring clusters. Still, it has to be noted that the durations of the update phases are proportional to the cluster size. The figure shows that a larger cluster size makes diffusion coarser because the information reaches one cluster at a time.

Figure 6. Comparison of the maximum traverse lengths of a packet on a monitoring status update

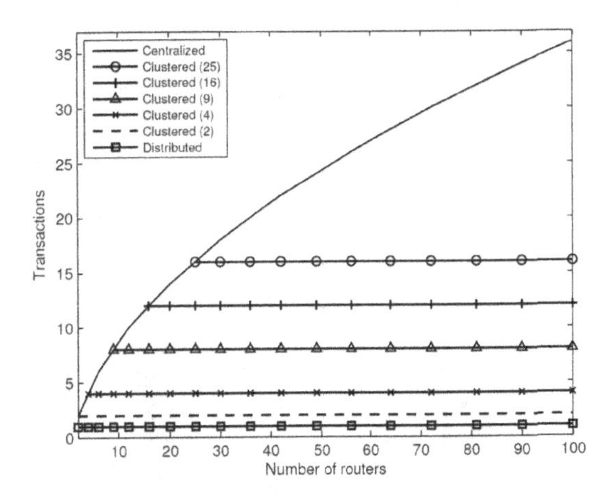

As Figure 7 shows, the shape of the monitoring cluster should be the same as the shape of the whole network. This guarantees that the data spreads evenly over the network and approaches each side of the network with the same rate. For example if the cluster size is two, the data diffuses to the other dimension with half the rate.

The area or the number of routers where the network status data diffuses during a certain time frame is represented with equation

$$D_{area} = \begin{cases} 0, p = 0 \\ C_{size}, p = 1 \\ C_{size} + \sum_{i=2}^{p} 4\left(i-1\right)C_{size}, p > 1 \end{cases}$$

where p is the number of update phases (See Figure 7). At the first update phase (p=1) the information diffuses to the local cluster. During the next phases the data diffuses to the neighboring clusters of the cluster or clusters which were updated at the previous phase. This additional area increases by the size of four clusters during every phase. This is represented with the sum term.

Figure 7. Diffusion of network status information

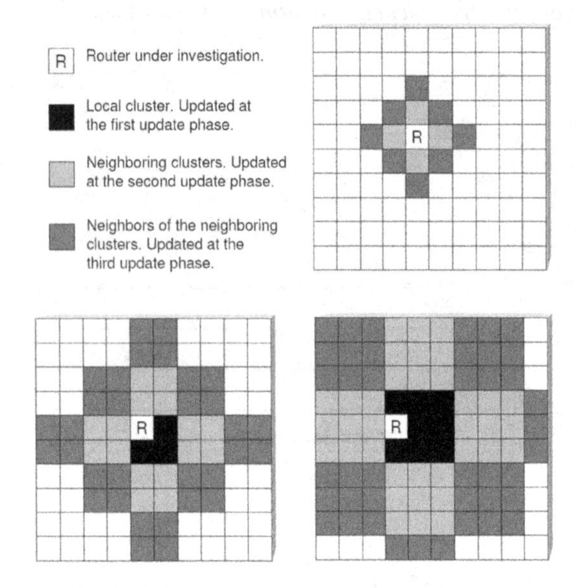

The duration of an update process can be calculated using equation

$$t = \left(p-1\right)\sqrt{C_{size}} + 4\left\lfloor \frac{\sqrt{C_{size}}}{2} \right\rfloor, p > 0$$

where the first term represents the time consumed to move data from one monitor to another while the second term depicts the time used to collect the data from routers and finally to deliver the data to routers on another cluster. The floor function gives the maximum hop count between a router and a monitor in a cluster.

Figure 8 shows how fast the information diffuses in the network. The distributed system outperforms other candidates. We can note that the structure with cluster size of nine is faster than the structure with cluster size of four. Respectively, the structure with cluster size of 25 spreads the data faster than the system with cluster size of 16. This is due to the more optimal structure of the clusters with sizes of 9 and 25. When C_{width} and C_{height} of a cluster are odd numbers the cluster monitor can be placed exactly in the middle of the cluster which minimized the theoretical average hop counts (H_{avg}) in a cluster. Figure 8 also shows that when a message needs to be delivered to the whole system immediately the distributed monitoring system accomplishes this with the highest rate. This may be useful if immediate system reconfiguration is required.

Cost of Monitoring Systems

Cost of a monitoring system implementation is related to the number and size of required probes and monitor logic and registers in the monitors. In every system, there is at least one probe for each router and registers for their own and their neighbors' statuses. The number of registers in a router is not related to the used monitoring structure so they are not taken into account in this analysis. The probes in the distributed monitoring system are simpler than in the clustered and centralized versions because they do not have to communicate over the network with the monitoring unit. The simple probes could be used also in clustered and centralized systems if there are dedicated communication resources between the probes and the monitors, and if the probe is connected to the router to which the monitor is

Figure 8. Diffusion of network status information as a function of time (size of a cluster in parentheses)

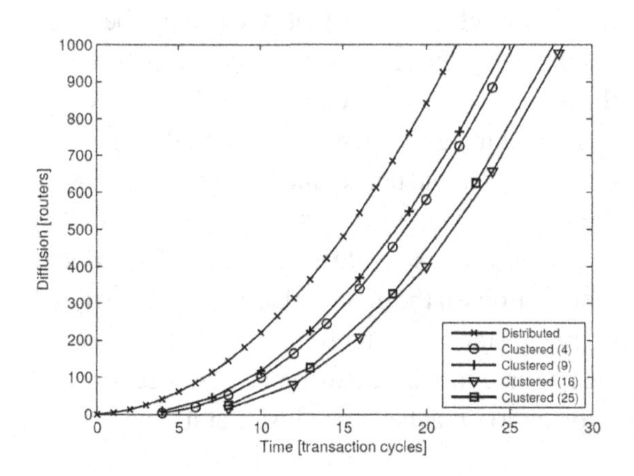

attached. Furthermore, monitors in distributed systems are simpler because they handle smaller amounts of data.

A monitor in a centralized structure needs registers to store the statuses of all the routers in the network. In clustered systems there have to be registers in monitors to store the statuses of the routers in a local cluster and the overall statuses of the neighboring clusters. A monitoring unit of the distributed monitoring system requires registers for the status of the local router and the statuses of the neighbors. In clustered and distributed systems the number of neighbor statuses to be stored is defined during the design process. Here we assume that the statuses of 12 neighbors are stored (closest neighbors and their neighbors). If the routing algorithm uses raw (i.e. unprocessed) neighbor status data, a monitor in distributed and clustered structures may use this from the router's memory and that way the demand for registers in monitors is decreased.

A cost comparison is presented in Table 1. Autonomous probes are probes which can communicate with a monitoring unit over the NoC while simple probes are directly connected to a monitor with dedicated communication resources. Cluster monitors are monitors that take care of several probes and routers. Those are used in centralized and clustered systems while simple monitors which just take care of a single probe and communicate with their neighboring monitors, are used in distributed monitoring systems. The required amount of memory in the monitors is defined as the number of registers where one register is able to store the status of one router. The number of required registers is listed in the table. The number of registers in distributed and clustered structures depends on how efficiently the registers of the local router can be utilized for the purposes of the monitor.

Table 1. Monitoring system cost comparison

	Distributed	Clustered	Centralized
Autonomous probes	0	$n-n/m$	$n-1$
Simple probes	n	n/m	1
Cluster monitors	0	n/m	1
Simple monitors	n	0	0
Registers in a monitor	$1 \ldots (q+1)$	$m \ldots (m+q)$	N

n = routers in the network
m = routers in a cluster
q = neighboring monitors/clusters whose overall statuses are stored to a monitor.

Exact cost comparison is impossible without circuit implementations but some assumptions of the component sizes and especially of their mutual ratios can be made. At first we define the units; the size of a simple probe: S_P, and the unit cost for monitors: S_M. An autonomous probe is a simple probe with capability to communicate over an NoC. The probe and the communication part of a probe are assumed to be roughly the same size which leads to an assumption that the size of an autonomous probe is $2S_P$. Another assumption is the ratio of the different monitor sizes. The approximated monitor sizes are related to the number of direct connections from a monitor to probes, routers and other monitors. S_M is equivalent to the cost of logic required for one connection, e.g., the size of a monitor in the middle of a distributed monitoring system is $6S_M$ because the monitor is connected to a probe, a router and four neighbor monitors. In addition we denote the size of a register with S_R.

By using the above assumptions we can carry out a cost comparison of monitoring systems for an NoC with 100 routers (Table 2). Note that there are no assumptions on the relations between units S_P, S_M and S_R which means that they are not comparable with each other.

The approximations in Table 2 show that the distributed implementation is the least complex in terms of probes. The number of registers depends on the used monitoring algorithm but is at

Table 2. Cost of 100-router NoC with different monitoring structures

	Distributed	Clustered (4)	Clustered (16)	Clustered (25)	Centralized
Probes	$100S_P$	$175S_P$	$191S_P$	$196S_P$	$199S_P$
Monitors	$560S_M$	$280S_M$	$224S_M$	$208S_M$	$200S_M$
Registers	$100S_R \ldots 1096S_R$	$100S_R \ldots 284S_R$	$100S_R \ldots 152S_R$	$100S_R \ldots 112S_R$	$100S_R$

Note: Size of a cluster in parentheses.

least equal to the number of routers and increases when the size of a cluster decreases. The analysis shows that when using distributed monitoring structure it is strongly recommended to integrate the router and the monitor and share the status registers among them. This can be done also when the cluster size is relatively small and raw monitoring data is stored to the status registers of the routers. In large clusters the neighbor cluster statuses are stored to the routers far away from the monitor, which makes the register sharing difficult. Table 2 also shows that the total cost of monitors decreases when the size of a cluster increases.

It has to be noted that in distributed implementations the computation is done in smaller parts by simple components working at the same speed with the larger monitors. That way the distributed system may reach significantly higher performance because the amount of data to be processed in a monitor is substantially smaller, or they can be optimized for low power consumption and area. Furthermore, the distributed monitoring system can be more fault tolerant because the computation is divided to the small monitoring units and each of them is much less critical in terms of overall system functionality than the larger monitors in clustered and centralized systems.

TRANSACTION LEVEL ANALYSIS

To justify the simplifications of our conceptual level simulations the performance of a 100-router NoC was analyzed with a *SystemC* model which utilizes *TLM 1.0* transaction level modeling library. Three NoC models with distributed, centralized and clustered monitoring structures were implemented with capabilities to model simultaneous traffic, network errors and congestion. The cluster size of the clustered system is 25, because it is suitable to analysis of all the essential aspects. The models utilize an adaptive routing algorithm and the monitoring system provides routers with information about the utilization levels of the twelve closest neighbors as well as the usability of the links between these neighbors. The update frequency of the monitoring data is equal in each of the systems. The routing algorithm is based on a dimension order routing (Dally & Towles, 2004) and it favors productive and minimal routes but is able to use non-productive directions in case of faulty links in the productive directions. Packets are dropped if they end to a situation where they cannot be routed or if the routing takes more than the defined maximum lifetime, which is defined so that reasonable performance can be achieved without substantial packet loss. U-turns are prohibited which can cause packet dropping but at the same time keeps the routing relatively straightforward. The delays of the components are defined statically which means that for instance the delay of a router or link is always the same.

The simulations were done with weighted traffic pattern so that 1/3 of the traffic is between neighboring cores, 1/3 between the neighbors of the neighbors and 1/3 between all the cores. The traffic pattern models a situation where the NoC is designed so that cores, which have a lot interaction, are situated close to each other. The results of the throughput simulations are presented in Figure 9 and the average transfer latencies in Figure 10. Throughput illustrates the number of delivered data packets excluding the monitoring packets. The system with distributed monitoring outperforms other candidates with a significant marginal in throughput. The differences are mainly caused by the different kind of monitoring traffic. Because the traffic pattern is weighted to low distance paths the monitoring packets paths in centralized and clustered systems are long related to the data packet paths and that way increase the number of transactions. The same performance difference can be seen in the latency simulations. The monitoring packets increase congestion in the network and that way slow it down. The presence of link errors decreases average latencies because the long distance packets are more probably dropped due to increasing error probability and the packet lifetime restrictions.

The age of the monitoring data differs in different systems because the transfer times between routers and monitors vary. In the distributed monitoring, the monitors are close to the routers, which minimizes the monitoring packet transfer delays and makes possible to keep the traffic status information up to date. On the other hand, in centralized systems the transfer delays can be significantly more than the average data packet latencies which makes the maintenance of up to date traffic status data difficult.

Figure 9. Throughput of data packets. A routing cycle is the duration of the routing procedure of a packet in a router. The cluster size in the clustered system is 25.

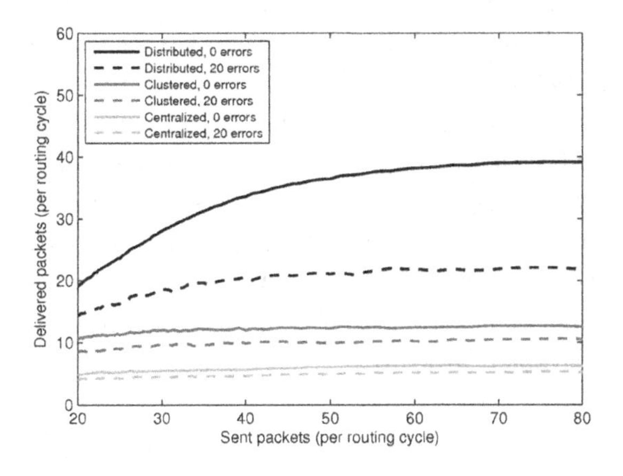

Figure 10. Average latency. The cluster size in the clustered system is 25.

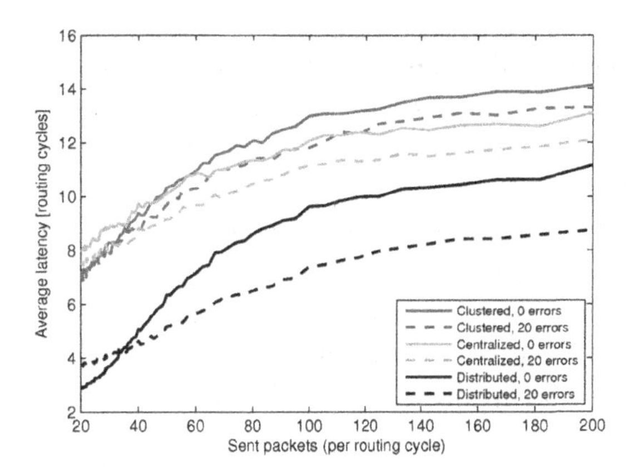

The distributed monitoring loads the system remarkably less than the other structures because there is no need to make any routing decisions while moving the monitoring packets. In the distributed structure the monitoring packets are just moved to neighbor routers which do not have to forward them. In clustered and centralized systems the monitoring packets have to be routed between routers and centralized or cluster monitors similarly than any other packet.

DISCUSSION AND FUTURE WORK

Our analysis shows that centralized control is slow and causes significant load to the system. There are tasks, especially related to traffic management, which can be implemented without centralized control. The intrusiveness of monitoring services is an essential factor in NoC design and is related to the selection between dedicated or shared resources. If the designer decides to use dedicated resources it could be reasonable to use centralized network monitoring to minimize the intrusiveness and the cost. Otherwise, if shared resources are used, at least network monitoring should be distributed to minimize the need for centralized control and the complexity of centralized monitors. In a large, complex system it may also be reasonable to use *hybrid monitoring structures*. A hybrid monitoring structure has both distributed and centralized monitoring services, possibly also clustered services, in the same system. Relatively simple and frequently repeating network monitoring tasks are carried out with distributed monitoring which optimizes the network resource utilization. Then centralized monitoring is used for system diagnostic purposes and for tasks which cannot be executed without centralized control. Network management is basically quite regional which means that most of the traffic related problems are solved using the resources near the problems origin. In most of the cases there is no need to know the situation in the other side of the network but the status of the neighbors, their neighbors and so on as far as reasonable. This decreases the need for centralized control and releases resources for more significant tasks, for instance load balancing, optimizations and reconfigurations.

In design of complex integrated circuits the focus should be kept on the computational features while the implementation of the communication infrastructure should be straightforward without time consuming redesign of essential structures. Carefully designed fully scalable distributed monitoring structures can be utilized in systems of different sizes without complicated customization processes. The structures should be developed towards general communication platforms which conform to different systems with varying performance and operational expectations.

Our conceptual analysis gives approximations of monitoring system performance and estimates of the costs, and the SystemC analysis supports the results. The future work includes development of the SystemC model and refinement of it with component VHDL implementations. The future work also includes design and development of monitoring and diagnostic services, traffic management and fault tolerant adaptive routing algorithms all of which can benefit from the monitoring structures analyzed in this paper. The research field will be extended towards both services and implementations. Realization of high performance communication requires intelligent adaptive routing algorithms which are able to efficiently utilize both the information on network traffic and faults to optimize the routing. One of the most essential challenges is to implement collection of traffic and fault information in a way that the routing algorithm can be provided with up to date information.

CONCLUSION

Distributed monitoring systems are flexible and scalable, which makes them suitable for NoCs. Partitioning the system into smaller parts also improves system level fault tolerance because a faulty component in a distributed monitoring system does not necessarily have a substantial influence to system functionality. The presented analysis shows that distributed monitoring systems load the network less than typical centralized systems. The analyzed metrics can be improved by clustering the system into smaller parts. However, clustered structures do not fully reach the performance of totally distributed monitoring systems. The main drawback of distributed monitoring is the lack of global status information which may be required in some reconfiguration tasks. The presented cost analysis shows that the proposed structure has a lower cost of probes but an increase in the monitoring unit and register costs. The outcome of the discussion is that in large and complex systems it might be beneficial to have both centralized and distributed structures

ACKNOWLEDGMENT

The authors would like to thank the Academy of Finland, the Nokia Foundation and the Finnish Foundation for Technology Promotion for financial support.

REFERENCES

Al Faruque, M., Ebi, T., & Henkel, J. (2008, November). ROAdNoC: Runtime Observability for an Adaptive Network on Chip Architecture. In *Proceedings of the IEEE/ACM International Conference on Computer-Aided Design (ICCAD 2008)* (pp. 543-548).

Benini, L., & De Micheli, G. (2002, January). Networks on Chips: a New SoC Paradigm. *IEEE Computer, 35*(1), 70–78.

Bertozzi, D., Benini, L., & De Micheli, G. (2005, June). Error Control Schemes for On-chip Communication Links: The Energy-reliability Tradeoff. *IEEE Transactions on Computer-Aided Design of Integrated Circuits and Systems, 24*(6), 818–831. doi:10.1109/TCAD.2005.847907

Ciordas, C., Basten, A., Radulescu, A., Goossens, K., & Meerbergen, J. V. (2005, October). An Event-based Monitoring Service for Networks on Chip. *ACM Transactions on Design Automation of Electronic Systems, 10*(4), 702–723. doi:10.1145/1109118.1109126

Ciordas, C., Goossens, K., Basten, T., Radulescu, A., & Boon, A. (2006, October). Transaction Monitoring in Networks on Chip: The On-chip Run-time Perspective. In *Proceedings of the International Symposium on Industrial Embedded Systems (IES '06).*

Ciordas, C., Goossens, K., Radulescu, A., & Basten, T. (2006, May). NoC Monitoring: Impact on the Design Flow. In *Proceedings of the IEEE International Symposium on Circuits and Systems (ISCAS 2006)* (pp. 1981-1984).

Dally, W. J., & Towles, B. (2001, June). Route Packets, not Wires: On-chip Interconnection Networks. In *Proceedings of the Design Automation Conference (DAC '01)* (pp. 684-689).

Dally, W. J., & Towles, B. (2004). *Principles and Practices of Interconnection Networks*. San Francisco: Morgan Kaufmann.

De Micheli, G., & Benini, L. (2006). *Networks on Chips*. San Francisco: Morgan Kaufmann.

Duato, J., Yalamanchili, S., & Ni, L. (2003). *Interconnection Networks: an Engineering Approach*. San Francisco: Morgan Kaufmann.

Fiorin, L., Palermo, G., & Silvano, C. (2009, April). MPSoCs Run-time Monitoring Through Networks-on-Chip. In *Proceedings of the Design, Automation & Test in Europe Conference and Exhibition (DATE '09)* (pp. 558-561).

Goossens, K., Dielissen, J., & Radulescu, A. (2005, September-October). Æthereal Network on Chip: Concepts, Architectures and Implementations. *IEEE Design & Test of Computers, 22*(5), 414–421. doi:10.1109/MDT.2005.99

Johnson, B. W. (1989). *Design and Analysis of Fault-Tolerant Digital Systems*. Boston: Addison-Wesley.

Lehtonen, T. (2009) *On Fault Tolerance Methods for Networks-on-Chip*. Unpublished doctoral dissertation, Turku Centre for Computer Science (TUCS), Turku, Finland.

Marescaux, T., Rångevall, A., Nollet, V., Bartic, A., & Corporaal, H. (2005, September). Distributed Congestion Control for Packet Switched Networks on Chip. In *Proceedings of the International Conference on Parallel Computing: Current & Future Issues of High-End Computing (ParCo 2005)* (pp. 761-768).

Mouhoub, R., & Hammami, O. (2006, October). NoC Monitoring Hardware Support for Fast NoC Design Space Exploration and Potential NoC Partial Dynamic Reconfiguration. In *Proceedings of the International Symposium on Industrial Embedded Systems (IES '06)*.

Murali, S., Theocharides, T., Vijaykrishnan, N., Irwin, M., Benini, L., & De Micheli, G. (2005, September-October). Analysis of Error Recovery Schemes for Networks on Chips. *IEEE Design & Test of Computers*, *22*(5), 434–442. doi:10.1109/MDT.2005.104

Nilsson, E., Millberg, M., Öberg, J., & Jantsch, A. (2003, March). Load Distribution with the Proximity Congestion Awareness in a Network on Chip. In *Proceedings of the Design Automation and Test Europe (DATE)* (pp. 1126-1127).

Nollet, V., Marescaux, T., & Verkest, D. (2004, June). Operating-system Controlled Network on Chip. In *Proceedings of the Design Automation Conference (DAC '04)* (pp. 256-259).

Nurmi, J., Tenhunen, H., Isoaho, J., & Jantsch, A. (Eds.). (2004). *Interconnect-Centric Design for Advanced SoC and NoC*. Dordrecht, The Netherlands: Kluwer Academic Publishers.

Rabaey, J. M. (1996). *Digital Integrated Circuits: A Design Perspective*. Upper Saddle River, NJ: Prentice Hall, Inc.

Rantala, V., Lehtonen, T., & Plosila, J. (2006, August). *Network on Chip Routing Algorithms* (Tech. Rep. No. 779). Turku, Finland: Turku Centre for Computer Science (TUCS).

Tagel, M., Ellervee, P., & Jervan, G. (2009, October). Scheduling Framework for Real-time Dependable NoC-based Systems. In *Proceedings of the International Symposium on System-on-Chip (SOC 2009)* (pp. 95-99).

Tianxu, Z., & Xuchao, D. (2003, November). Reliability Estimation Model of IC's Interconnect Based on Uniform Distribution of Defects on a Chip. In *Proceedings of the 18th IEEE International Symposium on Defect and Fault Tolerance in VLSI Systems* (pp. 11-17).

van den Brand, J., Ciordas, C., Goossens, K., & Basten, T. (2007, April). Congestion Controlled Best-effort Communication for Networks-on-Chip. In *Proceedings of the Design, Automation & Test in Europe Conference and Exhibition (DATE '07)* (pp. 1-6).

Wong, H.-S. P., Frank, D. J., Solomon, P. M., Wann, C. H. J., & Welser, J. J. (2005). Nanoscale CMOS. In Ionescu, A. M., & Banerjee, K. (Eds.), *Emerging Nanoelectronics: Life with and after CMOS* (*Vol. 1*, pp. 46–83). Norwell, MA: Kluwer Academic Publishers.

This work was previously published in the International Journal of Embedded and Real-Time Communication Systems, Volume 2, Issue 1, edited by Seppo Virtanen, pp. 49-67, copyright 2011 by IGI Publishing (an imprint of IGI Global).

Chapter 9
Optimized Communication Architecture of MPSoCs with a Hardware Scheduler:
A System-Level Analysis

Diandian Zhang
RWTH Aachen University, Germany

Han Zhang
RWTH Aachen University, Germany

Jeronimo Castrillon
RWTH Aachen University, Germany

Torsten Kempf
RWTH Aachen University, Germany

Bart Vanthournout
Synopsys Inc., Belgium

Gerd Ascheid
RWTH Aachen University, Germany

Rainer Leupers
RWTH Aachen University, Germany

ABSTRACT

Efficient runtime resource management in multi-processor systems-on-chip (MPSoCs) for achieving high performance and low energy consumption is one of the key challenges for system designers. OSIP, an operating system application-specific instruction-set processor, together with its well-defined programming model, provides a promising solution. It delivers high computational performance to deal with dynamic task scheduling and mapping. Being programmable, it can easily be adapted to different systems. However, the distributed computation among the different processing elements introduces complexity to the communication architecture, which tends to become the bottleneck of such systems. In this work, the authors highlight the vital importance of the communication architecture for OSIP-based systems and optimize the communication architecture. Furthermore, the effects of OSIP and the communication architecture are investigated jointly from the system point of view, based on a broad case study for a real life application (H.264) and a synthetic benchmark application.

DOI: 10.4018/978-1-4666-2776-5.ch009

INTRODUCTION

Heterogeneous multi-processor systems-on-chip (MPSoCs) are nowadays widely used in the embedded domain, such as wireless communication and multimedia, since they can provide efficient trade-offs between the computational power, the energy consumption and the flexibility of the system. One big challenge that comes with heterogeneous MPSoCs is system programming. This becomes even more critical, when taking runtime task scheduling and mapping into consideration. In large-scale systems, runtime scheduling and mapping are highly demanded from the performance and energy perspective, since it is very difficult to consider different dynamic effects at design time.

Various approaches have been proposed in academia and industry to address this problem. Generally, these approaches can be categorized into two groups: software solutions and hardware solutions. Software solutions such as TI OMAP (Texas Instruments, Inc., 2010) and Atmel D940 (Atmel Corporation, 2011) typically employ an operating system (OS), which runs on a RISC processor that dynamically distributes the workload to processing elements (PEs). These approaches are very flexible, but have low efficiency, especially when it comes to small tasks. The main reason is the high OS overhead in terms of power, memory footprint and performance.

This problem has been tackled by moving OS functionality to hardware accelerators, both for uni-core platforms (Nakano, Utama, Itabashi, Shiomi, & Imai, 1995; Kohout, Ganesh, & Jacob, 2003; Murtaza, Khan, Rafique, Bajwa, & Zaman, 2006; Nordström & Asplund, 2007) and for multi-core platforms (Park, Hong, & Chae, 2008; Seidel, 2006; Limberg et al., 2009; Lippett, 2004; Nácul, Regazzoni, & Lajolo, 2007; Pan & Wells, 2008). In the following, the hardware solutions for MPSoCs are further discussed.

The hardware OS kernel – HOSK introduced by Park, Hong, and Chae (2008) is a coprocessor that performs scheduling (fair and priority based) on a homogeneous cluster of simplified RISC processors. It features a low multi-threading overhead (less than 1% for 1-kcycle-tasks). However, a dedicated context controller should be included into the RISC processor to exchange context data between the processor and HOSK. This impedes its integration into traditional component-based design with off-the-shelf processors. Furthermore, to our best knowledge, no programming model exists for the HOSK-based MPSoCs.

In the work of Seidel (2006) and Limberg et al. (2009), a hardware scheduler called CoreManager is used to detect task dependency at runtime and schedule tasks. A programming model is provided along with CoreManager, following the synchronous data flow (SDF) model. Hence, this solution is only applied to a limited set of applications. High scheduling efficiency has been reported for CoreManager (60 cycles to schedule a task in average), which however is at the cost of high area overhead.

The approach of SystemWeaver (Lippett, 2004) focuses on the issue of task scheduling and mapping on heterogeneous MPSoCs, supported with a programming model. It has a slightly higher flexibility than HOSK and CoreManager by allowing the user to compose different basic scheduling primitives so as to implement complex scheduling decisions. However, its flexibility is still rather limited and the usage is rather difficult due to its design complexity.

In the SMP architecture introduced by Nácul, Regazzoni, and Lajolo (2007), a hardware RTOS is applied for scheduling a dual-ARM-system, based on the round-robin policy. In the system, for each ARM processor a hardware scheduler instance is required, which makes this approach difficult to scale to a large system without significantly increasing the area.

Hardware supported scheduling has also been considered in dynamically reconfigurable systems-on-chip (RSoCs) to improve the utilization of the reconfigurable components of the system. In the work presented by Pan and Wells (2008), a hardware unit is in charge of task scheduling for system reconfiguration and execution. Two types of tasks are distinguished: configuration tasks for reconfigurable logic cells and application software tasks. A priority-based policy is applied for the scheduling in this system.

In general, these pure hardware solutions provide high efficiency in dynamic scheduling in MPSoCs. However, they suffer from problems inherent to hardware such as low scalability, extensibility and usability, fixed scheduling properties and difficulties in system integration.

Based on the concept of application-specific instruction-set processors (ASIPs) (Ienne & Leupers, 2006), an operating system ASIP called OSIP has been introduced by Castrillon et al. (2009) to perform runtime task scheduling and mapping, which combines the advantages of both software and hardware solutions. On the one hand, special instructions are developed to efficiently support scheduling and mapping operations on task queues. On the other hand, coming with a C-compiler, OSIP is highly programmable, which provides the user with high flexibility to extend existing or add new scheduling features. Furthermore, the simple integration of OSIP as a standard memory-mapped component and a generic programming model consisting of well-defined software layers eases the usability of OSIP.

The work presented by Castrillon et al. (2009) has focused on the construction of OSIP-based MPSoCs and the development of the OSIP processor architecture, in which an ideal communication architecture is assumed in order to isolate the effect of OSIP. Unfortunately, one of the main development issues in today's MPSoC architectures is an efficient design of the communication architecture, as its impact on the overall performance is significant. In this work, an in-depth study of the communication architecture effect in OSIP-based systems is made.

Typical examples of on-chip interconnect are AMBA (ARM Ltd., 2010) and CoreConnect (IBM Corporation, 2010) for small systems or networks-on-chip (Benini & De Micheli, 2002; Jantsch & Tenhunen, 2003) for large-scale systems. In general, when a large number of PEs are connected in the system, high traffic load will be generated to the communication architecture. However, in OSIP-based systems, the big number of PEs also indicates high workload to OSIP. The question comes up: Which one would be the bottleneck of the system, OSIP or the communication architecture? For answering the question, different OSIP platforms with varied number of PEs and communication architectures as well as different schedulers are compared. Based on the comparison, OSIP-based MPSoCs are analyzed from a system point of view, jointly considering the effect of OSIP and the communication architecture on the performance.

The main contribution of this work is two-fold: (1) optimization of the communication architecture of OSIP-based systems and (2) thorough investigation of joint effects of the OSIP scheduling overhead and the communication overhead including synchronization.

The rest of the paper is organized as follows. First, a short introduction to OSIP-based MPSoCs is given. Next, an initial analysis of communication overhead based on a benchmark application from the multimedia domain (H.264) is presented. Following the analysis, three optimization steps for the communication architecture are introduced and evaluated. Afterwards, a detailed analysis of performance characteristics in OSIP-based systems is carried out. This analysis starts with the results of the H.264 application and is systematically investigated by using a synthetic benchmark application. Finally, a summary and an outlook are given.

OVERVIEW OF OSIP-BASED MPSOCS

In OSIP-based MPSoCs, an application-specific processor called OSIP is used as a programmable hardware scheduler to perform runtime task scheduling and mapping. It features special instructions to efficiently support scheduling and mapping algorithms, including instructions for task-level comparison, queue operations etc. In order to support different processing element classes and application requirements (e.g. hard/soft real time, best effort), the scheduling and mapping are performed in a hierarchical way, similar to the concept introduced by Goyal, Guo, and Vin (1996).

OSIP Software Integration

To simplify the process of MPSoC programming, OSIP-based systems are delivered with software layers, as shown in Figure 1. In addition to standard micro-kernels of the PEs for individual task scheduling, a software application programming interface (API) layer is inserted to support system-wide scheduling and mapping. It provides an abstraction of the actual low level communication primitives for multi-task management, e.g. task creation, task suspension or task deletion. These primitives are translated into commands and off-loaded to OSIP for dynamic task scheduling and mapping.

The PEs are grouped into several processing classes for performing runtime mapping decisions. These classes are user-defined and can be for example determined according to PE types such as RISC, DSP, hardware accelerator etc. In this way, a task can be mapped onto any PE in the corresponding class, which improves the utilization of PEs and greatly reduces the programming effort when extending the system. The superscripts of PEs in Figure 1 represent the different processing classes and the subscripts represent the individual PEs inside each class.

Furthermore, a firmware containing low level functions is provided to optimally exploit OSIP's dedicated hardware features. It comprises several basic scheduling algorithms such as round-robin, FIFO, priority-based and fair queue. Supported by a C-compiler, the user has the full flexibility to implement the own code to extend or add new scheduling features upon the provided firmware.

Figure 1. OSIP software layers

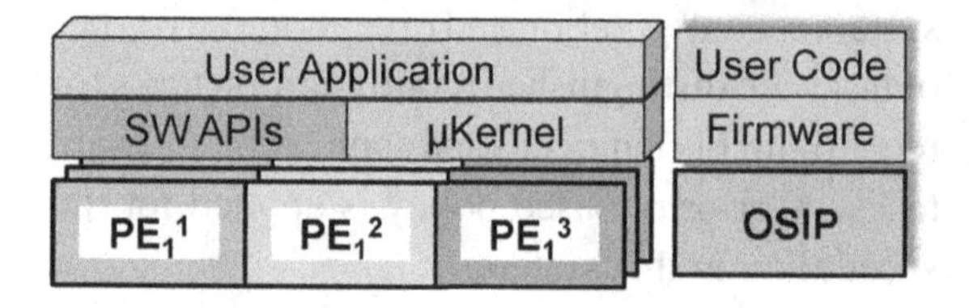

OSIP Hardware Integration

A typical hardware platform of OSIP-based MPSoCs is illustrated in Figure 2, which includes PEs of various types, a memory subsystem, peripherals and OSIP itself. From OSIP's point of view, the PEs are grouped into several classes. The communication between these components is via an arbitrary on-chip interconnect. In order to smoothly integrate into the system, OSIP is equipped with two interfaces:

- **Register Interface:** This interface enables OSIP to be accessed as a standard memory-mapped I/O, which makes the integration of OSIP very easy. Information is exchanged between the PEs and OSIP through this interface. Typical information includes the commands and arguments generated from the PEs to OSIP for multi-task management, the status (busy/idle) of OSIP and its return value. To synchronize the tasks, hardware semaphore registers are also implemented in this interface.

Figure 2. OSIP-based system

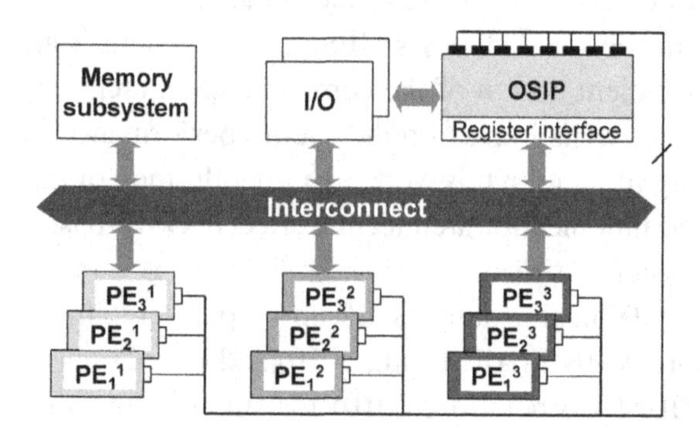

Since OSIP is a shared hardware resource, a special semaphore register is used in the register interface to guarantee that only one processor can get the access to OSIP. This is necessary for avoiding mixing the commands and arguments from different PEs. Only after getting this semaphore, a PE is allowed to continue sending commands. The other PEs will poll the semaphore, until it is freed by the owner PE.

- **Interrupt Interface:** Upon receiving a command from a PE, a corresponding handler is executed in OSIP. Depending on the commands and system state, interrupts might, but need not, be generated from OSIP to the PEs through dedicated interrupt lines in order to control platform-wide task execution. Typically, an interrupt from the OSIP triggers a new task execution on a corresponding PE. The current implementation of OSIP supports up to 32 interrupt lines.

INITIAL ANALYSIS OF COMMUNICATION OVERHEAD

Castrillon et al. (2009) have reported high efficiency of OSIP for dealing with dynamic scheduling and mapping. The evaluation was performed on a virtual platform created in SystemC using Synopsys Platform Architect (Synopsys, Inc., 2010a), which is further used as the baseline platform in this work.

The platform contains several ARM926EJS instruction-accurate processor models, the OSIP, a shared memory and several peripherals (input stream, virtual LCD and UART). OSIP is modeled at the cycle-accurate level. The components are connected with each other using an AMBA AHB bus. The ARM processors are bus masters, the other components the slaves. The whole system is driven by a 200MHz clock. As introduced above, in the work of Castrillon et al. (2009), the bus is bypassed to isolate the effect of OSIP in the system. In other words, no actual communication overhead exists in the system. This is further referred to as an ideal bus.

In order to get an impression of how a real communication architecture influences the system performance, an initial analysis of the communication overhead is made by enabling the AHB bus. The analysis results are shown in Figure 3, obtained with the H.264 video decoding application following the 2D-wave concept (Meenderinck, Azevedo, Alvarez, Juurlink, & Ramirez, 2008). By exploring the task parallelism, two processing classes are defined in the system for task mapping. One class contains all ARM processors, on which most of the algorithm kernels are mapped such as *inverse quantization*, *discrete cosine transform (DCT)* and *motion predication*. The *entropy decoding* is mapped onto another processing class, which contains only one ARM processor, since it is hard to parallelize.

As a common comparison criterion in multimedia applications, the frame rate of video decoding, given in frames per second (fps), is chosen to evaluate the performance of different OSIP platforms and especially to analyze the communication effect. The evaluated OSIP platforms are differentiated with respect to the number of ARM processors and different communication architectures.

In Figure 3, systems with an ideal bus and a real AHB bus are compared by varying the number of PEs from 1 to 11. A huge performance gap between both systems can be observed, which increases with more processors. While the measurement already shows a strong frame rate decrease by a factor of 2.3 in the single-ARM-system with the AHB bus, a larger slowdown factor of 3.8 is found in the 11-ARM-system. The increasing performance gap is not only due to the latency of the real bus, but mostly due to the bus contention which additionally introduces extremely high

Figure 3. Frame rate comparison: ideal bus vs. AHB

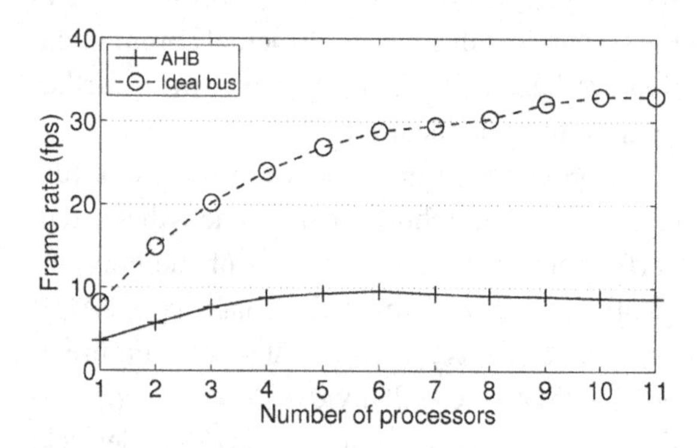

communication overhead. In fact, starting from the 6-ARM-system with the AHB bus, adding more processors even degrades the system performance. In this case, the benefit gained from the parallel execution of tasks is fully compensated by the additional communication overhead due to bus contention.

Figure 4 gives an inside view of the communication overhead in systems with the AHB bus by separating the execution time of individual processors into three components:

- **Active Time (t_A):** Which is the time that the PEs spend on executing the instructions.
- **Idle Time (t_I):** Which is the stall time, during which the PEs lie in a low-power state, waiting for interrupts from OSIP to get new tasks.
- **Communication Time (t_C):** Which is the time spent on the communication from the PEs to OSIP, the shared memory and I/Os.

For all systems, the communication time is the main contributor to the total execution time. In most cases, more than 60% of the execution time is spent on the communication, which is clearly the system bottleneck. In contrast, the active time continuously decreases, since each PE gets less tasks assigned with the increasing number of PEs. Therefore, they enter more frequently the idle state, which is illustrated by the trend of increasing idle time.

The impact of the communication architecture on the system performance can also be reflected by the OSIP state. Naturally, if more PEs are integrated into the system, requests are more frequently generated to OSIP. This would keep OSIP more often in the busy state. However, in Figure 5 this effect can only be found for the systems with the ideal bus, in which the busy time of OSIP increases steadily with more PEs. As a comparison, for the systems with a real AHB bus, OSIP stays mostly at an idle state, because most of time is spent on the communication. In fact, for systems with a large number of PEs, the percentage of busy time of OSIP is nearly unchanged. This indicates that the tasks are not assigned fast enough to the PEs due to the communication overhead, which results in a significant performance decrement.

Based on the analysis above, it can be concluded that an unoptimized communication architecture could have a disastrous effect on the

Figure 4. Time components from processor's view

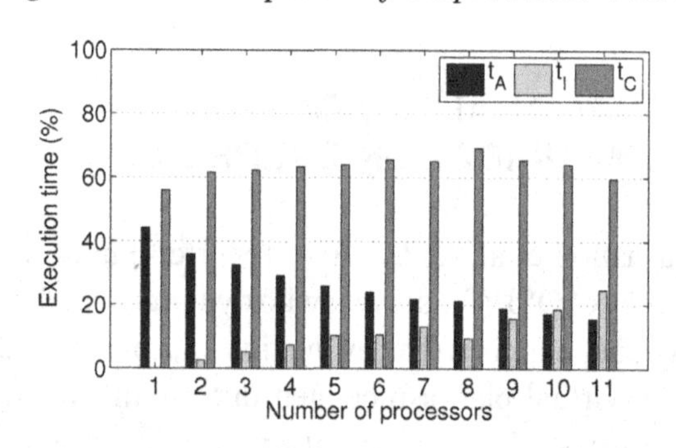

Figure 5. Influence of communication architecture on OSIP

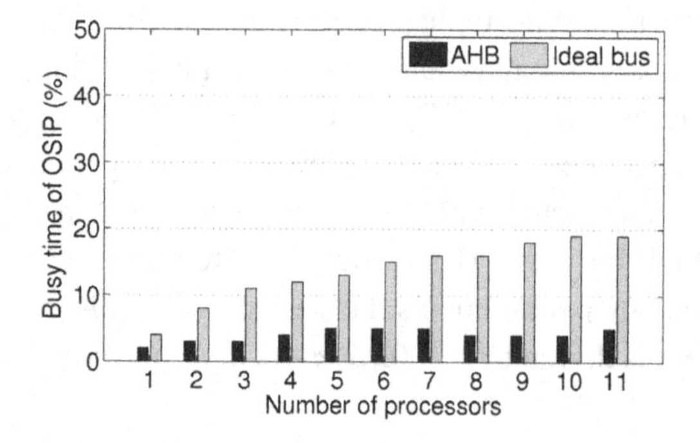

system performance and makes the system resources largely underutilized. Therefore, an optimized communication architecture is a must for efficient design and addressed in the following section.

OPTIMIZATION OF COMMUNICATION ARCHITECTURE

This section introduces three optimization steps for the communication architecture of OSIP-based systems, including a multi-layer AHB, a cache system with coherence control and write buffers.

Multi-Layer AHB

The advantages of a multi-layer AHB (ARM Ltd., 2010) against an AHB bus is that parallel accesses from the masters to the slaves of the bus are possible, as long as the addressed slaves are different. This fits well to the characteristics of the data communication in OSIP-based systems, in which three types of independent data communications exist:

- **Communication Between PEs and OSIP:** OSIP is accessed by PEs through the register interface as a standard peripheral connected to the bus. Information for system-wide scheduling, mapping and synchronization is exchanged in this interface. This communication type can be considered as a control-centric communication.
- **Communication Between PEs and Shared Memory:** Based on this communication, data are exchanged between different tasks running on different PEs. This communication type is data-centric.
- **Communication Between PEs and I/O:** This communication serves for reading an input data stream and outputting results. Normally this traffic contributes only to a minor part of the total communication.

By applying the multi-layer AHB, a lot of communication overhead due to the unnecessary bus contention caused in an AHB bus can be reduced, especially when the control-centric communication and the data-centric communication are well balanced. In the current simulation platform, each processor is connected to a bus layer, since all processors are treated as equal in this system. However, the area of this full multi-layer AHB for a large number of processors could be very big, which will be considered in future during the hardware overhead analysis.

Cache System

Compared to a multi-layer AHB, which mainly focuses on the improving the bus contention, a cache focuses on reducing bus access by exploiting the data locality, which reduces the communication cost due to the bus latency. More importantly, the reduced accesses to the bus potentially reduce the bus contention.

However, in multi-processor systems cache coherence becomes of vital importance. This problem has been thoroughly studied in the literature (Tomasevic & Milutinovic, 1994a, 1994b; Eggers & Katz, 1988; Loghi, Poncino, & Benini, 2006). In our work, a cache system based on the write-broadcast approach has been implemented.

As shown in Figure 6, the cache system contains two basic modules: a local cache module for each individual PE and a global cache coherence management unit (CCMU) ensuring the consistency between the local cache modules. The local cache module consists of a cache memory and a cache controller. Besides maintaining the cache memory, the local cache controller acts as a bridge to the CCMU with two additional ports, such that information exchanging is possible across all cache modules.

When reading the data, the cache system behaves like a normal single-cache system. The difference occurs when the data are written into the memory, in which case the CCMU is involved.

Figure 6. Cache system

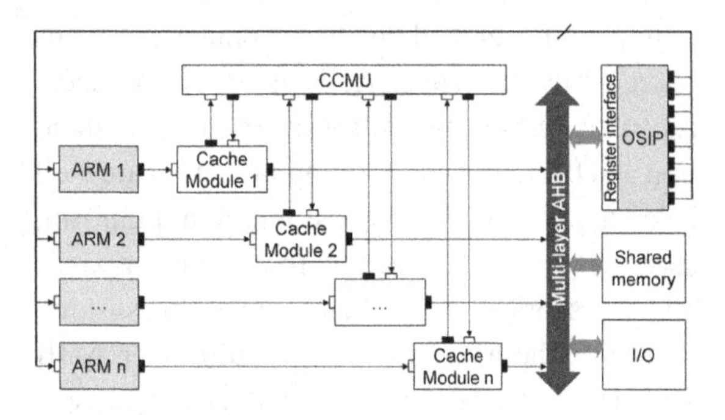

Whenever a cache controller captures a write request, it first sends the request to the bus to store the data into the shared memory. In addition, it forwards the request to the CCMU, which broadcasts the request to all cache modules. Upon receiving the broadcast from the CCMU, each cache module checks whether a local data copy at the required write address exists in the cache und updates the data correspondingly if it is the case. At the end, each cache controller sends a confirmation to the CCMU, which in turn generates a response over the initial cache module back to the PE to complete the write procedure. In this way, CCMU guarantees the consistency of the data stored in the caches and shared memory.

Note that the cache system mainly improves the data access to the shared memory. For reading the register interface of OSIP, the cache mechanism does not help much, because mostly the PEs poll a certain register for a desired value (e.g. to lock a semaphore), instead of just reading the register.

Write Buffers

Write buffers are used to improve the data writing. In fact, write buffers do not reduce the number of bus accesses, nor the bus contention, but they enable the parallelization of the communication and computation. The PEs offload the data transfer to the write buffers and then continue with the task execution, as if they were not aware of the bus traffic.

The implementation of the write buffers follows a similar principle as introduced by Sloss, Symes, and Wright (2004). In this approach, hardware FIFO queues are inserted between the cache modules and the bus. A write buffer confirms the write request of the PE before the request truly reaches the bus. Whenever there is a write request to the bus, it will be first appended to the FIFO queue, if the queue is not full, and a response to the PE will be sent immediately. Otherwise, the write buffer will wait, until at least one entry in the queue becomes free. In order to ensure the correctness of reading data, any read access to the bus will be delayed, until all requests in the write buffer have been processed. In other words, a read access empties the write buffer.

Similar to the caches, multiple write buffers also have the consistency problem. Suppose that PE_A writes a data to the shared memory, which shall be used by PE_B. It can happen that the data from PE_A still lies in the write buffer, when PE_B reads the shared memory and thus fetches the outdated value. However, in OSIP-based systems this race condition is prevented by the semaphore lock registers in the OSIP register interface. In the example, there is a data dependency between the tasks on PE_A and PE_B. A semaphore will be locked by PE_A, such that PE_B will not be allowed to fetch the data. After PE_A finishes the task, it frees the semaphore by writing a dedicated value. Since the write buffer is based on a FIFO queue, the write request to free the semaphore always follows the write requests before it. So, at the time when the semaphore register is written, the data in the memory has already been updated. Therefore, the race condition is avoided.

ANALYSIS OF OPTIMIZATIONS FOR COMMUNICATION ARCHITECTURE

To evaluate the three optimization steps for the communication architecture, the H.264 video decoding is chosen as the benchmarking application.

The setup of the evaluation system remains the same as introduced above, with the communication architecture enhanced by a multi-layer AHB, a cache system and write buffers. Following notations are made for the different optimization levels:

- **LV0:** AHB bus (unoptimized communication).
- **LV1:** Multi-layer AHB bus.
- **LV2:** Multi-layer AHB bus with cache.
- **LV3:** Multi-layer AHB bus with cache and write buffers.
- **LV4:** Ideal bus (no communication overhead).

If not stated otherwise, the cache size and the write buffer size are set to 4 Kbyte and 4 words respectively in the following discussions. The local cache module is implemented as a four-way set associative cache.

Performance Comparison

Figure 7 gives an overview of the effect of the different optimizations on the system performance by comparing the frame rate of H.264. The performance improvement is significant: From LV0 to LV3 in an 11-ARM-system, the speed-up of the frame rate reaches 2.5.

For this application, the improvement is mainly due to the cache system and write buffers. In the 11-ARM-system, 85.6% of the improvement is gained with these two optimization steps. In contrast, using a multi-layer AHB only shows a slight speed-up due to the dominating communication between the PEs and the shared memory. However, for balanced communications from the PEs to the slave components, more benefit would be expected from the multi-layer bus.

Naturally, the performance gap between the system with an ideal communication architecture (LV4) and the systems with a real communication architecture (LV0-LV3) gains advantage from the increasing number of PEs, as depicted in the figure. In latter systems, more bus contentions will happen when using more PEs, which greatly lowers the benefits obtained from the task processing parallelism. This is however not a limitation for the system with an ideal bus.

A detailed analysis of the communication overhead at different optimization levels is given in Figure 8, based on an 11-ARM-system. It can be observed that the communication time (t_C) is reduced significantly by a factor of 3.5 from LV0 to LV3. Meanwhile, the idle time of the PEs (t_I) is decreased by a factor of 2.6, which is an important side effect of the optimization for the communication architecture. By optimizing the communication, the execution time for the tasks is reduced, which in turn leads to shortened stall time of PEs for dependent tasks.

Figure 7. Frame rate at different optimization levels

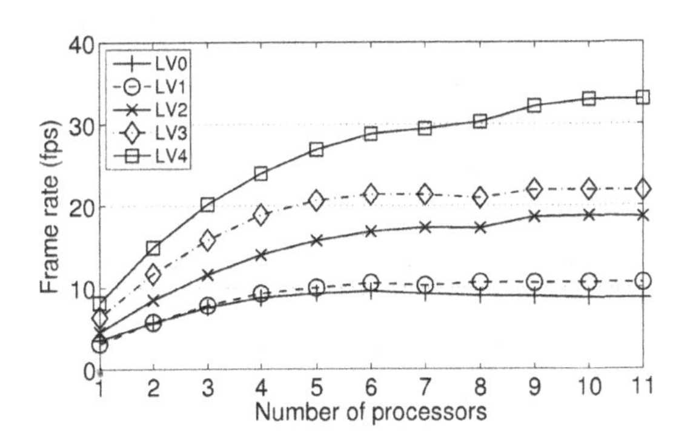

Figure 8. Time components at different optimization levels (11-ARM-system)

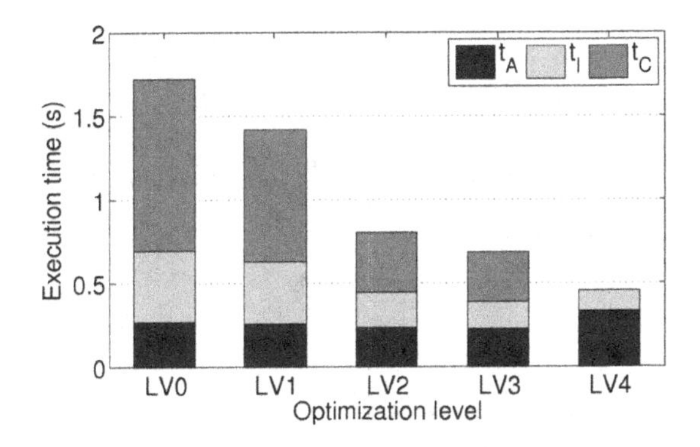

In comparison to t_C and t_I, the active time (t_A) of processors remains nearly unchanged, since each PE still gets the same amount of tasks assigned, independent of the optimization levels. An exception can be found in the system at LV4, which is a consequence of significantly more software API calls for polling the interface registers of OSIP due to the zero-latency in the bus. In this sense, the communication overhead is partially shifted to the execution of software APIs.

Effect of Cache Size and Write Buffer Size

Since the cache system and write buffers play an important role in the communication architecture, the effect of their sizes is analyzed, as presented in Figure 9. While changing the cache size has a high impact on the system performance, the write buffer size influences the performance only slightly, although whether or not using a write buffer does have a considerable impact. Using a large write buffer can reduce the write access time. But whenever there is a cache miss, the write buffer has to be emptied, which again compensates the reduced time.

The improvement by increasing the cache size is also limited to a specific range. As shown in the figure, beyond a size of 4 Kbyte, the impact of the cache size is rather low for the targeted application, since the cache hitrate is already high (95.5% with a 4-Kbyte-cache). A larger cache size is not able to improve the hitrate significantly.

Figure 9. Frame rate with different cache and write buffer sizes (LV3)

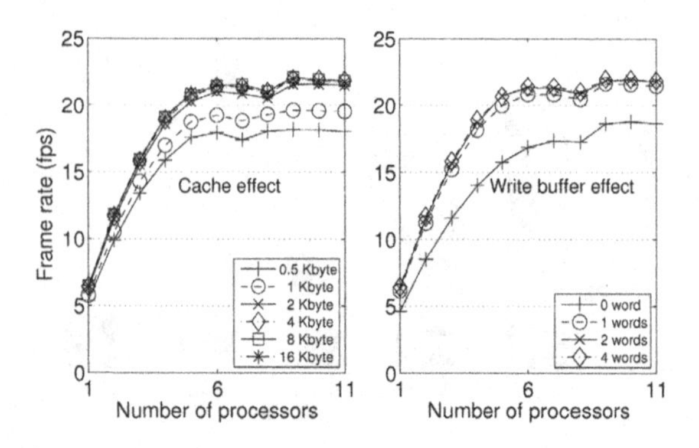

SYSTEM LEVEL ANALYSIS

The analysis above has solely focused on impact of the communication architecture on the system performance. In OSIP-based systems, the performance is also affected by other factors, among them the efficiency of OSIP being one major factor. When designing OSIP-based systems, both the communication architecture and OSIP have to be considered together. In this section, the joint effects of the communication architecture and OSIP are discussed based on the H.264 application and a synthetic benchmark application.

H.264 Benchmarking

To analyze the joint effects, two other implementations of the same OSIP functionality have been made for comparison:

- **UT-OSIP:** It is a pure untimed SystemC implementation, which implements the OSIP functionality within zero simulation time, hence can be regarded as an infinitely fast scheduler.
- **LT-OSIP:** It is a template RISC processor (LTRISC) provided with Synopsys Processor Designer (Synopsys, Inc., 2010b), to which the firmware of OSIP is ported. This processor is considered as a slow scheduler, since it is not specialized for the scheduling and mapping operations.

For having a better graphical presentation of the figures in the following analyses, the analyzed communication architectures are limited to LV0, LV3 and LV4.

Figure 10 presents the frame rates of H.264 in the systems with different OSIPs and communication architectures. Naturally, for all communication architectures, the system performance with LT-OSIP is lower than that with UT-OSIP and OSIP, as LT-OSIP is not an efficient scheduler. However, the performance gap between the systems with a fast scheduler (UT-OSIP/OSIP) and a slow scheduler (LT-OSIP) is gradually narrowed from LV4 over LV3 to LV0. Especially at LV0, a slow scheduler does not perform much worse than a fast scheduler from the system point of view. For example, in the 2-ARM-system, the performance of LT-OSIP-based system is very close to that of OSIP/UT-OSIP-based system. This indicates that a high-speed scheduler in OSIP-based systems will not be able to lead to a considerable improvement of the system performance, if the communication architecture is not properly designed. In this case, the efficiency of a fast scheduler is wasted to a large extent.

Once the communication architecture is highly optimized, the advantages of a fast scheduler appear clearly. Large performance difference can be found at LV3 and LV4, especially for a big number of PEs. In this case, a slow scheduler is becoming the bottleneck of the system. This can also be reflected by the fact that the performance of LT-OSIP-based systems increases only slightly from LV3 to LV4. Even an ideal communication architecture is not able to help in improving the performance much. In contrast, OSIP-based systems can still manage to achieve a significant speed-up, which shows its high potential of coping with task scheduling.

Figure 10. Joint impact of OSIP and the communication architecture

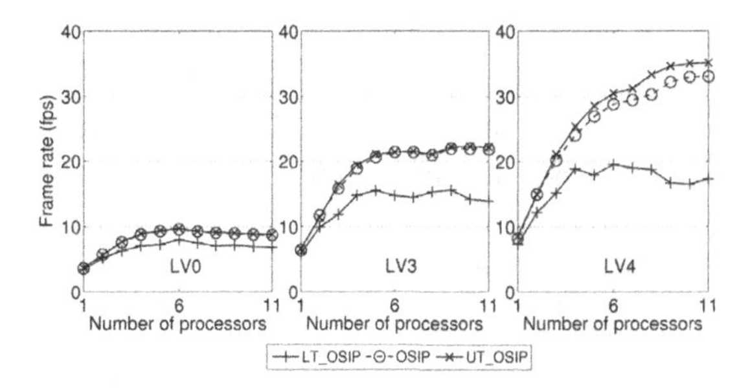

Furthermore, while a slight performance difference still exists between OSIP- and UT-OSIP-based systems at LV4, this difference can hardly be found in the systems with a real, but highly optimized communication architecture. This observation clearly shows the OSIP efficiency from the view of the practical use of OSIP in a real system.

Synthetic Benchmarking

The performance of OSIP-based systems greatly depends on both the scheduler efficiency and the communication architecture. Both are however in turn influenced by the task sizes (execution time of tasks on PEs). Different task sizes and also different number of tasks will cause different load to the scheduler and the communication architecture.

The case study based on H.264 is with fixed task sizes and fixed task number. The fundamental questions that come into mind are: a) How will such an OSIP-based system behave for other applications? b) How will the number and execution time of tasks affect the system? To answer these questions, a synthetic benchmark application is created and analyzed to get a broad view of the system performance behavior, in particular the joint effects of the communication architecture and OSIP.

The setup of the evaluation system remains the same as above. As illustrated in Figure 11, the synthetic application consists of three major tasks: *data producing* ($task_p$), *task generation* ($task_{tg}$) and *data consuming* ($task_c$). The first two tasks are mapped onto one ARM processor as a producer PE (*PPE*) and the last one is mapped onto the other ARM processors as a class of consumer PEs (*CPEs*). The *PPE* first produces a data block into the shared memory and then generates a set of $tasks_c$ for consuming the data. OSIP will schedule the *CPEs* to execute the $tasks_c$, which read and sum

Figure 11. Configuration of the synthetic benchmark application

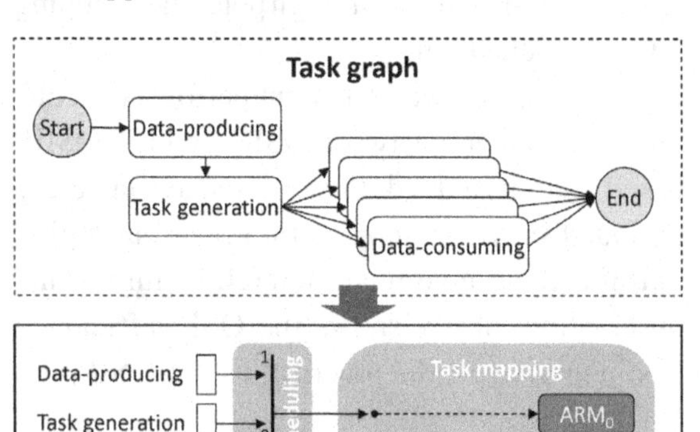

up the data from the shared memory and output the result to the I/O. A system with *n* consumer processors is referred to as *n-CPE-system*.

The following parameters are configurable in the benchmark application to influence the load of OSIP:

- **N_CPES ∈ {1, 3, 5, 7, 9, 11}:** Number of *CPEs*. Increasing N_CPEs will potentially create more requests to OSIP and more traffic, hence increases the load to OSIP and the communication architecture.
- **N_TASKS ∈ {11, 88, 165}:** Number of generated $tasks_c$. These numbers are chosen as a factor of 11, such that in an 11-CPE-system the tasks could be distributed to the *CPEs* in a balanced way.
- **N_ACCESSES ∈ {1, 6, 11}:** Frequency of accessing the same data. The size of the data block is fixed to 500 words, each 32 bits. The data access is configured in such a way that each word in the data block is read for multiple times, which is specified by N_ACCESSES. The function of this parameter is three-fold. First, changing N_ACCESSES will change the task size. An access frequency of 1, 6 and 11 represents a task size of 2.5 Kcycle, 15 Kcycle and 27.5 Kcycle, respectively. Note that these cycle numbers are obtained by running $task_c$ on the processor, with data located in the local memory. Certainly if the data are located in the shared memory, the cycle numbers will vary due to the communication overhead in the bus. However, they show qualitatively the different task sizes. Second, a larger N_ACCESSES indicates more data traffic. Third, the cache effect can be easily reflected by changing this parameter.

In addition, for including the cache coherence effect, the flow "data producing – data consuming" is iterated. In each iteration, the values in the data block are updated. Furthermore, the scheduling algorithm applied for the task queue is priority-based in order to stress the scheduler.

For evaluating OSIP-based systems, three representative application scenarios are derived from the benchmark application by configuring the above-mentioned parameters to produce different workloads to the scheduler.

- **Low Workload (N_ACCESSES = 11, N_TASKS = 11):** In this scenario, the size of $task_c$ is set to maximum within the scope of the benchmark parameters and the number of $tasks_c$ is set to minimum. For this type of applications, it takes a long time for a PE to finish one task before it requires a new task from OSIP. Therefore, the frequency of accessing the OSIP is low. Furthermore, it takes OSIP relatively less time to handle a request (to find the best candidate task from the task queue in this case), because the task queue is short. This scenario is further referred to as *best case scenario* in the following analyses.

- **High Workload (N_ACCESSES = 1, N_TASKS = 165):** This scenario is exactly the opposite case to the one above, in which the task size is set to minimum and the number of $tasks_c$ is set to maximum. In this case, the PEs are able to finish the tasks within a very short time, hence request very frequently new tasks from OSIP. The scheduling performed in OSIP also becomes much more intensive due to a much larger task queue. This scenario is further referred to as *worst case scenario*.
- **Moderate Workload (N_ACCESSES = 6, N_TASKS= 88):** In this scenario, both the task size and the number of tasks are set to medium. The load of OSIP is supposed to be at a moderate level in this case. This is further referred to as *average case scenario*.

Based on these scenarios, the behavior of the targeted evaluation systems is discussed in detail, considering the different schedulers, communication architectures and number of *CPEs*.

Best Case Scenario

The upper part of Figure 12 compares the execution time of the benchmark application in the best case scenario. In general, UT-OSIP- and OSIP-based systems have a very similar performance profile, while LT-OSIP-based systems have a reduced performance characteristic.

Similar to the observation made with the H.264 application, using an unoptimized communication architecture leads to a very small difference between the systems with different schedulers. This is due to the fact that the communication dominates the execution time of the application, which is the bottleneck of the system. By optimizing the communication architecture, the bottleneck starts to move to the other parts of the system.

With an optimized communication architecture, the effect of the different schedulers is still relatively small in small systems. In this case, the dominating factor is the execution time of the tasks. With continuously increased number of *CPEs*, soon LT-OSIP begins to reach its limitations in terms of scheduling tasks. Starting from 7-CPE-system, the performance gap between LT-OSIP-based systems and UT-OSIP-/OSIP-based systems becomes very large. In this case, a slow scheduler is not anymore able to efficiently distribute the various tasks to the PEs. This is also the reason for the limited performance improvement from LV0 to LV4 in LT-OSIP-based systems with a large number of *CPEs*, while in UT-OSIP/OSIP-based systems the execution time is significantly reduced.

Figure 12. Best case scenario (N_ACCESSES = 11, N_TASKS = 11)

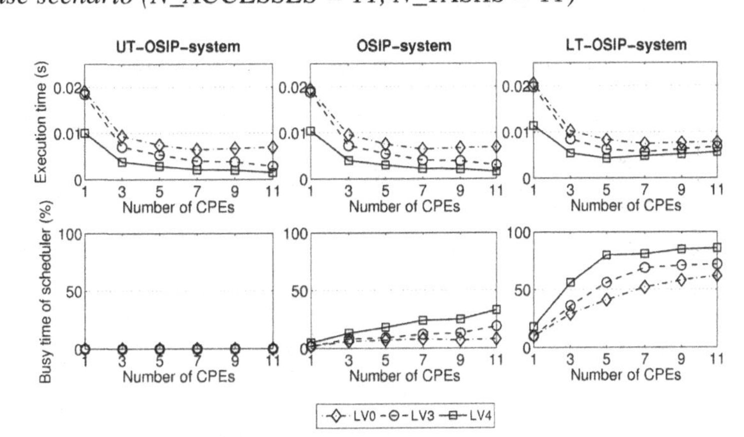

The scheduling efficiency can be well reflected by the busy state of the scheduler, which is illustrated in the lower part of Figure 12. Naturally, UT-OSIP is never busy, since it can perform any kind of scheduling within zero time. In comparison to UT-OSIP, OSIP and LT-OSIP in general become busier with increasing number of *CPEs*. Also further optimization in the communication architecture makes both schedulers more frequently enter the busy state, because the task execution time is shortened, which in turn increases the request frequency of *CPEs* to the scheduler. However, in comparison to LT-OSIP, OSIP is much less stressed. In the 11-CPE-system with a real but highly optimized communication architecture (LV3), the busy time of OSIP is below 20%, while LT-OSIP is for more than 70% of the time busy with the scheduling.

Worst Case Scenario

The execution time of the applications in Figure 13 demonstrates more peculiar system behaviors in the worst case scenario than in the best case scenario.

As indicated above, the worst case scenario provides a high load to the scheduler. In this scenario, using the LT-OSIP as the scheduler should not be considered. While it is still able to reduce the execution time from 1-CPE-system to 3-CPE-system, the system performance becomes disastrous when more *CPEs* are used. Even an ideal communication architecture is not able to really improve the system performance. In this case, not only the task scheduling is very inefficient, the time spent on the scheduling algorithm itself contributes to the major part of the total execution time of the application.

In contrast, UT-OSIP cannot be the bottleneck of the system. Therefore, significant improvement can be achieved by optimizing the communication architecture. However, it can also be observed that applying more than *3 CPEs* cannot help to continue improving the system performance even at LV4, where the communication architecture cannot be the system bottleneck, either.

The main reason lies in the task parallelism. In the application, the *PPE* creates $tasks_c$ based on the OSIP APIs. Although the time for executing the APIs is short, it often takes considerable time to get the semaphore register to lock the access to the scheduler, in order to send the commands including the arguments for creating tasks. This becomes especially critical in two cases: a) The task size is small. The *CPEs* will finish the tasks very fast and send commands to the scheduler for new tasks. Therefore, the possibility that more PEs compete for the same semaphore is greatly increased. b) The communication architecture is not well designed. It will take significantly more time for one processor to get the semaphore due to the bus contention. Even after getting the

Figure 13. Worst case scenario (N_ACCESSES = 1, N_TASKS = 165)

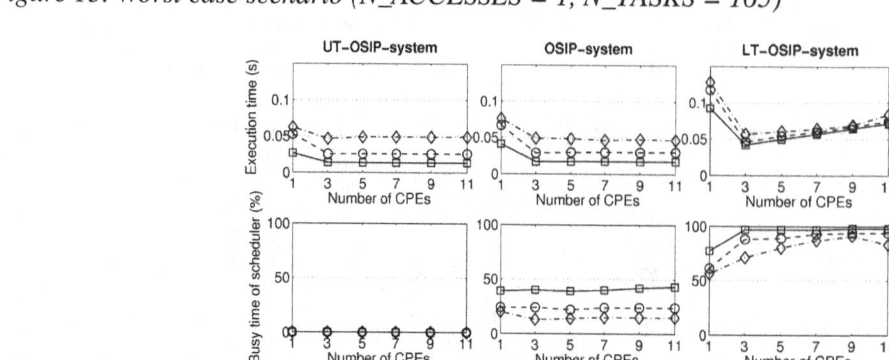

semaphore, sending the commands and arguments consumes much time, compared to the task size itself. All these can be considered as the synchronization overhead between the PEs and the scheduler. It leads to the fact that the *PPE* is not able to prepare the tasks fast enough, which are needed by the *CPEs*. In other words, the task parallelism cannot match the number of *CPEs*.

In the wide sense, the synchronization can also be regarded as one form of communication, which appears to be the bottleneck of the system in the worst case scenario. This is also the reason why in this case, OSIP-based systems still perform as well as UT-OSIP-based systems. In fact, mostly OSIP stays at an idle state in the system with a real communication architecture, as shown in the figure. Even in the largest system with the ideal bus, OSIP stays for 43% of the time at busy state.

Average Case Scenario

Often task partitioning is done in such a way that the task sizes are neither extremely large nor very small. Too large tasks would potentially reduce the parallelism degree of an application, while too small tasks would cause too much communication and synchronization overhead. Therefore, an average case scenario with moderate task sizes and number of tasks is studied.

The analysis results in Figure 14 show high similarity as in the best case scenario. Only in the systems with an unoptimized communication architecture, LT-OSIP-based systems still have comparable performance as the other two. In other cases, UT-OSIP- and OSIP-based systems, which have very similar performance, are much better than LT-OSIP-based systems.

To again highlight the conclusion that an unoptimized communication architecture can become the system bottleneck, adding more than 5 *CPEs* at LV0 even worsens the system performance. In this case, very high bus contention occurs, which in fact decreases the parallel task execution.

By comparing the average case scenario and worst case scenario for OSIP-based systems, one interesting phenomenon can be observed in the systems at LV3 and LV4. In the average case scenarios, OSIP even enters more frequently the busy state. The reason has been implicitly given above. In the worst case scenario, the task parallelism is in fact not high due to the synchronization between the PEs and OSIP. Therefore, the size of the task queue in OSIP in the worst case is on the contrary smaller than in the average case. This leads to a higher load of OSIP in the average case scenario. Still, in comparison to the task size, the scheduling overhead of OSIP is minor.

Figure 14. Average case scenario (N_ACCESSES = 6, N_TASKS = 88)

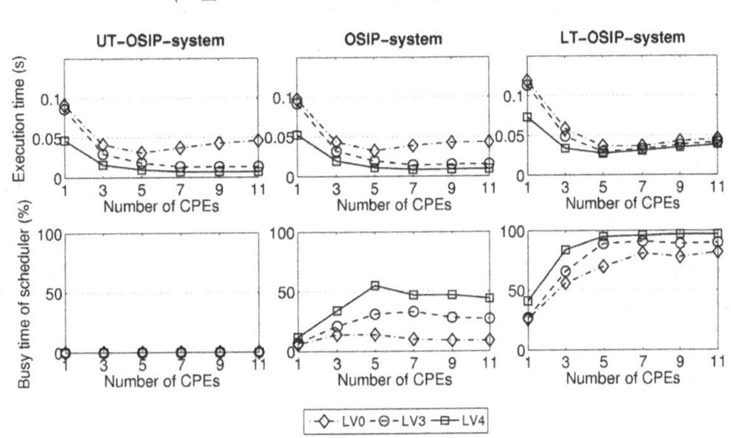

The behavior of OSIP-based systems is influenced by several factors including the scheduling efficiency, the communication and synchronization overhead and the task sizes. The bottleneck of the system can be shifted, depending on the different application scenarios. The case study above shows that OSIP has a very high efficiency to perform task scheduling and mapping at runtime for a realistic system from the system point of view. Furthermore, in order to make OSIP-based systems work in an optimal way, the communication architecture and the synchronization overhead need to be carefully considered during the system design.

CONCLUSION

In this work, we investigated the performance behavior of OSIP-based systems from a system point of view, considering both the OSIP efficiency and the communication overhead. The analysis based on a real-life application and a synthetic benchmark application gives a broad view of the joint effects caused by OSIP and the communication architecture. The results show a very high efficiency of OSIP for dynamic scheduling and mapping in MPSoCs. However, in order to fully utilize the OSIP-approach, not only an efficient OSIP processor, but also an optimized communication architecture is required. In addition, the synchronization overhead in OSIP-based systems should also be well considered to make the systems work in a proper way. More generally, designing MPSoCs based on a hardware scheduler must take special care of the balance between the efficiency of the scheduler and the communication architecture. Three approaches including a multi-layer bus, a cache system and write buffers were applied for the optimization of the communication architecture. The results showed a significant performance improvement by reducing the communication overhead.

In future, the cache system and write buffers will be implemented at RTL to analyze the timing and area overhead. Since the cache system based on the write-broadcast approach has a quite high area overhead, other cache coherence protocols will also be considered and compared in OSIP-based systems. Furthermore, sharing the bus layers of the multi-layer AHB among the processors will also be considered to trade off area against system performance.

REFERENCES

Ltd, A. R. M. (2010). AMBA on-chip connectivity. Retrieved from http://www.arm.com/products/system-ip/interconnect/

Atmel Corporation. (2011). DIOPSIS, D940. Retrieved from http://www.atmel.com/dyn/products/devices.asp?family_id=680

Benini, L., & De Micheli, G. (2002). Networks on chips: A new SoC paradigm. *IEEE Computer*, *35*(1), 70–78.

Castrillon, J., Zhang, D., Kempf, T., Vanthournout, B., Leupers, R., & Ascheid, G. (2009). Task management in MPSoCs: An ASIP approach. In Proceedings of the International Conference on Computer-Aided Design (pp. 587-594). New York, NY: ACM Press.

Eggers, S. J., & Katz, R. H. (1988). Evaluating the performance of four snooping cache coherency protocols (Tech. Rep. No. UCB/CSD-88-478). Berkeley, CA: University of California.

Goyal, P., Guo, X., & Vin, H. M. (1996). A hierarchical CPU scheduler for multimedia operating systems. In Proceedings of the Symposium on Operating Systems Design and Implementation (pp. 107-121). New York, NY: ACM Press.

Corporation, I. B. M. (2010). CoreConnect bus architecture. Retrieved from https://www-01.ibm.com/chips/techlib/techlib.nsf/productfamilies/CoreConnect_Bus_Architecture

Ienne, P., & Leupers, R. (Eds.). (2006). *Customizable embedded processors: Design technologies and applications (systems on silicon)*. San Francisco, CA: Morgan Kaufmann.

Jantsch, A., & Tenhunen, H. (Eds.). (2003). *Networks on Chip*. Boston, MA: Kluwer Academic.

Kohout, P., Ganesh, B., & Jacob, B. (2003). Hardware support for real-time operating systems. In Proceedings of the IEEE/ACM/IFIP International Conference on Hardware/Software Codesign and System Synthesis (pp. 45-51). New York, NY: ACM Press.

Limberg, T., Winter, M., Bimberg, M., Klemm, R., Tavares, M., Ahlendorf, H., et al. (2009). A heterogeneous MPSoC with hardware supported dynamic task scheduling for software defined radio. Paper presented at the DAC/ISSCC Student Design Contest, San Francisco, CA.

Lippett, M. (2004). An IP core based approach to the on-chip management of heterogeneous SoCs. In Proceedings of the IP Based SoC Design Forum & Exhibition.

Loghi, M., Poncino, M., & Benini, L. (2006). Cache coherence tradeoffs in shared-memory MPSoCs. *ACM Transactions on Embedded Computing Systems*, *5*(2), 383–407.. doi:10.1145/1151074.1151081

Meenderinck, C., Azevedo, A., Alvarez, M., Juurlink, B., & Ramirez, A. (2008). Parallel scalability of H.264. In Proceedings of the Workshop on Programmability Issues for Multi-Core Computers.

Murtaza, Z., Khan, S. A., Rafique, A., Bajwa, K. B., & Zaman, U. (2006). Silicon real time operating system for embedded DSPs. In Proceedings of the International Conference on Emerging Technologies (pp. 188-191). Washington, DC: IEEE Computer Society.

Nácul, A. C., Regazzoni, F., & Lajolo, M. (2007). Hardware scheduling support in SMP architectures. In Proceedings of the Conference on Design, Automation and Test in Europe (pp. 642-647).

Nakano, T., Utama, A., Itabashi, M., Shiomi, A., & Imai, M. (1995). Hardware implementation of a real-time operating system. In Proceedings of the TRON Project International Symposium (pp. 34-42). Washington, DC: IEEE Computer Society.

Nordström, S., & Asplund, L. (2007). Configurable hardware/software support for single processor real-time kernels. In Proceedings of the International Symposium on System-on-Chip (pp. 1-4).

Pan, Z., & Wells, B. E. (2008). Hardware supported task scheduling on dynamically reconfigurable SoC architectures. *IEEE Transactions on Very Large Scale Integration Systems*, *16*(11), 1465–1474..doi:10.1109/TVLSI.2008.2000974

Park, S., Hong, D.-s., & Chae, S.-I. (2008). A hardware operating system kernel for multiprocessor systems. *IEICE Electronics Express*, *5*(9), 296–302..doi:10.1587/elex.5.296

Seidel, H. (2006). *A task-level programmable processor*. Duisburg, Germany: WiKu-Verlag.

Sloss, A., Symes, D., & Wright, C. (2004). *ARM system developer's guide: Designing and optimizing system software*. San Francisco, CA: Morgan Kaufmann.

Synopsys, Inc. (2010a). Synopsys platform architect. Retrieved from http://www.synopsys.com/Tools/SLD/VirtualPrototyping/Pages/PlatformArchitect.aspx

Synopsys, Inc. (2010b). Synopsys processor designer. Retrieved from http://www.synopsys.com/Tools/SLD/ProcessorDev/Pages/default.aspx

Texas Instruments, Inc. (2010). OMAP. Retrieved from http://focus.ti.com/docs/prod/folders/print/omap3530.html

Tomasevic, M., & Milutinovic, V. (1994a). Hardware approaches to cache coherence in shared-memory multiprocessors, part 1. *IEEE Micro*, *14*(5), 52–59. doi:10.1109/MM.1994.363067

This work was previously published in the International Journal of Embedded and Real-Time Communication Systems, Volume 2, Issue 3, edited by Seppo Virtanen, pp. 1-20, copyright 2011 by IGI Publishing (an imprint of IGI Global).

Chapter 10
Implementation and Evaluation of Skip-Links:
A Dynamically Reconfiguring Topology for Energy-Efficient NoCs

Simon J. Hollis
University of Bristol, UK

Chris Jackson
University of Bristol, UK

ABSTRACT

The Skip-link architecture dynamically reconfigures Network-on-Chip (NoC) topologies in order to reduce the overall switching activity in many-core systems. The proposed architecture allows the creation of long-range Skip-links at runtime to reduce the logical distance between frequently communicating nodes. This offers a number of advantages over existing methods of creating optimised topologies already present in research, such as the Reconfigurable NoC (ReNoC) architecture and static Long-Range Link (LRL) insertion. This architecture monitors traffic behaviour and optimises the mesh topology without prior analysis of communications behaviour, and is thus applicable to all applications. The technique described here does not utilise a master node, and each router acts independently. The architecture is thus scalable to future many-core networks. The authors evaluate the performance using a cycle-accurate simulator with synthetic traffic patterns and compare the results to a mesh architecture, demonstrating logical hop count reductions of 12-17%. Coupled with this, up to a doubling in critical load is observed, and the potential for 10% energy reductions on a 16×16 node network.

INTRODUCTION

Network-on-Chips (NoCs) provide a general purpose communications fabric for multi-core architectures that allow functional blocks to communicate with each other. These networks are intended to provide a scalable, power and space efficient means to support growth in component density and many-core designs (Dally & Towles, 2001). Modern Chip Multi-Processors (CMPs) exploit NoCs to efficiently connect many processing elements (PEs) and other devices.

DOI: 10.4018/978-1-4666-2776-5.ch010

System performance is dependent on low-latency, high-bandwidth communications between cores. As the number of cores integrated in a single chip increases, greater demands are placed on the communication network. Thus, the design of the inter-connection fabric and not the PE is now the bottleneck in computation. Power and area are central considerations in the design of the interconnect. This has led to a wide range of literature, with topics including power analysis (Banerjee, Mullins, & Moore, 2007), router design (Michelogiannakis & Dally, 2009) and routing algorithms such as Opt-y (Schwiebert & Jayasimha, 1993) and IVAL (Towles, Dally, & Boyd, 2003). However, the main issue today is the delivery of energy-efficient and scalable interconnects: the focus of this paper.

NoC Design

An ideal NoC provides an area-efficient low-power network that exhibits good latency characteristics across a range of traffic patterns. A NoC can be designed for a specific application, affecting the choice of topology, routing algorithm and router design. These specialised networks can offer a performance advantage over a regular architecture, but are inflexible and costly to design. They require static analysis of the target application's inter-process communications behaviour and offer poor performance for general purpose computation.

In Asanovic et al. (2006) Dwarfs were introduced, where applications are classified in terms of their computation and communication patterns. For example, some applications have fixed, predictable communication between specific cores, while others have sporadic communication between all cores. In Asanovic et al. (2006) the shortcomings of fixed-topology on-chip interconnects are discussed and it is suggested that many applications have communication patterns for which an optimal topology can be configured. Therefore, the design of a flexible, common NoC topology is of paramount importance in the design of efficient next-generation systems. For example, Shalf et al. (2005) consider reconfigurable hybrid interconnects for Ultra-scale applications. They conclude that many Ultra-scale applications will benefit from a reconfigurable interconnection fabric. In this paper, we demonstrate how this can be achieved in a general sense, using the mesh topology as a basis. We will introduce dynamically-reconfigurable Skip-links to enable on-the-fly adaptation of the network topology to traffic behavior.

Energy in NoCs

As shown by Moore and Greenfield (2008), the energy cost of communicating a single 32-bit word off-chip is now 1300 times greater than the cost of a single ALU operation. With the increase in the use of parallel algorithms and multiprocessing, communication and not computation is becoming the power and performance bottleneck. NoC routers dissipate both dynamic and static power, in particular when switching and reading/writing flits to buffers. Inter-router links also consume a significant amount of energy. The energy consumption of routers is discussed by Hu and Marculescu (2005) in the context of a model proposed by Ye et al. (2002), and can be expressed in terms of the router energy E_R (buffering and switching) and link energy E_L requirements. In this model, the total energy required to send a single bit from node n_i to n_j is

$$E^{n_i,n_j} = N_{hops} \times E_R + (N_{hops} - 1) \times E_L \qquad (1.1)$$

where N_{hops} is the number of routers the bit traverses between nodes ni and nj. Equation (1) clearly shows that the energy efficiency of a NoC can be increased by reducing the value of E_R (the energy expenditure when switching); or E_L, the energy expended when linking flits from one router to the next; or N_{hops}. Reductions in E_R have been the subject of much work, such as the clock modifications suggested by Mullins (2006) and 'on/off' links by Peh and Soteriou (2004). A number of other low-power NoC implementation methods are discussed by Lee, Lee, and Yoo (2006).

This paper, however, considers the reduction of $Nhops$. A number of methods exist that permit such a saving. The topology of a NoC is an import consideration as it dictates the choice of routing algorithm and router design, as it determines the lengths of the possible paths across the network. If the topology of the architecture maps well to an application (by ensuring the short paths between frequently communicating processes), then the network will operate efficiently.

It is possible to generate application specific process-to-core mappings, so that frequently communicating processes are located on physically close cores, reducing the number of hops packets must take between them. In turn, this reduces the switching activity and energy expenditure required to route the packets. These mappings may be created dynamically at runtime or statically before execution. For examples of each see Nollet, Marescaux, Verkest, Mignolet, and Vernalde (2004) and Chen, Li, and Kandemir (2007). However, a major drawback of these approaches is the need for program predictability at compile time, along with the time complexity of the method.

Conversely, it is possible to optimise the topological locality of processes. The physical location of each process is maintained while the logical location is reconfigured, by reducing the number of hops between the processes. This allows a reduction in the switching activity required to route packets between processes, without having to locate those processes on physically closer cores. Although the same link distance will have to be traversed by any traffic, significant energy savings can be made due to reduced router activity.

We presented an analysis of the potential for savings in Hollis and Jackson (2009), where it was shown that if 70 – 80% of traffic leaving an output port has taken a 0° turn, then a net energy saving can be achieved by configuring the topology to let packets skip the router altogether. Our work on Skip-links builds on this intuition to offer savings to traffic, irrespective of its pattern or amenability to static analysis.

IRREGULAR AND RECONFIGURABLE TOPOLOGIES

To achieve the logical proximity discussed above, Stensgaard and Sparsø presented a NoC architecture called ReNoC that allows the network topology to be statically reconfigured (Stensgaard & Sparsø, 2008). An extra layer of topology switches surrounding each router makes it possible to route flits around (or Skip) switches so that the energy cost of switching and buffering the flit is not incurred. Combined, these allow for a network that is capable of forming point-to-point communications links between non-adjacent routers whilst maintaining the flexibility of the underlying mesh topology. The authors demonstrate that a ReNoC network configured with an application specific mapping has a 56% energy saving when compared to a network with a static mesh. Only a 10% increase in area of the NoC was required for the topology switches. The topology of ReNoC is configured on power-up or before new applications are assigned to cores, requiring a static and compute-intensive task graph to be computed that determines the bandwidth of communication between each process and assigns point-to-point links accordingly.

Express Virtual Channels (EVCs) were introduced in Kumar, Peh, Kundu, and Jha (2007). EVCs are a novel method of flow control that allows packets to virtually bypass individual routers. This is achieved by statically connecting a virtual channel at one router to a channel at another more than a single hop away and giving these EVCs priority over the standard VCs. The presented results show a 38% reduction in energy use and up to 84% reduction in packet latency.

Ogras and Marculescu have also described a method, Long-range links (LRL) that creates an optimal irregular mesh topology for the network (Ogras & Marculescu, 2006). First, an analysis of the frequency and magnitude of communication between processes in a specific application is performed. This identifies the optimum locations to insert LRL into the topology. An NoC is then synthesised with the optimum LRLs superimposed, adding an extra port to each router connected to an extra link. Critical load increases of up to 24.8% were achieved for an 8 × 8 network, as well as an average packet latency reduction of 76.9%. In addition, switch energy consumption was reduced by 7% on average, though link energy increased slightly, resulting in an overall reduction of 5%.

ROUTER SKIPPING

We propose a development of the ReNoC architecture (Stensgaard & Sparsø, 2008), such that topology switches can be adjusted at runtime. Instead of using statically derived task-graphs to reset the network topology prior to the execution of specific applications, the Skip-link architecture allows the network topology to be dynamically configured based on current traffic patterns—Skiplinks are inserted at runtime. Hence this technique has wider applicability than static analysis, including to applications for which the communication behaviour is not predictable or has not been computed.

A Skip-link *refers to a single router where a single input has been directly connected to its directionally-opposite output (Figure 1). This ensures that any flit arriving at the input will bypass the router.*

Figures 2 and 3 show an intuitive understanding of the benefits of Skip-linking. In Figure 2, we see what happens when multiple traffic streams are injected into a Network-on-Chip. As traffic is absorbed into routers, in this example arranged along the sides of the network, router buffer space is occupied.

Skip-links are designed to reduce overall buffer utilisation in these situations, and Figure 3 shows how this can occur. Here, Skip-links cause routers 1–6 to be skipped, and so router buffer use drops to zero for these locations.

A Skip-chain *refers to two or more adjacent routers that are skipped along the same input such that the Skip-links are effectively chained together (Figure 4). Single Skip-links can be thought of as Long Range Links (LRLs) of length = 2 and Skip-chains of length > 2. As with ReNoC, we disable an output port on each router that is skipped. An example of a router being skipped is shown in Figure 1. When a Skip-link is inserted, the logical distance between some source-destination pairs is reduced; delivering hop savings; but the logical distance between some other pairs will be increased, producing hop penalties. We only allow a Skip-link to be placed if, when summed over all traffic flows, the savings outweigh the penalties. The analysis of when this situation occurs is the focus of the remainder of the paper.*

Figure 1. The structure of a Skip-link enabled router (based on the ReNoC router (Stensgaard & Sparsø, 2008). Only the East/West channels are displayed. Enabled paths are in solid; disabled paths are dashed. A maximum of one direction per router may activate a Skip-link at any one time.

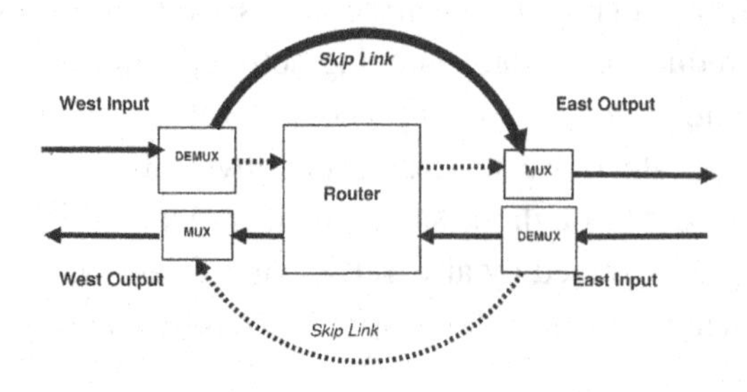

Figure 2. Total router input buffer use in a network with simple traffic flows

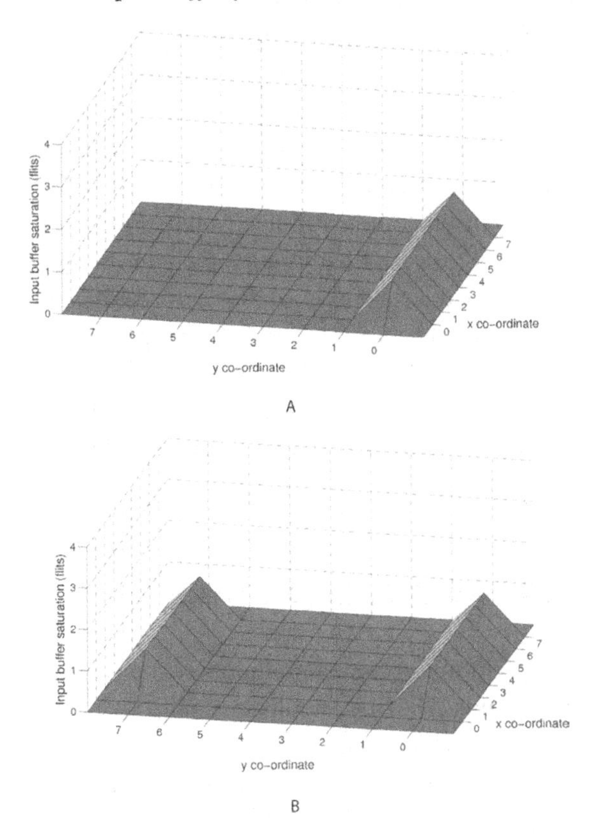

Figure 3. Total router buffer use in a network with simple traffic flows and Skip-links

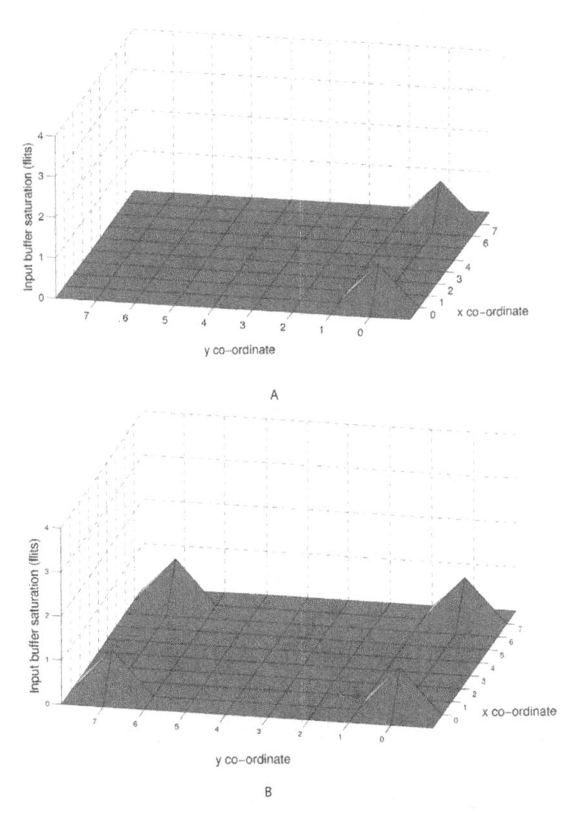

Placement Constraints and Routing

From a routing point of view, it is necessary to constrain the placement of Skip-chains for two reasons. First, to prevent the topology from becoming too deformed and creating source-destination pairs for which there is no path. Second, if Skip-chains are placed too close to each other, a packet may have to route around several disabled ports on its path, incurring many extra hops. If more hops are being spent on routing around disabled ports than are being saved by taking a Skip-link, then all benefits of the architecture are lost (Hollis & Jackson, 2009). Only 0° turn Skip-links are permitted so that the complexity of routing is reduced. The placement constraints, visualised in Figure 4, are:

1. Any router with an output port in a non-default state (either disabled or sending packets more than one hop) must not be adjacent to any other router with an output port in a non-default state.
2. An edge node may not be Skip-linked (though they can connect to a skipped node) - represented by 'X' in Figure 4. Furthermore, a Skip-link is only enabled in one direction.

While these constraints limit the performance gains that are possible, they are necessary to ensure the technique is applicable to all possible traffic patterns. As stated previously, a deadlock-free routing algorithm has been developed to support the irregular mesh topologies and ensure routability even as topology reconfigurations occur on the fly. The precise details of the algorithm are not

Figure 4. The second placement constraint for new Skip-chain creation in the presence of an existing chain. New Skip-links may not be created within the dashed region once the illustrated chain has been established, though more Skip-links may be added to the chain.

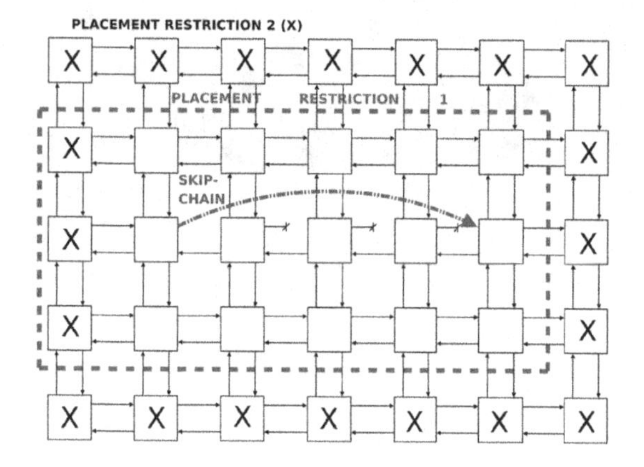

discussed in this paper, but it behaves identically to the Opt-y (Schwiebert & Jayasimha, 1993) algorithm in the absence of any irregularity.

Topology Status Tables

Each node on the network maintains a table that describes the state of the output ports of all directly connected nodes. This table is used to ensure placement restrictions are obeyed and to aid routing decisions. A node is aware of the status of its own output ports, as well as the output ports of any adjacent routers. This limits the extent to which data must be transmitted and maintains scalability. We must also record the status of input ports. However, the input port status can be represented with a single bit, whereas the output port status requires an integer. All values can therefore be compactly represented in hardware.

To store local output port status information requires $2n$ entries, where n is the network dimensionality. Therefore to store information on each neighbouring node's output ports in a 2-dimensional mesh requires $4n2$ entries. This is a total of $2n + 4n2$ entries, plus the same number of bits. The state of each output port is represented by an integer: 0 indicates that the port is disabled and 1 that the port is enabled as in a regular mesh configuration. When an output port is connected to a Skip-link chain, the length of the Skip-chain in standard links l is used to represent the status. Precisely the same number of entries are required for input ports, but a 0 indicates that a standard link connects to it, a 1 indicates that a Skip-link connects to it (the length of the chain does not matter). Consequently, whenever a Skip-chain is initiated, each node along the Skip-chain, as well as the node to which it connects must update each of its immediate neighbours so that their tables are accurate. This is infrequent, thus the overhead is low.

PLACEMENT ALGORITHMS

Our Skip-link technique requires two placement algorithms. The first establishes a Skip-link when suitable traffic conditions have been detected. The second monitors traffic around an existing Skip-link to determine if it should be removed. Together, these algorithms ensure that the network dynamically reconfigures to optimise the topology. The algorithms are executed at individual nodes. There exists no master node or central record of the global state of the network topology, ensuring that the technique is scalable to support future many-core NoCs.

A node may only establish a Skip-link over itself, based on the traffic it routes. The removal of a Skip-link can be triggered by two nodes; the skipped node and the node that routes packets along the Skip-chain. Skip-chains are formed by neighbouring routers with adjoining Skip-link configurations. This is a powerful concept, as each router acts independently to benefit the network. The emergent behaviour of the routers, interacting only via topology status table updates, increases the benefits of the Skip-link architecture. For example, consider the case where a single traffic flow exists along the length of a row or column. Each router along the flow's path will establish a Skip-link over itself, resulting in a Skip-chain leading from the source node to the destination. Thus, for long distance traffic, our approach converges to optimal behaviour, whilst using only local information. Using only local information also decreases the latency between the detection of suitable conditions and placement. In addition, we ensure that the architecture is not dependant on the distribution of global state information. The process of topology reconfiguration, the placement and removal of Skip-links is transparent to both by the application and user. The reconfiguration is continual in that the turns at each router are constantly monitored and the topology may change repeatedly throughout the execution of a single application. Importantly, each router acts independently, with very limited communication solely to each adjacent router. This communication occurs only when a router places or removes a Skip-link. Hence, this approach will scale to large networks.

Skip-Link Placement

We now describe the algorithm used to detect suitable traffic conditions for Skiplink placement. The activity of each router is observed in order to calculate the fraction f of flits exiting each output port that have turned 0°. To prevent oscillation between path initialisation and tear-down, the algorithm must be resistant to short-term changes in otherwise constant traffic flows. Hence the fraction of 0° turn traffic over a number of packets p must be considered. Exponential averaging is commonly used for predicting the magnitude of the next CPU burst of a thread based on a history of recent bursts. We apply this technique to Skip-link placement and removal. Instead of considering periods of time, we consider b bins of p packets, and aim to predict f. The formula can be expanded to consider n bins of observed behaviour:

$$f_{n+1} = \alpha \times (t_n + (1-\alpha) \times t_{n-1} + (1-\alpha)^2 \times t_{n-2} + \ldots + (1-\alpha)^n \times t_0) + (1-\alpha)^{n+1} \times f_n \quad (2)$$

where t represents observed values and f predicted values. The weighting of immediate past usage against the total previous history is determined by $0 < \alpha \leq 1$, where $\alpha = 1$ considers the magnitude of the last burst only. Once f has been predicted, we compare it to the value of the creation threshold, which was investigated in our previous work (Hollis & Jackson, 2009). If it is greater than or equal to the threshold value, then a Skip-link is placed. We showed that for a mesh, if $f > 0.7$ then Skip-link establishment is positive overall, but the precise value is dependent on actual traffic behaviour. We perform a detailed evaluation of these values at the end of this section.

Skip-Link Removal

As with Skip-link placement, each router independently determines when its Skiplink should be removed. However, it is also possible for the router connecting to a Skip-link to cause the skipped router to return to the default state. We refer to these two types of removal as local and head of chain removal respectively. The exponential averaging technique is applied to react to the behaviour of the traffic, using the equalities given.

Local Skip-Link Removal

A router removes its local Skip-link when

$$2 \times N_P \geq N_S \quad (3)$$

We use Np to represent the number of packets that must take a longer route due to the Skip-link and Ns for the number of packets using a Skip-link to reduce hop count. When a packet is skipped over a router, the router is strobed so that it is aware a hop saving of 1 has occurred due to the presence of the Skip-link, incrementing Ns. When the packet cannot utilise the disabled output port, then Np is incremented.

Head-of-Chain Skip-Link Removal

Any Skip-link, which may be the head of a Skip-chain, can only be reached from the output port of a single router. When a packet needs to travel in the direction of the Skip-link, this router can forward packets along the link or route them around the link. Where the presence of the Skip-link or chain results in an increased hop count for a packet, Np is incremented. Note that there is a constant penalty of two hops for these packets. When the hop penalties exceed the savings gained by forwarding packets into the Skip-link, then the head of the Skip-chain must be removed — if:

$$2 \times N_P \geq (l - 1) \times N_S \quad (4)$$

The above equation considers the penalty for a longer route (always two hops) and the saving achieved by using a Skip-chain (one less than the length l of the Skip-chain).

SKIP-LINK PLACEMENT PARAMETER EVALUATION

We now evaluate the effect of different values of the three Skip-link placement parameters (α, f, p).

In Sect. 4.1, we presented the Skip-link placement algorithm and stated that a 'good' placement occurs when $f > 0.7$. This original statement came from an analysis of synthetic traffic patterns, and so we wanted to see whether this was still true for these, more realistic, statistical traffic patterns. We wished also to determine the effect of varying the history averaging function α and the granularity of history bins p. To do this, we ran a detailed sweep over the valid space of these variables, using our previously-developed NoC simulator, in combination with traffic traces generated using a statistical traffic model developed by Soteriou et al. (2006).

Methodology

To perform our evaluation of the threshold parameters, we implemented a NoC simulator, capable of processing statistical traffic flows, based on the work by Soteriou et al. (2006). This work allows traffic to be reproduced that represents a range of real application traces, and a table of these is displayed in their original paper.

The traffic model contains three key parameters, (H, ρ, σ) which allow the construction of flows resembling carefully calibrated applications. For our evaluation here, we chose a set of five applications that covered the majority of the potential (H, ρ, σ) space, and thus represent a good cross-section of potential traffic that may be encountered in a Skip-linked system.

We then simulated these five applications across the possible ranges of Skip-link parameters (α, f, p). The results are discussed below.

Threshold Parameter Results

In general, the results were surprising, since far less variation in Skip-link performance was observed than expected over the full exploration range. In particular, it appears that the quality of Skip-link placement is almost entirely insensitive to changes in α, as will be described below.

History Decay Factor (α)

Skip-link placement decision relies on our prediction of future traffic via an exponentially-decaying history log. In this setup the value of α determines how quickly the significance of previous history samples decays — large α causes a slow decay; small α causes the smallest decay.

It would seem that α is important, since paying attention to more history will aid in identifying long term 0° turns, and less history will give more adaption to more dynamic traffic patterns.

However, the results we have been able to generate dispel this intuition. The value of α is so insignificant, compared to the f and p parameters that were it to be graphed on the same scale, no change in physical or logical hop count would be visible across its entire range. Numerical analysis of our simulations verified this, and α so insignificant that a poor choice would only add 2% to the overall hop count (based on the largest variation across the applications of 'swim' with $f = 0.85$, $p = 310$).

Thus, to an approximation, any value of α is acceptable for Skip-link implementation, however our results show that ($\alpha = 0.3$), is optimal for the majority of applications, and we recommend that Skip-link deployments use this value initially. In keeping with this, the remainder of this paper's threshold evaluation is conducted with respect to a fixed value of $\alpha = 0.3$.

0° Turn Threshold (f)

The threshold value f defines whether or not a given history prediction indicates that a Skip-link should be placed in a given location or not. It is clear, therefore, that a bad choice of f will result in a poorly-performing Skip-link system. The main question then revolves around how significant changes in f are to the functioning of Skip-links. In Figures 5 and 6, we show how variations in p and f affect the physical and logical hop counts of traffic in the 'qsort' application. If we focus on the variations with f, we immediately see two things: first, it is obvious that a larger f delivers a lower physical hop count, but a larger logical one; second, the total variation in hop counts is small at around 2% for physical hops and 9% for logical hops.

Since overall system performance is reduced by more logical hops, and improved by fewer physical hops, we seek a balance between the two curves and, given the dominance of the logical hops; it appears that a value of $0.70 \leq f \leq 0.85$ is optimal. This assertion is substantiated by additional data we have for all five benchmark applications, across a range of α values, but we do not have space to include all these data here, so ask the reader to believe they follow similar trends (with small shifts in the acceptable range of f +/-0.05). Thus, these data substantiate our earlier assertion that $f > 0.70$ is a good threshold, and adds an upper-bound that $f \leq 0.85$ is most optimal.

Bin Size (p)

The bin size parameter p decides what time window traffic sampling will occur over. A selection of too short a time will result in potential instability in the Skip-link placements, since they may be placed when only a short amount of data is detected mak-

Figure 5. Physical hops for qsort (α=0.3)

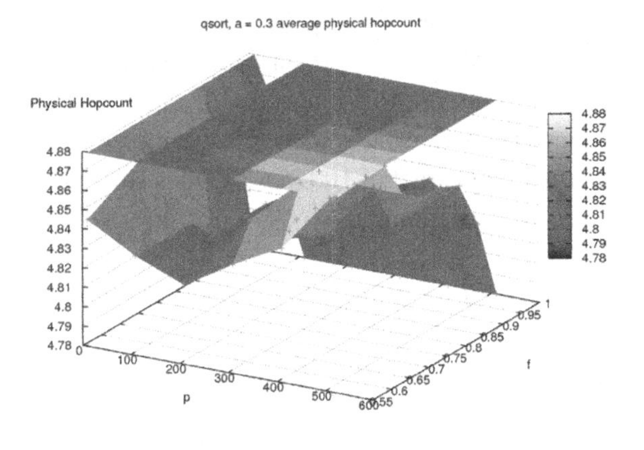

Figure 6. Logical hops for qsort (α=0.3)

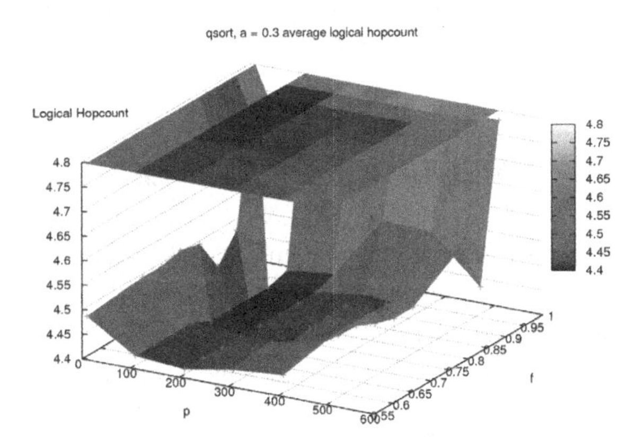

ing 0° turns. This traffic may quickly die away, resulting in the Skip-link being torn down shortly afterwards, with the potential for this behaviour to oscillate if the traffic returns for a short period.

On the other hand, too large a value of p makes Skip-link placement sluggish and slow to react to changes in traffic patterns. Since Skip-links have maximum advantage when they properly mirror the traffic flowing through the network, being too insensitive to traffic variations is potentially as damaging at too fast an adaption. Thus, a middle ground is sought, whereby p is similar to the timescale of traffic variation. By its very nature, this makes p tightly coupled to the particular traffic running over the network, in a way not found for f. Therefore, we now present an analysis of p across all our sample applications.

We display the graphs for all five of our test applications in Figure 5 through 14. It can be seen that the optimal value of p changes across the applications, with 'qsort' noticeably having the lowest value of ~ 100, and 'equake' the highest of ~ 200, but this range is still rather small. Therefore, we are once again able to conclude that a small range, that of $100 \leq p \leq 200$ is sufficient to correctly capture all types of traffic.

Summary

Following the investigation above, we can conclude the following parameter ranges for Skip-links, that are optimal for the majority of traffic flows, and can be recommended for common deployment: $\alpha = 0.3$, $0.70 \leq f \leq 0.85$, $100 \leq p \leq 200$.

Figure 7. Physical hops for equake (α=0.3)

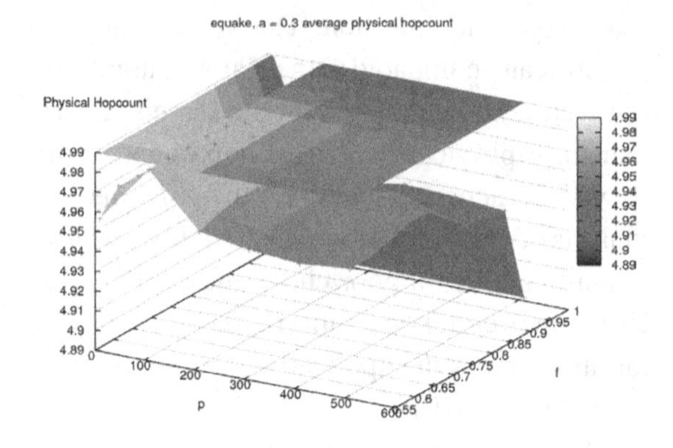

Figure 8. Logical hops for equake (α=0.3)

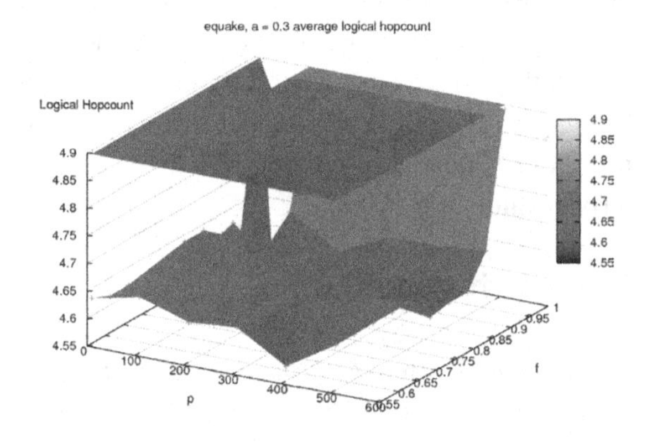

Figure 9. Physical hops for 8bencode (α=0.3)

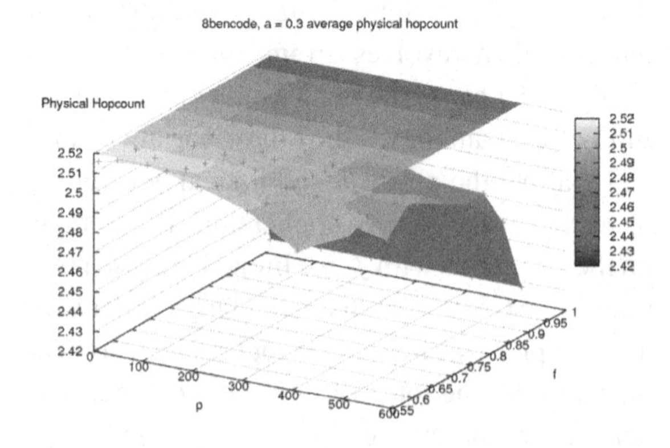

IMPLEMENTATION OF SKIP-LINKS

We now discuss the implementation details of Skip-links, focusing on how Skiplinks can be created in a manner that ensures no data is lost, and that the two merging paths through a router node (through the router, or along the Skip-link) can be supported. A basic assumption underlying the applicability of Skip-links is the ability to create a fork/merge structure like that shown in Figure 1, where a router block can be augmented with a MUX/DEMUX pair on each output and input,

Figure 10. Logical hops for 8bencode (α=0.3)

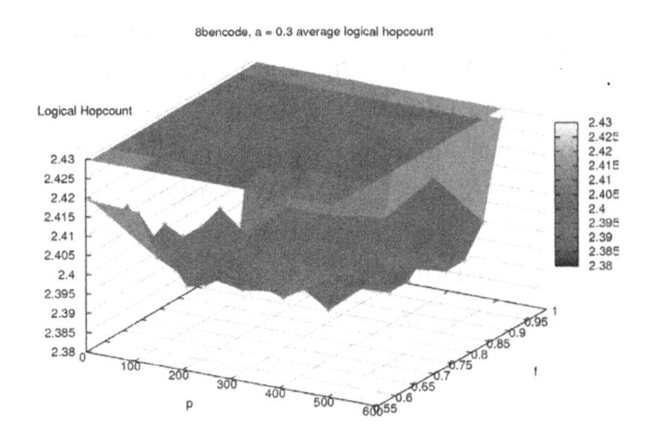

Figure 11. Physical hops for swim (α=0.3)

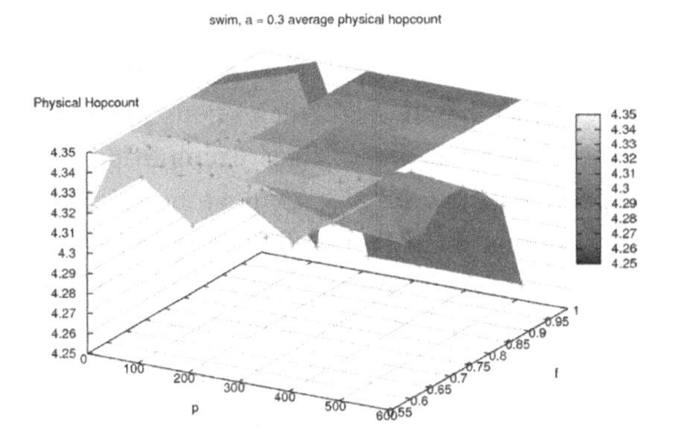

Figure 12. Logical hops for swim (α=0.3)

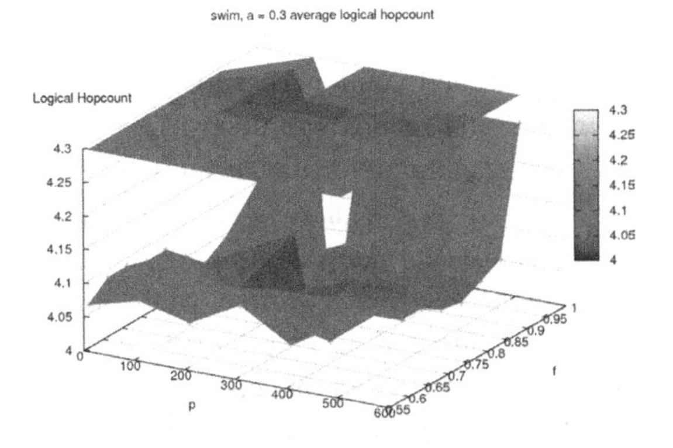

respectively. The implementation of such logic is straightforward, but so far in this paper we have not addressed the subtleties that arise when dynamically switching ongoing data flows from one path to another. We do this now.

Figure 13. Physical hops for gzip (α=0.3)

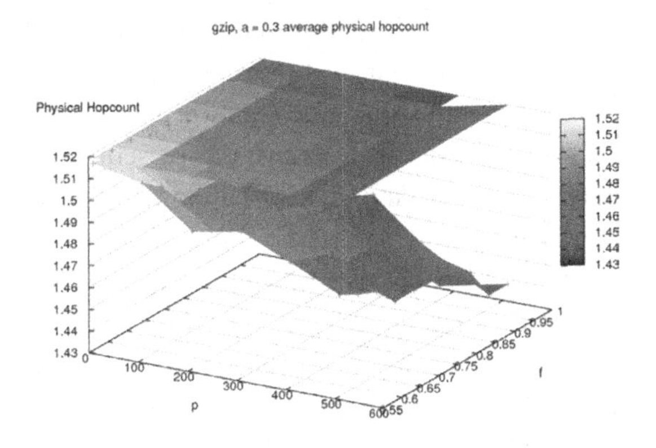

Figure 14. Logical hops for gzip (α=0.3)

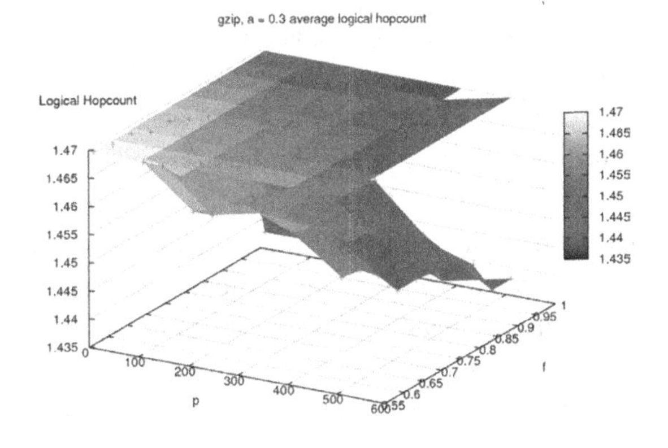

Ensuring Flit Ordering

At an application level, data transmissions are broken up into units called packets, and entire packets are sent and received by applications. In an NoC, packet transmissions are further broken up into basic units of transmission, called flits. A given transmission may comprise multiple flits or even multiple packets. Most network implementations and applications expect that the order of transmission is preserved between the injecting and receiving router node.

In a system with a non-uniform topology, such as one supporting Skip-links, this issue becomes more problematic as there may be multiple minimal (optimal) path through the network that differ depending on whether or not Skip-links are currently deployed along a route. In this situation, packet-based routers using our Opt-bypass adaptive routing algorithm may return different paths for different packets or even different flits between the same source/destination pairs. Thus, it is possible for flits to arrive out-of-sequence. Unless the underlying network supports ordering guarantees, out-of-order arrival can cause problems with erroneous data transfers or traffic QoS guarantee violation.

Take, for example, the Skip-link setup of Figure 1, where there are two possible paths for data to take to the output port. When the topology is altered by establishing a Skip-link over a router, a single output port on that router must be disabled. This can cause a local synchronisation problem in that a packet may be divided between the original path through the router and the new Skip-link path. That is, the Skiplink was initialised when only some of the packet's flits had entered and were switched by the router. The Skip-link path includes no buffering or switching elements, giving it significantly less latency than the path through the router. Therefore, it is possible that the flits that did not enter the router will reach the output before those that did. Thus, it is possible for old data to be overtaken by new ones arriving at the input.

Clearly this presents both an ordering and arbitration problem as the packet's flits could now be transmitted out of order and flits on each path may arrive simultaneously. In addition, the input buffers of the router may hold flits that need to use the now disabled output. Such synchronisation issues do not exist when disabling a Skip-link, as this action can be delayed until a tail flit travels through the Skip-link input, ensuring that a packet is not divided. Neighbouring routers are only informed of the router's output ports returning to normal when the Skip-link is removed.

Along with the ordering problem, we have the issue of ensuring data safety (i.e. all data will be delivered, and none lost) in the presence of links that may be brought up or go down autonomously and unpredictably.

In the following sections, we discuss how flit ordering and safety may be completed for two different logic implementation scenarios, based on the Globally- Asynchronous, Locally Synchronous (GALS) paradigm (Krstic, Grass, Gurkaynak, & Vivet, 2007). We will make use of asynchronous logic (Sparso & Furber, 2001) on our control path to ensure the safety and ordering properties we desire.

Asynchronous Logic Implementations

Asynchronous Signaling

The issues above can be solved by the introduction of signalling based on Asynchronous logic. The value of such an approach is exemplified with the popularity of Globally Asynchronous Locally Synchronous (GALS) based NoC systems. The use of local synchrony allows standard IP blocks and standard-cell design flows to be retained, easing design and verification, whilst the addition of asynchrony on the inter-block data path gives benefits such as clock-skew agnosticism and actual-case delay performance (Krstic et al., 2007).

In GALS systems, dual-rail signalling (Sparso & Furber, 2001) is often deployed, due to its relative simplicity and additional benefit of inserting a non-transitioning guard wire alongside each transitioning one, helping to guard against crosstalk. 1-of-N and M-of-N signalling schemes (Sparso & Furber, 2001) are also popular, as illustrated by the contemporary XMOS device, which utilises 1-of-5 encoding for its chip-to-chip signaling (http://www.xmos.com/products).

In this discussion, however, we will consider the use of the most commonly implemented bundled-data paradigm, where data can be transmitted without the need for additional wires or extra formatting. This approach is compatible with synchronous logic data paths with minimal overhead, and no need for data format conversion at all. Therefore, in a NoC system, where data path latency is critical, if forms the best choice for implementation, We refer the reader to the above references for full details of asynchronous techniques, but for the following discussion it is useful to know that, under 4-phase bundled-data asynchronous signalling (Sparso & Furber, 2001), the following transmission protocol is maintained:

1. Data is placed on the data wires in normal binary format;
2. A request (req) wire is transitioned by the sender to signal that new data is available for the received;
3. Upon reception of the request, the receiver latches in data and causes a;
4. Transition of an acknowledgment (ack) control wire to notify the sender of successful reception;
5. A phase by which first the req and then the ack wires are returned to their idle state.

The above *sequence, known as handshaking* results in *the transmission of a data word, and is repeated for each subsequent word.*

*Whilst ou*r analysis is based on this asynchronous paradigm, the system also works as a drop-in replacement for a synchronous data path with conventional flow control, where the req signals correspond to data clock signals, and the ack signals correspond to credit incrementing signal.

Timing and Safety

We are now in a position to consider how our GALS paradigm may be used to ensure our data ordering and safety properties.

We have identified two methods with which to mitigate the synchronisation problem. The first allows the network to establish a Skip-link with low latency, but with the need for addition of resynchronising logic. The second solution requires an analogue component (a MUTEX), but does not require a resynchroniser for each router.

The first solution works correctly with any router, provided the router has the ability to indicate via a flushed signal that its output queue is empty in a given direction (and no flits in its input queues wish to use that output in the future). This solution requires this signal, but has no timing requirements on the router, so can be used safely in a wide variety of situations.

Our second solution assumes the router is a 'black box', which can be parameterised, but additional signals, such as the flushed wire cannot be added. It is particularly suitable as a drop-in replacement, or where routers are protected IP blocks with no source designs available. This solution, however does require that the maximum inter-flit delay $\tau iFlit$ on a router's output port be characterised.

The two designs have different applicability, but as we shall show, both deliver data- and ordering-safe delivery of flits in the presence of dynamically-reconfigured Skip-link paths. We start now with an introduction of our first solution.

Timing Agnostic Solution

In our 'timing agnostic solution', we assume one piece of support from the routers deployed in the NoC: the ability to notify our data path when their output buffers are empty. This is particularly useful for our situation, since it provides us a solution to the flit ordering problem our solution is to wait until a router's output port in a given direction has completely emptied before activating the Skip-linking path. Since flits entering a router traverse in order, those traversing a Skip-link path also do (it contains no state, so re-ordering is impossible), and all Skip-link path flits traverse after all router flits, then ordering is preserved on Skip-link path setup.

Skip-Link Placement

Assume a condition suitable to Skip-linking has already been detected. Initially any packet currently holding the relevant output channel is allowed to complete switching. The complete steps are listed below.

1. Any packet currently holding the relevant output channel is permitted to complete switching. Each neighbouring router is sent a single flit indicating that this output port is disabled.
2. The Skip-linked router stops accepting new flits on its input ports (by setting its buffer full flags). Its input buffer is flushed (by switching the flits to outputs) so that no flits in the router are waiting for the now disabled output.
3. The Skip-link path is enabled.
4. Router resumes accepting flits, but with the new topology, where one input and one output port is disabled (Figure 1).

A simple MERGE element (Sparso & Furber, 2001) is used to select between router output and Skip-link data. A circuit diagram is shown in Figure 16. A signal from the router (Flush1) is ANDed with the req2a signal from the DEMUX. This ensures that the merge unit does not receive requests from the Skip-link path until the router has switched out all flits waiting on the output. Since the MERGE element provides state and handshaking the simple gating provided by the AND element does not affect the correctness of the 4-phase protocol. The internal control circuitry is shown as Figure 17. The data path is shown as Figure 15, and it is clear that this very simple solution will have a low overhead on the data path, inserting at most one gate delay on the data path.

Figure 15. Timing agnostic data path

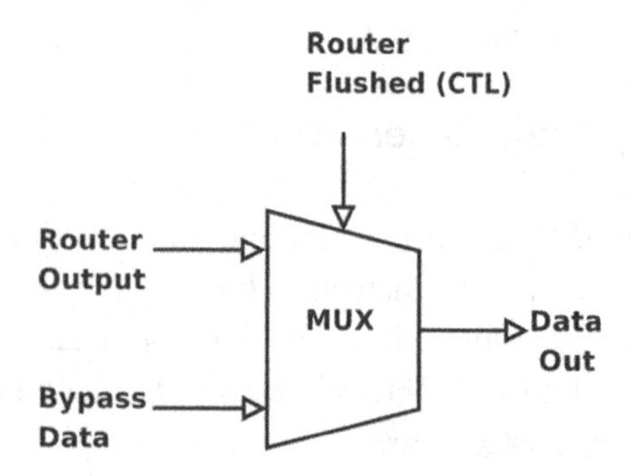

Figure 16. Timing agnostic control path

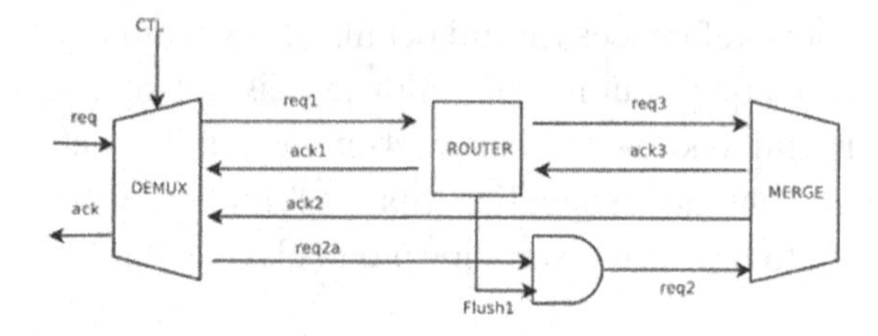

Skip-Link Teardown

Tearing down a Skip-link path is easier, since no flits can be stored anywhere in the system. Therefore, the input port can be changed over arbitrarily, as long as any flit (and there can be only one due to the absence of state) currently in the Skip-link path is allowed to complete transmission on the output port.

Advantages of this Approach

This implementation is completely Delay Insensitive (DI) (Sparso & Furber, 2001), which means that it is capable of waiting an arbitrary amount of time between Skip-link creation being decided, and the router's buffers becoming empty. DI circuits are the most robust asynchronous ones, and ensure data safety by definition.

Disadvantages of this Approach

Router IP source code is needed to be able to implement the flush signal. Further, flushing must occur before the input side is enabled, so an additional delay penalty is experienced.

Performance Evaluation

When synthesised with the FreePDK 45nm library (Stine et al., 2007), the timing agnostic solution used a total of 430μm2of area and 48μW of power, having 133ps of delay on the worst-case path, approximately 4-FO4 in this technology.

Figure 17. Timing agnostic control logic

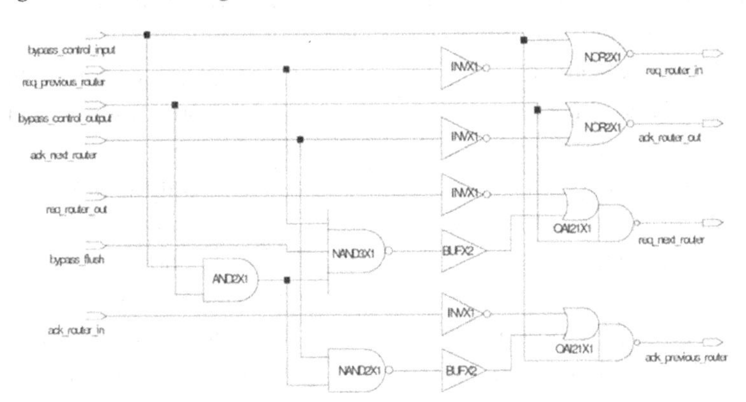

Router Agnostic Solution

It is important to minimise the delay between the detection of traffic patterns that would benefit from Skip-linking and the deployment of Skip-link paths. Any latency in initialising the Skip-link is a wasted opportunity to decrease switching activity. With the addition of a blocking data-merge unit it is possible to initialise a Skip-link immediately whilst avoiding potential ordering and synchronisation problems. However, to maintain ordering, before the router can be Skip-linked it must empty the relevant output buffer, as well as all input buffers to ensure no flits are waiting for the disabled port.

Our 'router agnostic' solution aims to tackle this issue by using a priority arbiter with hysteresis to ensure correct flit priority, and thus ordering. In the case where the router has no more data to send, Skip-linked traffic is allowed to proceed as soon as the arbiter can create a grant. Thus, this implementation is more responsive to newly Skip-linked traffic. A discussion of its implementation now follows.

The router output and Skip-link output connect to a conventional multiplexer as shown in Figure 18. The output select signals for the MUX are generated by an arbiter that receives the request signals from both the Skip-link (req2) and router outputs (req1a). Before the data waiting in the Skip-link output can be used, it must be ensured that there is no data left on the router output. Potentially a Skip-link can be initialised when a packet is being switched by the router across the links that form the Skip-link. As explained previously, this can lead to flits reaching the output link in the wrong order. To ensure this cannot occur, the router's output buffer must be emptied before accepting the Skip-link output. Therefore the arbiter must only grant access to the Skip-link output when there is no request being transmitted by the router output (i.e. all flits have been flushed). The circuit shown in Figure 18 ensures that all requests from the router are given priority over those from the Skip-link. The delay element is calibrated to the

Figure 18. `Router agnostic' control + data path

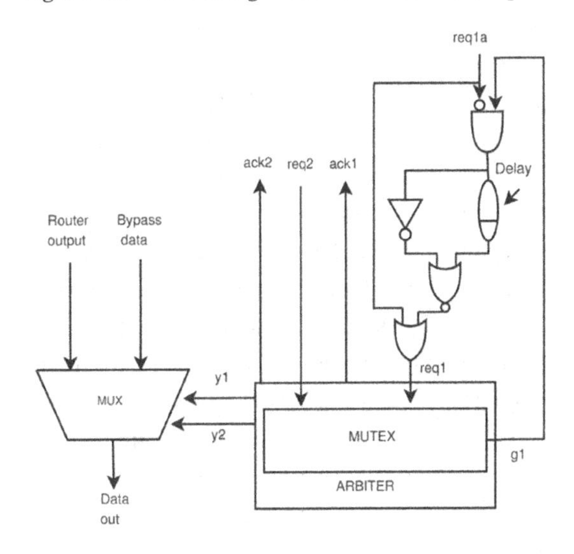

maximum cycle time of the router (*t*iFlit), and holds the request (req1) high until the router has completed flushing. There is a latency penalty when the router has completed as the request line will be held high by the delay element even though the router no longer has any output. The following describes the operation of the circuit that holds the request signal from the router:

1. A flit is waiting to be output from the router, so req1a is high. The arbiter grants MUX access to the router, acknowledges the output (ack1) and so req1a resets.
2. A Skip-link is initialised and so req2 goes high. However, the delay element maintains a high output and so req1 is held.
3. As the arbiter granted access to the router initially and req1 has not dropped low, access to the MUX is still granted to the router.
4. If the router still has output, then go back to step 1. Otherwise, access is granted to the router for the period of the delay element. When this period has passed, req1 goes low and finally the Skip-link output is granted access to the MUX.

Simulated Performance

To ensure operational correctness of our designs, and also to be able to characterise their performance, we carried out a number of simulations using the latest Cadence® analogue toolset, when applied to a contemporary 90nm technology (unfortunately, we did not have an analogue 45nm library available with which to perform a like-for-like comparison with the timing-agnostic design). We designed laid out and extracted values for our router agnostic units. We present an analysis of their correctness, latency, area overhead and energy consumption below. All slew measurements are taken between the 20–80% of vdd point, and timing for full swing is taken as the appropriate 20% of vdd or 80% of vdd as appropriate.

Functional Correctness

A demonstration of correctness of the router agnostic solution comprises many parts, including function correctness and the validation of the intended operating mode of elements like the delay-based grant extension.

Figure 19 shows the output from a simulation of our router-agnostic circuit, illustrating the role of the delay in extending the grant's active time beyond that indicated by the request input. In this simulation, req2 was held low, whilst req1 was pulsed. It can be clearly seen that the gnta line is held high significantly longer than would be normal if the device's output was influenced only the profile of the request input req1. This is one validation that the circuit is functioning as desired. It can also be seen that the falling edge of the grant (gnta) line closely mirrors in time and shape that of the output from the internal delay. This further demonstrates the correctness of operation, which allows a designer to tune the delay to match the operational characteristics of the router to which it is attached.

MUTEX Unit

The performance of the MUTEX unit is critical to the operation and performance of our router-agnostic solution. With a MUTEX, we not only require correct functional behaviour, as described in (Sparso & Furber, 2001), but since its performance is on the critical path of our design, we wish it to have a low forward latency, so that the MUX element controlling the datapath may be quickly switched.

As such, the req→gnt latency is critical. In a MUTEX element, however, this varies dependent on the relative arrival time of the two req inputs: the closer they are together, the longer the forward latency, following an inverse exponential curve (Kinniment, 2007); the further apart, the faster, up to some minimal limit.

Figure 19. Extension of the grant signal gnt1 (solid) based on the output from the delay element (dashed). The input request signal req1 is shown for reference (dotted).

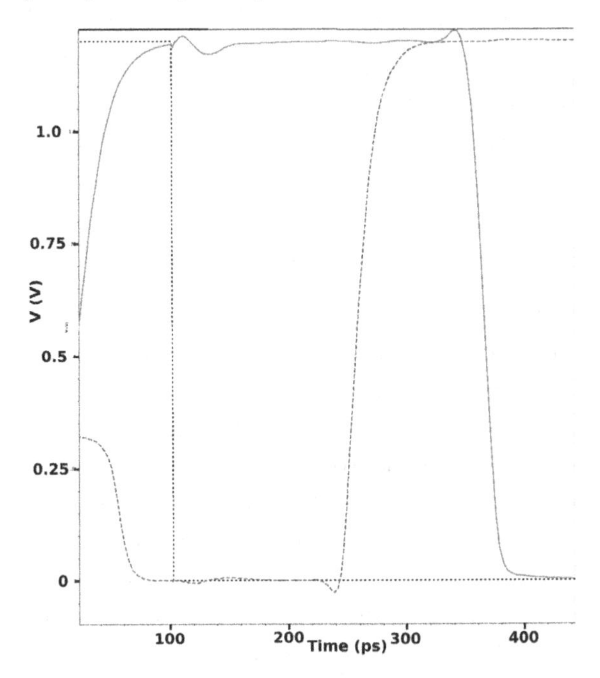

If a MUTEX is designed entirely symmetrically, then the output latency (corresponding to the amount of time it is internally metastable) can become very long if coincident inputs are applied. This phenomenon would have an adverse (if exceedingly rare) impact on our data flows. To get over this problem, and also to implement the prioritisation of the router request signal, as explained earlier, we laid out the MUTEX following (Sparso & Furber, 2001) in a slightly asymmetric manner.

To characterise MUTEX behaviour, we carried out a sweep through the space in which it may potentially become internally metastable, with impact on its resolution time. The MUTEX element had step inputs and was loaded by a single FO4 capacitance. The results are plotted in Figure 20, where the various curves show the voltage at the gnta output when the arrival times of the req1 & req2 input signals are offset by various amounts from 0–40ps (from right to left, respectively). An additional curve with 1ns separation (left-most) is also included to show the minimum transition time (since with this separation, internal metastability is near impossible). We see the expected effect: the closer the request signals arrive at the multiplexer, the longer it takes to resolve, from 115ps for the 1ns spacing, to 155ps for the coincident inputs (80% of vdd transition times). In our technology, this corresponds to a range of 3.4 . . . 4.6 FO4 delays.

Adding in the remainder of the logic, characterised for a 45nm process, adds a worst-case of 71ps. Thus, the total delay of the system ranges from 115ps (gnt2→req2, with no other wires asserted, to 226ps for (gnt1→req1, with gnt2 simultaneously asserted. The latter situation is rather unlikely to occur in practice, so we assume 115ps as an average case delay for our system. Total area for the combined system is estimated at $31\mu m2$, with the system consuming 1254nW of power.

Figure 20. Coincident requests and resolution at the MUTEX unit for various request arrival time differences

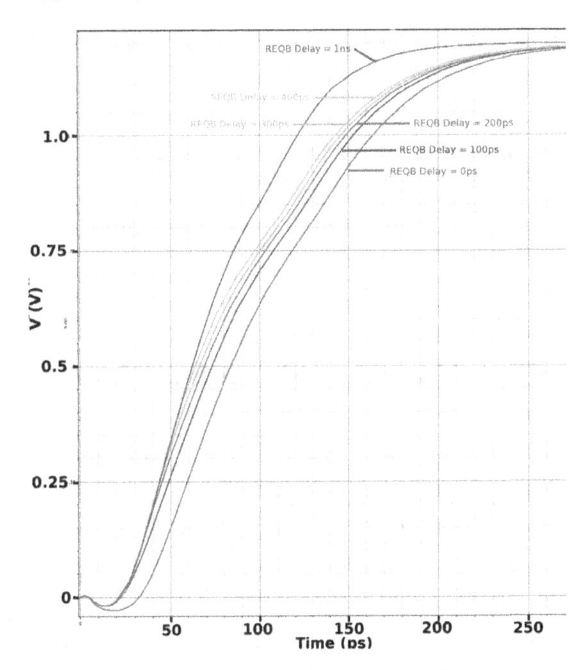

Comparison of the Two Solutions

Table 1 shows a comparison of the two solutions, where we also include information for an implementation of a 32-bit data path in 45nm. Three main observations can be made:

1. The router agnostic solution has a slightly smaller average delay, compared to the timing agnostic one at 45nm. This is due to the shorter logical path, of just two gates on when requests rise on the input; for the timing agnostic solution, there are three gates, and one is a slower 3-input NAND gate.
2. They are broadly comparable in area and power use.
3. The data path dominates both implementations massively for all resource values.

Given the massive dominance of the data path delay and resource use, it is unlikely to matter which control path solution is chosen, from a performance point of view. Therefore, the one chosen for a particular Skip-link implementation should be based solely on the characteristics desired. If an all-digital flow is desired, the timing agnostic design should be chosen, and the additional router flush signal implemented; if the router is protected IP, then the router agnostic design should be chosen. Overall, the data path limits the maximum operating frequency at 1.4GHz, but we would expect that, with hand optimisation, this could be reduced significantly, enabling operation at up to 3GHz.

SKIP-LINK TRAFFIC PERFORMANCE

Now that we have established a set of optimal Skip-link creation parameters and defined their implementation, we are in a position to evaluate their performance properly. In this section, we evaluate the performance of Skip-links by simulating a k-ary 2-mesh network with a range of synthetic traffic patterns. To evaluate the Skip-link architecture we reused our cycle-accurate simulator and generated traffic traces using the 2D tornado and transpose permutations (Dally & Towles, 2003). For transpose traffic we considered 8×8 and 16×16 node networks (this permutation requires the number of nodes to be an even power of two). The transpose permutation is representative of traffic encountered during matrix transpose operations. For tornado traffic 12×12 node networks were also considered. The tornado permutation was designed as an adversary for torus networks and stresses a network significantly more than the oft used uniform random traffic permutation. Our simulator incorporates fully-detailed router and physical link models, and we simulated six Virtual Channels per link, in each direction. A router with empty buffers is capable of processing a flit per channel per cycle. To produce our latency figures, our wire interconnects were assumed to be pipelined, with each hop requiring a single cycle. In this paper, we do not consider the routing algorithm needed by our dynamically-changing topology, however one has been developed: Opt-Bypass (Jackson & Hollis, 2011), based on Opt-y (Schwiebert & Jayasimha, 1993). This was incorporated into our simulations to ensure routability.

Table 1. Implementation resource use (45nm; † 90nm)*

Design	Delay (ps)	Area (μm²)	Power (nW)
Timing agnostic	133*	28*	1300*
Router agnostic	115†	31*	1254*
Data path (32-bit)	702*	411*	46900*

Hop Count Changes

We extracted figures from our simulations for the number of times each flit passed through a router (a Logical (Log.) hop) or traversed a physical wire segment (a Physical [Phy.] hop). Table 2 shows the results of averaging all flits in a run for both a Static Mesh (SM) and one augmented with Skip-links (SLM). The configuration number denotes the number of routers simulated.

We see that SLM consistently provides a logical hop saving of between 12– 17%, signifying that proportion of routers along its path are skipped on average. The figure also grows with network size, as expected, since the longer-distance routes provided by our approach are more likely in larger networks.

These observations not only prove our concept's value, but also demonstrate significant energy savings, as we will discuss shortly. Neither scheme can overcome the need to cover physical distance between source and destination nodes, and so the physical distance cannot be reduced by SLM, and we see this is the case. SLM actually increases slightly (by 1–4%) the average physical distance travelled by flits. This is due to the fact some traffic must be routed around links disabled due to the skipping mechanism. However, this overhead is very small, compared to the logical hop saving.

Energy Consumption

To evaluate the energy savings of our approach, we combined the hop counts in Table 2 with data taken from ReNoC (Stensgaard & Sparsø, 2008), which gives values for router and wire energy consumption in a typical 45nm ASIC design. We assumed an inter-router wire length of 0.5mm, which is consistent with a design aim of fitting the 256 simulated cores on a 1cm × 1cm die. We understand that this is only a single point in a very wide design space, but given the space limitations of this paper, it is not possible to present a full design-space exploration. However, we believe that our figures are indicative of the savings that could realistically be made in a wide range of NoC realisations.

Figure 21 shows the average per-flit end-to-end energy consumption with the traffic injection rate routed under both SM and SLM for 64 and 256 core systems. All results are plotted until saturation, with the absolute values scaled. They show similar pictures: due to the different average hop path lengths of the different sized networks, SLM traffic consumes less overall energy than SM traffic; tornado traffic consumes more energy than transpose traffic; SM-routed traffic has a very steady energy consumption graph with injection rate; SLM routed traffic has a more unpredictable energy consumption pattern.

We now examine each of these observations in turn. First, SLM traffic always consumes less overall energy than SM traffic. This is explained by noting that the weighting given in Stensgaard and Sparsø (2008) roughly translates to three wire segments consuming the same amount of energy as a router in our configuration. Therefore, the extra physical hop increases of SLM are easily outweighed by its logical hop savings, resulting in a net energy reduction over SM.

Table 2. Difference in Logical (Log.) and Physical (Phy.) hop counts between a Static Mesh (SM) and a Skip-link augmented Mesh (SLM)

Configuration	SM	SM Log.	SLM Phy.	Δ Log.	Δ Phy.
64way Tornado	4.97	4.38	5.13	-12%	+3.1%
144Way Tornado	8.84	7.69	8.96	-13%	+1.3%
256Way Tornado	12.26	10.21	12.48	-17%	+1.8%
64Way Transpose	5.99	5.20	6.10	-13%	+1.8%
256Way Transpose	11.33	9.36	11.60	-17%	+3.6%

Figure 21. Per-flit energy consumptions between the static mesh (SM) and a Skip-linked mesh (SLM)

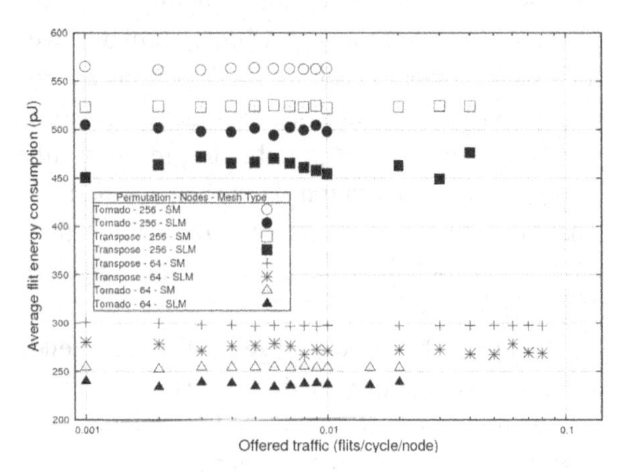

Figure 22. Latency vs. injection rate: tornado and transpose traffic; static mesh (SM) and a Skip-linked mesh (SLM)

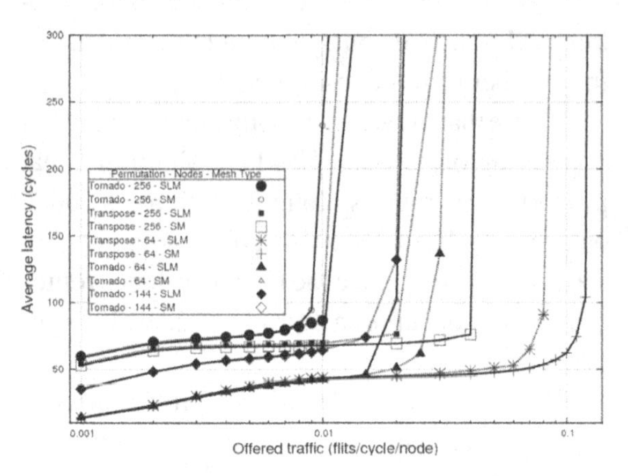

Next, tornado traffic is more energy-intensive than transpose traffic, since under tornado, all nodes must generate traffic, whereas for transpose some nodes sit idle. As can also be seen for larger network sizes, the steady-state behaviour of tornado also sends flits further on average (roughly half the network size in each dimension). This delivers a saving of approximately 10% of energy on a 256 node network, compared to around a 6% saving for a 64 node network. For transpose traffic, this increases to a 14% saving on a 256 node network, down to around 10% for 64 nodes.

Finally, the energy consumption pattern of SM is steady with injection rate since the route cannot change, and so the path (and hence energy requirement) is always the same. For SLM, this is not true, and topologies and routes can be reconfigured during runs to support more optimal execution. Therefore, there is a (bounded) variance in the routes taken between runs, and related energy consumption.

Latency and Critical Load

We also measured the average flit latency of traffic during our simulations, as it varied with traffic injection rate. Figure 22 shows the results for tornado and transpose traffic.

All latency measurements show the expected horizontal asymptote slowly rising from zero-load (Dally & Towles, 2003), then sloping upward to infinity as the network reaches saturation throughput. We see that larger networks saturate at lower injection rates, as expected for the permutations used (since larger networks produce a larger total traffic load for a given injection rate).

Exact latencies for various configurations can clearly be seen in Figure 22, but we draw the reader's attention to the generally positive impact of Skip-links on the critical load of the network, as defined by the saturation point of the graphs, when they begin to tend strongly to a vertical asymptote. For a 256 node network with Tornado traffic, Skip-linked, saturation occurs at 0.01 flits/cycle/node, compared with 0.009 for the standard mesh; this increases to 0.03 compared to 0.01, with the smaller 64 node network. For transpose, large networks both behave much better, and perform better without Skip-links with 256 non-Skip-linked nodes accepting an injection rate of 0.04 flits/cycle/node before saturating, compared to 0.02 flits/cycle/node for Skip-linked ones. As networks shrink, performance is again increased, and a 64 node standard mesh network can handle over 0.1 flits/cycle/node, but only 0.07 flits/cycle/node if Skip-linked.

So, in general, with tornado traffic, SLM produces lower latencies than SM, and is also capable of shifting the saturation point slightly to the right. For transpose, the opposite effect is observed, and this is consistent with its ability to better optimise traffic that travels long distances on average. Thus, when considering only latency, Skip-linking only benefits tornado traffic, and not transpose.

CONCLUSION

We have defined a new NoC architecture to support runtime topology changes. As part of this work, we have developed the notion of a dynamic placement algorithm where each router independently determines when to place a Skip-link. Emergent behaviour can lead to the formation of Skip-chains, formed by adjacent Skip-links that further optimise the topology. Each router is also capable of restoring the regular mesh topology, should a placed Skip-link no longer be beneficial.

When combined with an already-developed deadlock-free routing algorithm, we are able to show that our system reacts well to two well-known adversarial traffic patterns, tornado and transpose. It offers lower logical hop count and energy consumption for both permutations, while exhibiting lower latency and a higher traffic saturation point for tornado traffic. It operates in a distributed manner, without the need for any a priori static analysis or hinting.

Logical hop counts are reduced and both energy consumption and latency results are better for a Skip-link enabled topology than for a standard mesh. These results exemplify the advantages of a dynamically-reconfiguring topology. They also act as a proof of concept for a system that automatically places Skip-links and chains in near-optimal locations. Our results illustrate the usefulness of dynamic reconfiguration, and we believe that these advantages will lead to future systems adopting this principle.

Finally, we presented two potential implementations for Skip-links: one that follows a completely digital flow, and can be used with synchronous or asynchronous logic systems; the other a purely asynchronous system that requires an analogue MUTEX element to be designed and utilised. Both systems were simulated, their performance extracted and displayed, and both we shown to be robust and high performance.

REFERENCES

Asanovic, K., Bodik, R., Catanzaro, B. C., Gebis, J. J., Husbands, P., Keutzer, K., et al. (2006). *The landscape of parallel computing research: A view from Berkeley* (Tech. Rep. No. UCB/EECS-2006-183). Berkeley, CA: University of California.

Banerjee, A., Mullins, R., & Moore, S. (2007, May). A power and energy exploration of network-on-chip architectures. In *Proceedings of the First International Symposium on Networks-on-Chip* (pp. 163-172). Washington, DC: IEEE Computer Society.

Chen, G., Li, F., & Kandemir, M. (2007). Compiler-directed application mapping for Noc based chip multiprocessors. *SIGPLAN Notes*, *42*(7), 155–157. doi:10.1145/1273444.1254796

Dally, W. J., & Towles, B. (2001). Route packets, not wires: On-chip interconnection networks. In *Proceedings of the 38th Conference on Design Automation* (pp. 684-689). New York, NY: ACM Press.

Dally, W. J., & Towles, B. (2003). *Principles and practices of interconnection networks*. San Francisco, CA: Morgan Kaufmann.

Hollis, S. J., & Jackson, C. (2009, September). When does network-on-chip bypassing make sense? In *Proceedings of the 22nd IEEE International SOCC Conference* (pp. 143-146). Washington, DC: IEEE Computer Society.

Hu, J., & Marculescu, R. (2005). Energy- and performance-aware mapping for regular Noc architectures. *IEEE Transactions on Computer-Aided Design of Integrated Circuits and Systems, 24*(4).

Jackson, C., & Hollis, S. J. (2011). A deadlock-free routing algorithm for dynamically reconfigurable networks-on-chip. *Microprocessors and Microsystems, 35*(2), 139–151. doi:10.1016/j.micpro.2010.09.004

Kinniment, D. J. (2007). *Synchronization and arbitration in digital systems*. Chichester, UK: John Wiley & Sons. doi:10.1002/9780470517147

Krstic, M., Grass, E., Gurkaynak, F. K., & Vivet, P. (2007). Globally asynchronous, locally synchronous circuits: Overview and outlook. *IEEE Design & Test of Computers, 24*, 430–441. doi:10.1109/MDT.2007.164

Kumar, A., Peh, L.-S., Kundu, P., & Jha, N. K. (2007). Express virtual channels: Towards the ideal interconnection fabric. *SIGARCH Computer Architecture News, 35*(2), 150–161. doi:10.1145/1273440.1250681

Lee, K., Lee, S.-J., & Yoo, H.-J. (2006). Low-power network-on-chip for high-performance soc design. *IEEE Transactions on Very Large Scale Integration Systems, 14*, 148–160. doi:10.1109/TVLSI.2005.863753

Michelogiannakis, G., & Dally, W. J. (2009). Router designs for elastic buffer on-chip networks. In *Proceedings of the Conference on High Performance Computing Networking, Storage and Analysis* (pp. 1-10). New York, NY: ACM Press.

Moore, S., & Greenfield, D. (2008). The next resource war: Computation vs. communication. In *Proceedings of the International Workshop on System Level Interconnect Prediction* (pp. 81-86). New York, NY: ACM Press.

Mullins, R. (2006, Nov.). Minimising dynamic power consumption in on-chip networks. In *Proceedings of the International Symposium on System-on-Chip* (pp. 1-4).

Nollet, V., Marescaux, T., Verkest, D., Mignolet, J.-Y., & Vernalde, S. (2004). Operating-system controlled network on chip. In *Proceedings of the 41st Annual Conference on Design Automation* (pp. 256-259). New York, NY: ACM Press.

Ogras, U. Y., & Marculescu, R. (2006). It's a small world after all: Noc performance optimization via long-range link insertion. *IEEE Transactions on Very Large Scale Integration Systems, 14*(7), 693–706. doi:10.1109/TVLSI.2006.878263

Schwiebert, L., & Jayasimha, D. N. (1993). Optimal fully adaptive wormhole routing for meshes. In *Proceedings of the ACM/IEEE Conference on Supercomputing* (pp. 782-791). New York, NY: ACM Press.

Shalf, J., Kamil, S., Oliker, L., & Skinner, D. (2005). Analyzing ultra-scale application communication requirements for a reconfigurable hybrid interconnect. In *Proceedings of the ACM/IEEE Conference on Supercomputing* (p. 17). Washington, DC: IEEE Computer Society.

Soteriou, V., & Peh, L.-S. (2004, October). Design-space exploration of power- aware on/off interconnection networks. In *Proceedings of the 22nd International Conference on Computer Design* (pp. 510-517). Washington, DC: IEEE Computer Society.

Soteriou, V., Wang, H., & Peh, L.-S. (2006). A statistical traffic model for on-chip interconnection networks. In *Proceedings of the 14th IEEE International Symposium on Modeling, Analysis, And Simulation* (pp. 104-116). Washington, DC: IEEE Computer Society.

Sparso, J., & Furber, S. (2001). *Principles of asynchronous circuit design - a systems perspective*. Dordrecht, The Netherlands: Kluwer Academic.

Stensgaard, M. B., & Sparsø, J. (2008). Renoc: A network-on-chip architecture with reconfigurable topology. In *Proceedings of the Second ACM/IEEE International Symposium on Networks-on-Chip* (pp. 55–64). Washington, DC: IEEE Computer Society.

Stine, J. E., Castellanos, I., Wood, M., Henson, J., Love, F., Davis, W. R., et al. (2007). Freepdk: An open-source variation-aware design kit. In *Proceedings of the IEEE International Conference on Microelectronic Systems Education* (pp. 173-174). Washington, DC: IEEE Computer Society.

Towles, B., Dally, W. J., & Boyd, S. (2003). Throughput-centric routing algorithm design. In *Proceedings of the Fifteenth Annual ACM Symposium on Parallel Algorithms and Architectures* (pp. 200-209). New York, NY: ACM Press.

Ye, T., Benini, L., & De Micheli, G. (2002). Analysis of power consumption on switch fabrics in network routers. In *Proceedings of the 39th Design Automation Conference* (pp. 524-529). New York, NY: ACM Press.

This work was previously published in the International Journal of Embedded and Real-Time Communication Systems, Volume 2, Issue 3, edited by Seppo Virtanen, pp. 21-49, copyright 2011 by IGI Publishing (an imprint of IGI Global).

Chapter 11
Self-Calibrating Source Synchronous Communication for Delay Variation Tolerant GALS Network-on-Chip Design

Alessandro Strano
University of Ferrara, Italy

Carles Hernández
Universidad Politécnica de Valencia, Spain

Federico Silla
Universidad Politécnica de Valencia, Spain

Davide Bertozzi
University of Ferrara, Italy

ABSTRACT

Source synchronous links for use in multi-synchronous networks-on-chip (NoCs) are becoming the most vulnerable points for correct network operation and must be safeguarded against intra-link delay variations and signal misalignments. The intricacy of matching link net attributes during placement and routing and the growing role of process parameter variations in nanoscale silicon technologies are the root causes for this. This article addresses the challenge of designing a process variation and layout mismatch tolerant link for synchronizer-based GALS NoCs by implementing a self-calibration mechanism. A variation detector senses the variability-induced misalignment between data lines with themselves and with the transmitter clock routed with data in source synchronous links. A suitable delayed replica of the transmitter clock is then selected for safe sampling of misaligned data. The manuscript proves robustness of the link in isolation with respect to a detector-less link, but also assesses integration issues with the downstream synchronizer and switch architecture, proving the benefits in a realistic experimental setting for cost-effective NoCs.

DOI: 10.4018/978-1-4666-2776-5.ch011

voltages and frequencies. In this context, multi-synchronous designs, making use of synchronizers and source-synchronous communication for clock domain crossing, turn out to be a more flexible and readily available alternative to embody the GALS concept into industry-relevant designs rather than clockless handshaking.

In source synchronous links, data is routed together with a strobe signal (it might be the transmitter clock itself) which enables safe sampling at the receiver side regardless of the clock phase and/or frequency ratio between the communicating clock domains. However, synchronization interfaces now become the true weak-point of the system that needs to be safeguarded against timing failures. In fact, source synchronous links are generally designed under the assumption that there will be no or very limited routing skew between data lines and the transmitter clock, an assumption that might be easily impaired in nanoscale CMOS processes. The reason for this is twofold. On one hand, significant length, resistance and load deviations among different wires of a link should be expected even when advanced bus routing features of place-and-route tools are used. On the other hand, process parameter variations impact various device characteristics, such as effective gate length, oxide thickness and transistor threshold voltages, which may in turn lead to significant variations in power consumption and to timing violations. These effects impair the functionality of the link by reducing the data stability window and by causing the uncertainty on the precise sampling time over the clock period.

Although reducing delay variations between wires of a GALS NoC link would be the ideal requirement for advanced NoC implementation styles, it is virtually impossible to completely predict nominal delay and routing skew deviations occurring in the manufacturing process. Therefore, NoC link design should consider self-calibration, which is the approach taken by this manuscript. Instead of relying on the worst-case characterization of design parameters, self calibrating systems determine autonomously the boundary of correct behavior, and set design parameters accordingly.

In this work we apply the self-calibration design principle to source synchronous links for use in synchronizer-based GALS NoCs. The basic idea is to use a variability detector at the receiver side of a source synchronous link with the capability of sensing the offset between data wires with each other and with the transmitter clock routed along with them. Based on misalignment quantification, the circuit selects the earliest delayed replica of the transmitter clock which can safely sample input data in the synchronization interface. The detector is meant for use during system reset, during which each GALS link is supposed to perform a self-calibration procedure. Repeated at every system bootstrap, the procedure can ensure robustness against wear-out effects.

RELATED WORK

Many recent works analyze the impact of process variations on the performance of integrated circuits, providing data on how parameter variations impact the maximum design frequency (Bowman et al., 2002) or variability models that characterize variations in microarchitecture (Sarangi et al., 2008; Bonesi et al., 2008). However, these studies do not consider the implications of variations in the interconnect infrastructure. Although (Nicopoulos et al., 2010) is a step forward in this direction, this study neglects the impact of manufacturing deviations on NoC links. Unfortunately, this impact is not negligible (Mondal et al., 2007; Hernández et al., 2010; Hassan et al., 2009). On one hand, although there are examples of repeater-less NoC self-calibrating links (Jose et al., 2005), they typically undergo repeater insertion. Therefore, they suffer from Lgate variations and dopant fluctuations in the transistors building up repeater stages, and also suffer from the variability introduced by the chemical metal planarization process (Mondal et al., 2007).

Process variation in NoC links causes that links in the network feature different delays, despite that they were initially designed to be identical (Hernández et al., 2010). Thus, some links will not be able to switch at the intended frequency, thus reducing overall performance. This will be the case when the NoC operating speed is constrained by link delay (Gilabert et al., 2009). A more destructive effect concerns the delay differentiation between the individual wires of a NoC link. In the case of GALS links, bridging mesochronous or fully asynchronous clock domains, this delay differentiation increases the probability of sampling failure at the receiver end (Loi et al., 2008; Ludovici et al., 2010).

Architecture and circuit designers have to deal both with the random and with the systematic components of link delay variability. The systematic component can be partially addressed by making all wires undergo the same physical routing conditions, like in Kakoee et al. (2010). Unfortunately, this approach retains sensitivity to random variations. One common choice is to add worst-case guard bands to critical paths (Borkar et al., 2004). However, this comes at the expense of a high reduction in performance. A more refined technique is using statistical delay calculation tools in order to reduce design margins and improve design speed (Synopsys, 2007). This probabilistic framework avoids the pessimistic timing estimation introduced by the classical static timing analysis (Orshansky & Keutzer, 2002). Other kinds of solutions are based on tolerating infrequent run-time timing violations, where delay failures are tolerated at the cost of performance (Ernst et al., 2003; Murali et al., 2006). On the other hand, there exist several proposals that perform post-silicon detection and compensation. In these techniques the circuit is designed for the typical case and variability compensation is performed post-silicon at some cost (performance, power) (Bonesi et al., 2008; Paci et al., 2009). An effective technique to design differently than worst-case is to use self-calibrating links. The work in Worm et al. (2005) dynamically scales down the voltage swing, while ensuring data integrity. In Merdadoni et al. (2008) the voltage swing is adapted to the link delay budget during a calibration phase. The Stars system uses non intrusive monitors to measure the delay across the longest path in a device and adjust clock frequency accordingly (Diamos et al., 2007). Finally, in Höppner et al. (2010) data is serialized into several parallel high speed lanes and the delay deviation of each lane is compensated by means of a digitally controlled delay cell.

Source synchronous communication has recently emerged as a potential alternative to clockless handshaking for clock domain crossing in GALS NoCs (Panades et al., 2008). In contrast to well-consolidated application domains such as off-chip memory sub-systems and networking as well as I/O (Collins & Nikel, 1999), its application to an on-chip setting is challenged by the uncertainty of the manufacturing process discussed so far. In particular, there is consensus among industrial designers that naively implementing source synchronous links in an on-chip setting is very likely to lead to timing violations. Typical solutions to mitigate this concern include configurable delay elements on data and/or clock wires (Yu & Baas, 2006), adaptive delay tuning (Elrabaa, 2006), clock recovery schemes (Kihara et al., 2003; Moore et al., 2000), multiple registers on the data bus (Dally & Poulton, 1998), or alternating sample edges (Tran et al., 2009). In all cases, proposed schemes are either overly costly since inspired by off-chip networking, or assume timing constraints that in real-life are unpredictable until tape-out and testing, or assume matched delays between similar components, which is hard to achieve in a variability-dominated environment. With respect to previous work, this paper for the first time applies the self-calibration design principle to source synchronous links for use in GALS NoCs. Differently than shared busses, flow control issues need to be more closely accounted for.

TARGET GALS PLATFORM AND MOTIVATION

Without lack of generality, the synchronizer-based GALS network addressed in this work consists of implementing the on-chip network and the networked IP cores as disjoint clock domains, and therefore to place circuitry to reliably and efficiently move data across asynchronous clock boundaries between NoC switches and connected network interfaces. Figure 1 depicts our target GALS platform.

Unfortunately, the network ends up spanning the entire chip and might be difficult to clock due to the growing chip sizes, clock rates, wire delays and parameter variations. We find that mesochronous synchronization can relieve the burden of chip-wide clock tree distribution while requiring simpler and more compact synchronization interfaces than dual-clock FIFOs. Hierarchical clock tree synthesis is an effective way of inferring mesochronous links, as already experimented in Panades et al. (2008).

In our architecture, the upstream links of both dual-clock FIFOs and mesochronous synchronizers are source synchronous: data is routed together with the transmitter clock, which is used at the receiver end to safely sample input data in the receiver (mesochronous or asynchronous) clock domain. Minimizing routing skew between data wires and transmitter clock is a challenging task for state-of-the-art physical routing tools (Kakoee et al., 2010), as current support of physical routing tools for bundled routing proves not always capable of matching wire characteristics (e.g., resistance, length, delay), especially over long links. Moreover, the reduced signal slope over long links causes a larger delay of downstream logic cells, thus resulting in a delayed generation of the sampling control signal.

Even assuming ideal routing, the onset of significant random process variation effects makes it very difficult to meet the routing skew constraints. As a consequence, a sampling failure may occur at the receiver side of a source synchronous link. This scenario calls for an augmented GALS link architecture for robustness to layout mismatches and process variations

ARCHITECTURE OF THE VARIATION DETECTOR

This section presents a novel architecture circuit (the *variation detector*) guaranteeing the reliability of NoC source synchronous interfaces under high circuit and wire delay variability. Implicitly, the circuit copes with mismatches of layout characteristics as well (such as unmatched link net attributes).

The variation detector we propose should be placed in front of the regular synchronizer (such as dual-clock FIFOs or input latches of mesochronous synchronizers) and aims at restoring the required alignment between data lines and the transmitter clock for safe sampling. The operating principle is the following. At first, the amount of misalignment between the wires of the source synchronous link (data plus clock) is sensed. Then, the receiver synchronizer is fed with a clock signal version that guarantees correct sampling. The clock signal is a delayed replica of the transmitter clock. A schematic of the proposed architecture is illustrated in Figure 2. It is composed of a parametric number of brute force synchronizers sampling the output of an AND block. The AND gate receives the incoming data as input generating a high value as soon as all the data input bits are high. The source synchronous clock signal (routed along with the data) crosses a set of delay chains introducing an incremental amount of delay with respect to the nominal source clock. The delayed clocks feed 2-flop (or more) brute force synchronizers that sample data at different time instants within the clock period. Therefore, at least one of them will be able to sample input data during the stability window of these latter.

Figure 1. GALS target platform of this work

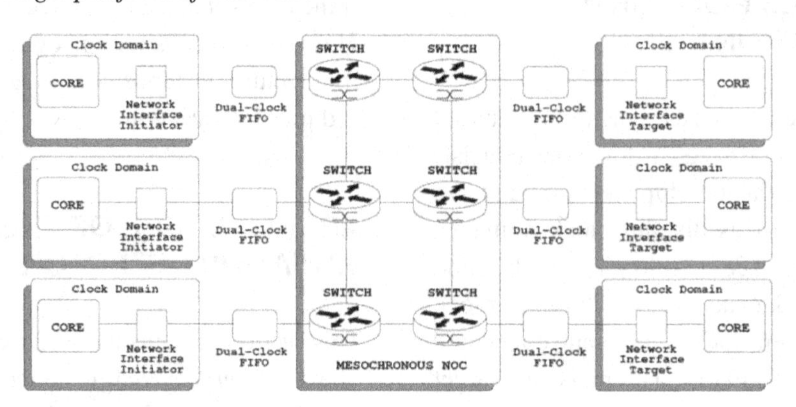

In order to preserve the synchronism property between reset and clock signal, the reset is bundled with the data as well and it crosses the delay chains together with the clock. The number of delay chains depends on the number of brute force synchronizers in order to guarantee to every synchronizer block a clock/reset with a different and equally interspaced offset. Moreover, it depends on the amount of expected routing skew between data and clock and between data lines with themselves: the shorter the data stability window, the higher the number of brute force synchronizers needed to correctly sample the data at least once.

Finally, the outputs of the brute force synchronizers drive the multiplexer control signals and select through it the delayed clock signal replica required to feed the receiving mesochronous synchronizer with a safe strobe signal. The selection of the safe clock signal takes place during the NoC reset phase and it represents a key step in the proposed architecture. It is detailed in the following section ("*Operating principle*"). However, it is worth recalling that once a suitable strobe signal is selected during the reset process (namely, at every system bootstrap), then the selection stays the same throughout the entire use cycle of the system, thus not generating an overhead for normal system operation. Used over time, the proposed architecture allows the network to deal also with wear-out effects that might prolong gate and wire delays in an unpredictable way.

Two implementation issues should be observed. On one hand, by increasing the number of flip flops in the brute force synchronizers, the only implication would be the prolonging of the reset phase. This course of action might be taken so to be able to resolve metastability in future technology nodes, where the resolution time constant of synchronizers is expected to degrade (Beer et al., 2010). On the other hand, the logical AND gate introduces a delay that actually induces further routing skew between data and clock wires. However, this effect is mitigated by practical considerations.

First, the AND gate could be synthesized for maximum performance, thus minimizing delay at the cost of area. Second, when the data stability window is short, design guidelines in the "*Design guidelines*" section will indicate that a higher number of brute force synchronizers is needed.

Figure 2. Variation detector architecture

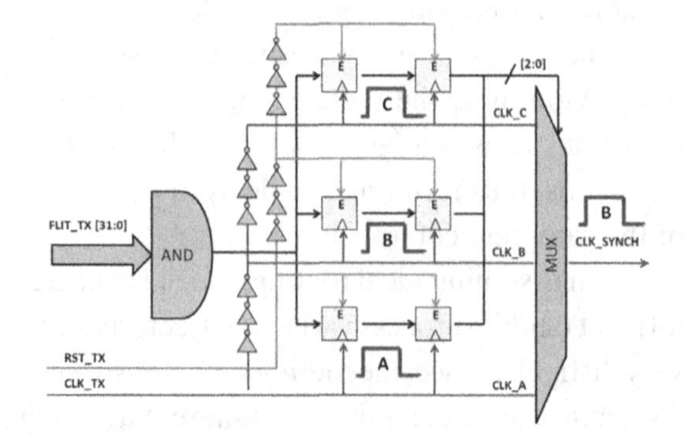

Therefore, the selected clock signal replica will typically sample data close to the middle of the stability window. Finally, it is possible to engineer the delay lines in such a way to mask the AND gate delay. In fact, a clock signal *clk_A'* might be derived by delaying *clk_A* by an amount of time equal to the AND gate delay. Then, *clk_A* would be sent to the multiplexer, while *clk_A'* would feed the first brute force synchronizer stage. The same would then be done for the later stages. In practice, for the sake of clock signal selection, the clock signals would be rigidly delayed by the same delay of the AND gate. This latter solution has been experimented in this paper and makes the variability detector operate correctly with 5 synchronization stages at 1 GHz with only 50% data stability window (with respect to the clock period) in 65nm CMOS process.

Operating Principle

The architecture of the variation detector provides the safe clock signal before the downstream synchronizer starts its operation. To achieve this result, the proposed architecture comes into play during the NoC reset phase. Furthermore, to guarantee the synchronization between the incoming data and the source synchronous clock signal, the circuit must detect the exact arrival time of every data bit in the GALS link. This is possible by sensing the incoming bits when they are switching from a high logic value to a low logic value or vice versa. As a result, the proposed architecture can properly sense routing skews when it receives two consecutive data patterns so that every data bit of the first transaction is negated with respect to the data bit of the other one. These two transactions can be purposely generated from the output buffer of the upstream switch during the reset process. To meet this goal, we designed an output buffer in the xpipesLite architecture (Stergiou et al., 2005) capable of driving, during the reset phase, a sequence of low-logic value bits (00..0) followed by a sequence of high-logic value bits (11..1). To note that different patterns could be implemented in order to better cope with the crosstalk effect. In fact, the AND block can be always set in compliance with the expected patterns (i.e., the high/low bits of the first/second pattern can be negated at the AND input).

Figure 3 reports the post-layout reset phase waveforms of a variation detector composed of 3 delay lines and 3 brute force synchronizers, like the one in Figure 2. We assume in this figure that input data is sampled on *positive* clock edges. As represented in the figure, once the first sequence of zeros is driven at the variation detector input port, the output of the AND module is set to a low logic value (*and_out*). In the meantime, the source synchronous clock signal (*clk_tx*) is propagated through the delay lines and it generates 3 replicated clock signals with a different offset (*clk_A, clk_B, clk_C*). Until the incoming sequence has low value, the brute force synchronizers sample the low AND output and they set to zero the control signal driving the multiplexer (see *ctr_A, ctr_B, ctr_C* in Figure 3). In this configuration the multiplexer output is permanently at a low logic value by default and the downstream mesochronous synchronizer does not receive any clock signal.

Then, the transmitter starts to drive the sequence of ones and the data bits (*flit_tx*) start to switch at the variation detector input port. Finally, when the incoming data is stable and every data bit has switched to the high logic value, the AND output assumes a high logic value. Although the AND output feeds all the brute force synchronizers, only some of them are able to sample the high value at the AND output. In fact, the brute force synchronizers driven by clock signals with an early positive edge will not reveal the high value; on the contrary, the brute force synchronizers driven by the clock signal with a late positive edge will sample it and will set the multiplexer control signal. As a result, the control signal collects all the information about the result of the AND output sampling and this latter information is exploited to distinguish between the safe clock

Figure 3. Variation detector reset phase

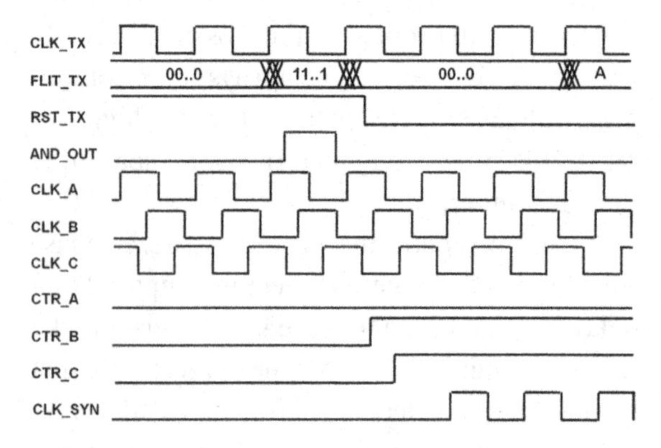

signal and the potentially unsafe clock signals (early clock signals). In Figure 3, the *ctr_B* and the *ctr_C* signals are set to the high logic value (i.e., *CTR(2:0)=110*) and this means that the *clk_B* and *clk_C* are safe: the positive edges of these latter clock signals occur after all data bits have switched to the high logic value (i.e., the data is stable).

Since it is the first of the safe clock signals to be selected to cross the multiplexer (i.e., the first high bit of *CTR(2:0)* enables the respective input port of the multiplexer), in our example, the *clk_B* is the clock signal driving the receiver synchronizer (*clk_synch*). It should be noted that a source synchronous synchronizer in a similar scenario could not properly work if driven by the nominal source clock signal (*clk_tx*). In fact, as shown in Figure 3, the positive edge of the source clock signal occurs close to the switching window of input data, hence resulting in a failure of the hold constraint.

The sequence of 1s is driven by the transmitter during the last clock cycle of the reset phase. As result, once the control signal in the variation detector is correctly set, the reset switches to the low value and the variation detector circuit interrupts its operation by freezing the safe clock signal at the multiplexer output. Since the detector circuit works only during the reset phase, the power consumption overhead is marginal over the system lifetime.

As a general comment, the architecture of the variability detector is quite simple and its operation in clear in principle. The main challenge consists however of assessing its timing robustness with respect to layout mismatches and to process variation-induced delay variability of link wires and of detector logic gates as well. An extensive validation effort of the architecture follows hereafter.

DELAY VARIABILITY ROBUSTNESS

In order to test the capabilities of our design, we set up a Verilog testbench able to drive the data lines and the transmitter clock signal to a mesochronous synchronizer in the receiver clock domain. The mesochronous synchronizer is the same as in Ludovici et al. (2010), however the considerations that follow are synchronizer architecture independent. Therefore, a dual-clock FIFO in place of the mesochronous synchronizer would not make any difference for the results that follow.

Variability detector and companion synchronizer have been synthesized, placed and routed with a commercial toolflow on a 65nm STMicroelectronics technology.

Misalignment Between Data and Clock

In order to study the robustness of the communication when the ideal alignment between data and transmitter clock is not respected, we inject an increasing negative/positive skew into the clock signal wire. Therefore, we compare the performance of a baseline source synchronous architecture with the one achieved by a source synchronous architecture augmented with the proposed variation detector.

Figure 4(a) reports the timing margin derived from a post-layout analysis of a source synchronous mesochronous synchronizer synthesized at 1 GHz without the variation detector. Setup and hold times have been experimentally measured by driving the mesochronous synchronizer under test with a transmitter clock affected by an increasing routing skew and by monitoring the relative waveforms. *FF_Time* refers to the minimum values required by the technology library for correct sampling.

First of all, we observe that the setup time increases and the hold time decreases linearly with the increase of the clock skew. When the positive clock skew reaches 20% of the clock period, the hold time violates the minimum timing margin and a failure is detected in the mesochronous synchronizer. On the contrary, a setup violation is verified when injecting a negative clock skew higher than 45%. As a conclusion, the baseline mesochronous synchronizer is able to support a skew between clock and data within a range of -45% and +20% of the clock period.

Overall, 35% of the clock period is unsafe for sampling and a positive clock edge, loosely synchronous with the data, can only occur during the remaining 65% of the clock period. This experiment was carried out considering an incoming data stable for 75% of the clock period at the mesochronous synchronizer input port (Figure 5). This accounts for the delay variability between data lines and for the non-null settling time usually found in post-layout link switching.

In order to perform a fair comparison, a similar experiment, shown in Figure 4(b), was performed for a mesochronous synchronizer with the variation detector (composed of 5 delay lines and 5 brute force synchronizers) connected in front of it. We can notice a significant difference of the results due to the insertion of the variation detector. The setup margin is strictly symmetric with

Figure 4. Timing margin of the baseline mesochronous synchronizer (a) and the mesochronous synchronizer with a variation detector in front of it (b)

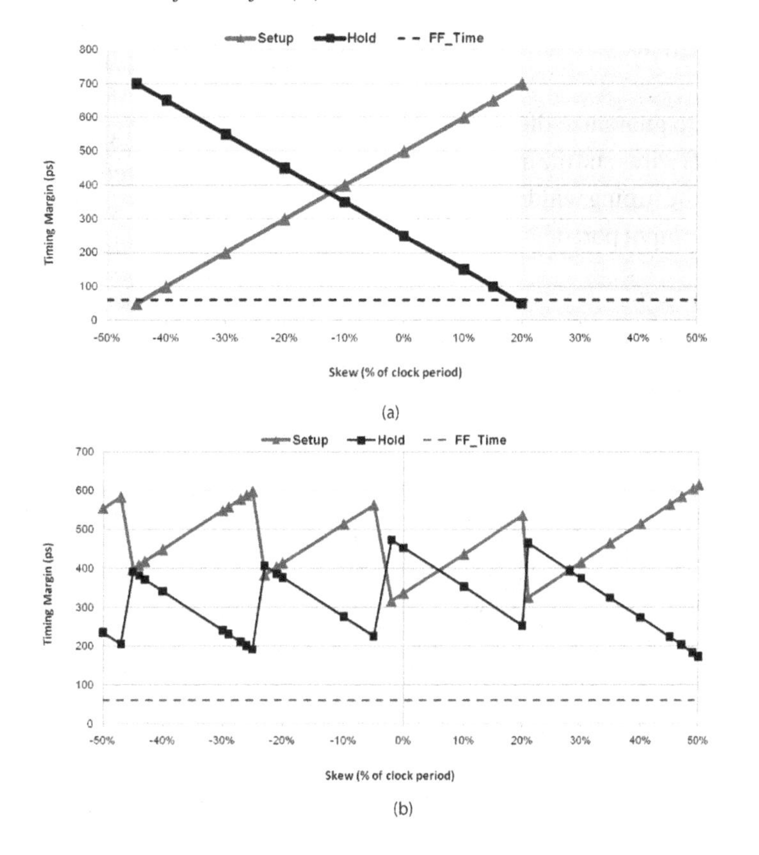

Figure 5. Settling time of the incoming data

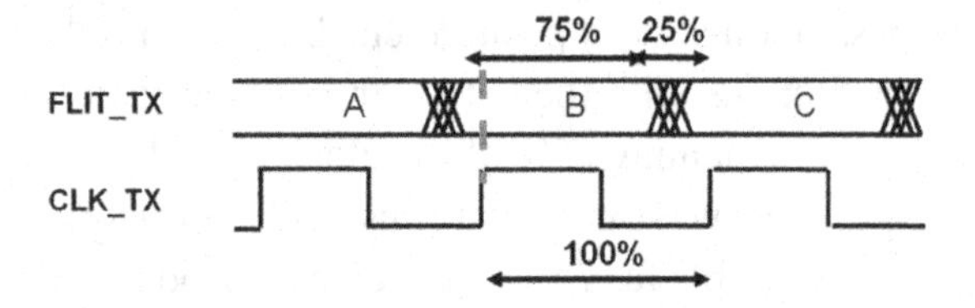

respect to the hold margin. Moreover, the timing margin is composed of the periodic repetitions of the same triangular-shaped trend.

As before, at the increase of the clock skew corresponds a decrease of the hold margin although, in this case, as soon as the hold time degradation becomes significant and dangerous for correct circuit operation the detector re-establishes safe margins. In conclusion, the proposed architecture guarantees a safe communication in a mesochronous link in every clock delay scenario as opposed to the baseline detector-less GALS link architecture.

Misalignment between Wires of the Data Link

Although it is essential to guarantee the in-phase property between the data wires and the transmitter clock signal, the size of the timing window during which data is stable at the input port of the receiver synchronizer is also an important parameter for reliable communication. In particular, we measured timing margins of the baseline source synchronous architecture vs. the detector-augmented architecture when varying this latter parameter.

Since the considered timing window is longer when the data bits arrive simultaneously, this parameter is tightly dependent on the amount of process variations affecting the upstream link. Therefore, we injected an increasing amount of delay variability in the wires carrying the data bits and we monitored the behavior of the synchronizer under test.

Figure 6 reports the timing margin derived from a post-layout analysis of a mesochronous synchronizer synthesized at 1GHz. The timing window when input data is stable is identified by the two horizontal lines crossing the y-axis in -50% and +50%. Y-axis reports the position of the data sampling edge inside (i.e., as a percentage of) the data stability window. 0% means sampling in the middle of the window, while sampling close to the upper bound (50%) incurs setup time violations; on the contrary, a sampling close to the lower bound (-50%) can incur hold time violations.

The figure also compares the results in the presence of a different amount of delay variability in the data link wires, i.e., different sizes of the data stability window. The analyzed timing

Figure 6. Timing margin in the presence of process variation in the data link wires

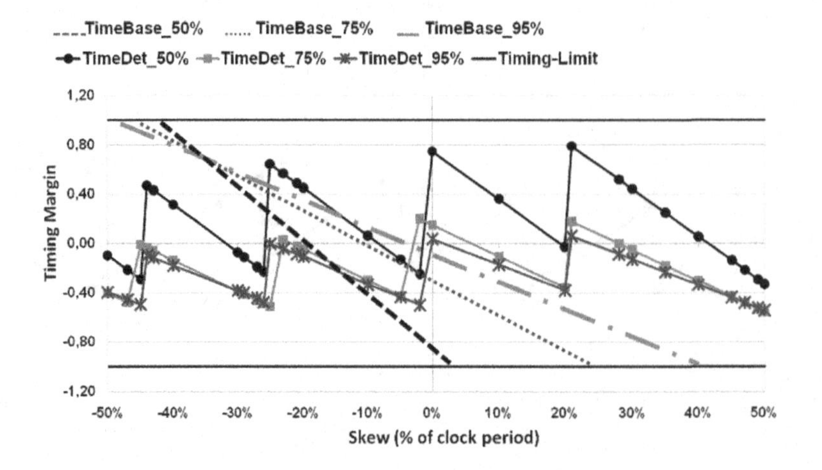

windows range from 95% to the 50% of the clock period size. As expected, in the configuration without the variation detector (*TimeBase 95%, TimeBase 75%, TimeBase 50%*), at the increase of the positive/negative clock skew, the data sampling time moves closer to the lower/higher bound of the window until an hold/setup violation occurs. Moreover, the higher the delay variability in the data wires the lower the probability of sampling in the safe timing window.

A similar experiment was carried out by reporting the timing margin of a mesochronous synchronizer augmented with the variation detector (*TimeDet 95%, TimeDet 75%, TimeDet 50%*). As result, the data sampling occurs inside the safe timing window in every experimented scenario. We can notice how the sampling time moves closer to the unsafe lower/upper bound as the timing window size shrinks. In every case, the proposed architecture does not incur timing violations even when the data is stable for only 50% of the clock cycle.

Random Process Variability Affecting the Detector

While the robustness to link delay variability is already implicit in the above experimental results, the robustness of the detector to the process delay variability of its own logic cells still has to be demonstrated. For this purpose, we characterized the circuit behavior by modeling the random variations for technologies ranging from 65nm to 16nm, based on the methodology presented in Hernández et al. (2010). Therefore, this section shows the effect of process variations affecting the detector on the correct operation of this latter.

In Figure 7, we measured the timing margin of the proposed architecture in the presence of best, nominal and worst random process variation conditions. We considered a pessimistic timing window, when the data is unstable at the synchronizer input port for 50% of the clock period. The timing margins are derived from a post-layout analysis performed at 1GHz.

As reported in the figure, the variation detector proves robust to the random delay variability achieving similar margins in the different process variation scenarios. Similar tests have been performed by applying the variability model considering even the wider variances of parameter spread foreseen by the ITRS for the 45nm, 32nm, 22nm and 16nm technology nodes, but are omitted for lack of space. Again, the result proved the relative insensitivity by construction of our detector to the amount of injected delay variability.

Architecture Integration

Since a mesochronous synchronizer is designed to support a given amount of skew between two mesochronous domains, it is relevant to analyze how link delay variability affects the skew tolerance of the synchronizer and whether the integration of the variation detector into the synchronizer degrades its final skew robustness figures. At this point, the actual synchronizer architecture comes into play. This manuscript builds on the latest advances in synchronizer design for cost-effective on-chip networks.

In fact, there are two main schemes to instantiate a mesochronous synchronizer in a source synchronous link. The basic scheme consists of placing the synchronizer in front of the downstream switch input port, but as an external block to it.

Figure 7. Random variations affecting the delay of detector logic cells

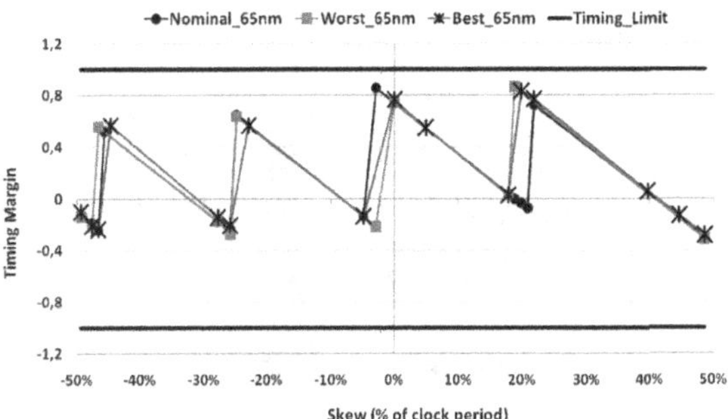

A link taking this approach follows a *Loosely Coupled Synchronization* scheme (Ludovici et al., 2009) and the synchronizer used in this context might be similar to the one graphically illustrated in Figure 8.

The circuit receives as its inputs a bundle of NoC wires representing the regular NoC link, carrying data and/or flow control commands, and a copy of the clock signal of the sender. The circuit is composed by a front-end and a back-end. The front-end is driven by the incoming clock signal, and of parallel latches in a rotating fashion, based on a counter. The back-end of the circuit leverages the local clock of the downstream switch, and samples data from one of the latches in the front-end thanks to multiplexing logic which is also based on a counter. The rationale is to temporarily store incoming information in one of the front-end latches, using the incoming clock wire to avoid any timing problem related to the clock phase offset.

Once the information stored in the latch is stable, it can be read by the target clock domain and sampled by a regular flip-flop. In practice, such a sampling is performed by the input buffer of the downstream switch. Anyway, the loosely coupled design introduces a severe overhead in the communication. A key principle for reducing the latency, area and power overhead of switches with loosely coupled synchronizers is to co-design the synchronizer with the downstream switch input port. The co-optimization consists of using

Figure 8. Loosely coupled synchronizer architecture

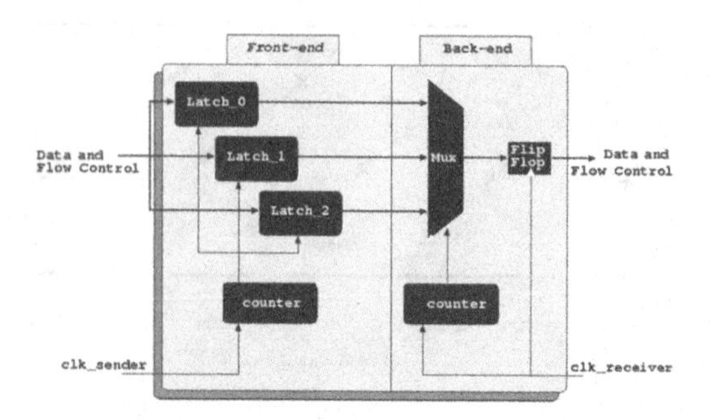

the buffering resources of the switch input stage not only for retiming and flow control as in the synchronous architecture, but also for synchronization purposes. This would allow to completely removing the switch input buffer and to replace it with the synchronizer itself and the ultimate consequence is that the mesochronous synchronizer becomes the actual switch input stage.

The basic architecture of the new switch input port is illustrated in Figure 9 and denoted *Tightly Coupled Synchronizer* (Ludovici et al., 2009). In this case, the ultimate sampling of the data flowing out of the synchronizer multiplexer and going through the switch internal logic will occur at the switch output port. On the contrary, in the loosely coupled synchronizer, the synchronized data is directly sampled by a regular flip-flop. As a consequence, the two architectures are affected by different timing constraints and exploit a distinct phase offset budget.

In particular, the skew tolerance of the two architecture schemes depends on the relative alignment of data arrival time at latch outputs, multiplexer selection window and sampling edge in the receiver clock domain.

Anyway, due to the worst case timing path between the tightly coupled synchronizer output and the switch output port, a switch with tight coupling of the synchronizer with the switch incurs in a degradation of the timing margin (which is actually the skew tolerance of the synchronizer). This is a clear trade-off that is at the disposal of the designer, who selects the suitable design point based on his requirements.

The variability detector is integrated with the synchronizer based on the scheme in Figure 10. This scheme refers to a loose coupling with the NoC, but the same holds also for a tight coupling. The output of the variability detector feeds a counter which in turn generates the latch enable signal for the synchronizer.

Figure 9. Tightly coupled synchronizer architecture

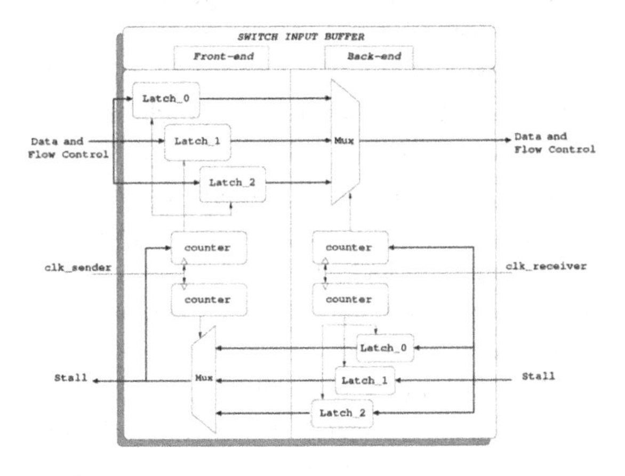

Impact of Variability Detector on Synchronizer Timing Margins

In order to evaluate the skew tolerance in the case of link delay variations in the source synchronous link, we injected an increasing negative/positive skew into the transmitted clock signal wire, thus inducing a routing skew with the data wires. A the same time, we applied a different clock phase offset between the transmitter clock (*clock_TX*) and the receiver clock (*clock_RX*), to assess evolution of the timing margin. This experiment is performed for both the detector-augmented and the detector-less synchronization architecture where a tightly coupled and a loosely coupled synchronizer are alternatively instantiated. Ultimately, we aim at assessing to which extent the detector impacts the skew tolerance of the two synchronization schemes in a noisy environment.

In Figure 11 the results for the Loosely Coupled Synchronizer are reported. The y-axis reports negative and positive values of the skew, expressed as percentage of the clock period, while the x-axis reports the misalignment of the transmitted clock signal wire with respect to data wires.

As expected, the baseline detector-less synchronizer is not able to work in every variability scenario. It is able to properly absorb a skew between two mesochronous domains only when the skew between the transmitted clock and data is within a range of -38% and +3% of the clock period. In this latter range the architecture supports 100% of negative skew and between 60% and 90% of positive skew. A positive skew means that the clock at the transmitter is delayed with respect to the one at the receiver.

On the contrary, by applying the proposed variation detector to the synchronizer, the architecture guarantees an acceptable skew tolerance in every process variation scenario. In particular it achieves similar performance to the baseline architecture when the skew between the transmitted clock and data is within a range of -38% and +3% while it

Figure 10. Integration of the variation detector with the loosely coupled synchronizer

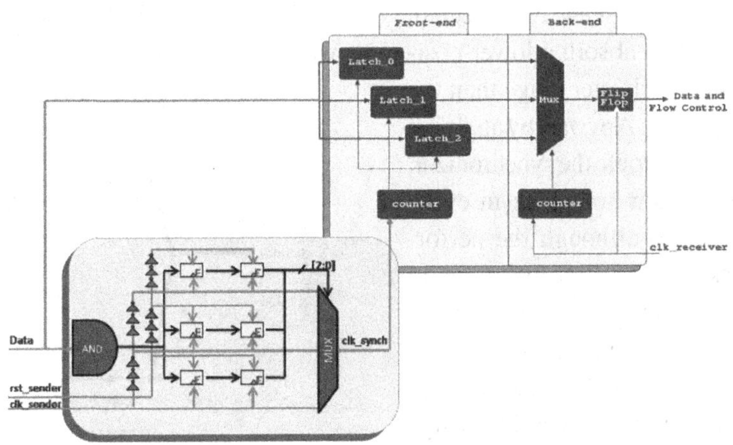

Figure 11. Skew tolerance as a function of the misalignment between data and transmitter clock in the loosely coupled synchronizer augmented with the variability detector

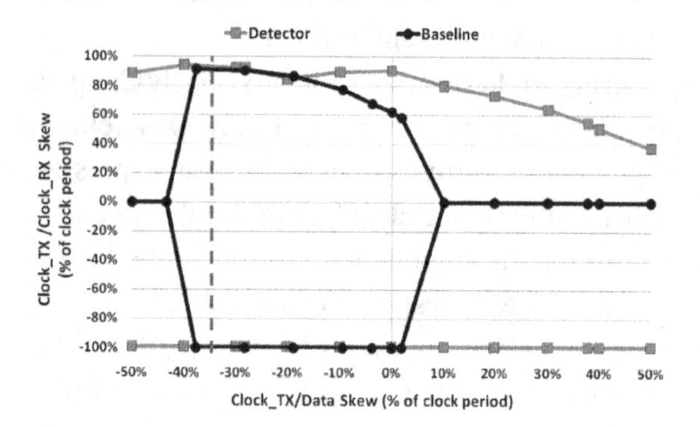

supports 100% of negative skew and between 40% and 90% of positive skew in the other scenarios. As a result, in those cases where the detector-less architecture works fine, the detector-augmented one does not do a worse job, while at the same time prolonging the feasibility space.

Finally, Figure 12 reports the results for the Tightly Coupled Synchronizer. As expected, the tightly coupled synchronizer is not able to work in every variability scenario as the loosely coupled counterpart. In particular, the tightly coupled synchronizer is able to properly absorb a skew between two mesochronous domains only when the skew between the transmitted clock and data is within a range of -28% and +3% of the clock period. In this latter range the tightly coupled architecture supports 100% of negative skew but only 40% and 60% of positive skew. As expected, its timing margin allows to absorb a lower positive skew and to work in a shorter range than the loosely coupled counterpart. Anyway, by applying the proposed variation detector to the synchronizer, it is still guaranteed a skew tolerance in every process variation scenario although the performance for positive skews is clearly lower due to the severe timing constraints. Especially, similar performance to the stand-alone tightly coupled architecture are achieved when the skew between the transmitted clock and data is within a range of -28% and +3% while it supports still 100% of negative skew and until 50% of positive skew in the other scenarios.

In practice, without the variation detector, skew tolerance of the mesochronous synchronizer at the receiver end is seriously impacted both in the loosely coupled and in the tightly coupled scenario by even small misalignments between data and transmitter clock. Vice versa, the variation detector enables a smoother degradation of the skew tolerance and its extension to operating conditions featuring large routing skew between data and clock. Ultimately, having the variability detector can help better cope with the layout constraints and manufacturing process uncertainties, while marginally impacting skew tolerance in the best cases.

Area Overhead

According to our post-synthesis area figures, our variability detector consumes approximately one third of the area of the compact loosely coupled mesochronous synchronizer in Ludovici et al. (2009) that we have used throughout this manuscript for the experiments (Figure 8). The detector with 5 brute-force synchronizers consumes 440 um2 as opposed to the 1150 um2 of the mesochronous

Figure 12. Skew tolerance as a function of the misalignment between data and transmitter clock in the tightly coupled synchronizer augmented with the variability detector

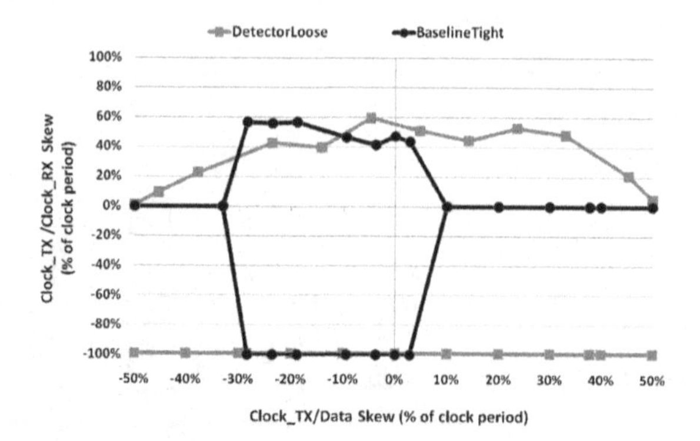

synchronizer. When considering a 5x5 switch commonly found in a 2D mesh and synthesized for max performance, the variation detector takes less than 1% of total switch area.

Design Guidelines

As a final contribution, the variation detector can be designed to match well-defined operating conditions at the minimum implementation cost. Indeed, the proposed architecture can be parameterized as a function of the target performance and of the timing margins expected for the source synchronous interface. Next, design equations follow for this purpose.

In a high variability environment, the alignment between the sampling edge of the clock signal and the data can be unpredictable. As a result, sampling can occur regardless of the stability or settling timing window of data. In order to ensure the reliability of the source-synchronous communication, the detector should always guarantee a safe clock version with a sampling edge falling in the data stability window. To meet this requirement, a minimum number of clock versions should be designed to sweep the entire clock period in such a way that stable data can be safely sampled at least once regardless of its position within the clock period.

As a consequence, to increase the reliability of the proposed architecture and to minimize the number of required stages, the number of clock versions and the delay introduced by each delay line must be constrained. In particular, the length of the clock period to be swept and the length of the stability data window are the two main parameters to be taken into account to properly design a reliable variation detector. Then, we can define the safe timing window (*Swd*) where the output of the AND block is stable at a high value (i.e., all the bits of the incoming data arrived at the receiver end) and the metastability risk is avoided, as follows:

$$Swd = Tstability - Tsetup - Thold \quad (1)$$

where *Tstability* represents both the timing window where the AND output is high and the window where the incoming data is stable. Moreover, *Tsetup* and *Thold* correspond to the minimum timing margins required by the technology library for correct sampling.

Intuitively, to ensure that at least a clock version will safely sample, we can suppose to split the clock period into equal parts of *Swd* length and generate a rising clock edge in each sub-period. The following formula can be exploited to determine the minimum amount of variation detector stages:

$$(Tclk/Swd) < NumStage \quad (2)$$

It should be observed that *Tclk* is the source clock period and the delay chains are supposed to be composed by the same number of delay elements to guarantee an equal offset between the falling edge of the clock versions. Clearly, *NumStage* represents the minimum number of stages for a reliable communication but a greater number of stages can be adopted without affecting the correct operating principle of the system.

In the baseline architecture, described in the *"Operating Principle"* section, the multiplexer selects the first safe clock between the available versions. However, it is possible to enable a circuit variant exploiting the multiplexer control signals to further increase the robustness of the variation detector. In fact, an unlucky condition can occur. A clock version can violate the hold/setup time constraint of its brute-force stage. As a consequence, the resulting output can fall into a metastability state and it can erroneously set to 1 the multiplexer control bit. Since this clock version will be the first to set the multiplexer, it will be allowed to cross the multiplexer. As a result it will provoke potential metastability concerns in the source synchronous interface since its rising edge will be premature for the safe sampling of the data. In this case, the real safe clock version was not considered since it was the second one to set the multiplexer.

Anyway, we can force a *Swd* length constraint more severe than in formula 1 to avoid the metastability concerns. In fact, we can require to have two safe clock versions for every *Swd* scenario and as a result, we can set the multiplexer to allow always the second clock version instead of the first one to cross its logic. In this case, the clock version violating the brute-force hold/setup time constraint will be not considered since it will be the first to set the multiplexer. To conclude, formula 1 should be extended as follows for increased robustness to metastability concerns:

$$Swd = (Tstability - Tsetup - Thold)/2 \quad (3)$$

$$Swd > Max[Tsetup; Thold] \quad (4)$$

To note that metastability robustness is guaranteed when both the conditions (Formula 3 and 4) are respected. The violation of Formula 4 condition can bring two consecutive clock versions to violate the brute-force hold/setup time and set erroneously the multiplexer (the worst case scenario). In practice, we have to avoid that sampling occurs at such a fine granularity that two consecutive sampling edges occur in the metastability window.

In order to quantify the number of required brute-force and delay line stages for real-life design problems, Table 1 shows some of the possible scenarios as a function of target frequency and data instability timing window. It should be observed that in Table 1 formula 3 was applied and a *Tsetup* and *Thold* of 100ps was considered.

Interestingly, the *Swd > Max[Tsetup; Thold]* condition is not verified when working at 1.25Ghz with 500ps of settling time. In fact, this can be considered an infeasible scenario where the safe sampling data window lasts just 12% of the clock period. As a final remark, throughout the manuscript safe timing margins with a 5 stage variation detector working at 1Ghz were achieved since the *Tsetup/Thold* constraints of the used technology library were clearly lower than the pessimistic constraints considered in Table 1.

CONCLUSION

This manuscript has presented a novel variation detector architecture for use in source synchronous links in front of the receiver synchronizer (either a mesochronous synchronizer or a dual-clock FIFO). The detector was in particular conceived to deal both with layout mismatches and with link delay variability and builds on the principle of self-calibration, i.e., adaptation of the transmitter clock signal phase to in-situ actual operating conditions. We have proved the ability of detector-augmented GALS links to cope with large signal misalignments and small data stability windows, unlike baseline GALS links. Above all, the variation detector has proven robust to the delay variability of its logic cells. Overall,

Table 1. Number of stages of the variability detector based on the frequency and the settling time requirements

	Target Frequency				
Settling Time	**1.25Ghz**	**1Ghz**	**750Mhz**	**500Mhz**	**250Mhz**
500ps	X	7	5	4	3
300ps	6	4	4	3	3
100ps	4	3	3	3	3

analyzing the comprehensive metric of skew tolerance for a mesochronous synchronizer has revealed that the insertion of the variation detector has advantages in most cases and for the main synchronization architectures of practical interest (loosely vs. tightly coupled with the NoC). Even for non-variability-dominated links, it can better cope with the layout constraints of the link.

REFERENCES

Asenov, A., Kaya, S., & Davies, J. H. (2002). Intrinsic threshold voltage fluctuations in decanano MOSFETs due to local oxide thickness variations. *IEEE Transactions on Electron Devices*, *49*(1), 112–119. doi:10.1109/16.974757

Beer, S., Ginosar, R., Priel, M., Dobkin, R., & Kolodny, A. (2010). The devolution of synchronizers. In *Proceedings of the IEEE Symposium on Asynchronous Circuits and Systems* (pp. 94-103).

Bonesi, S., Bertozzi, D., Benini, L., & Macii, E. (2008). Process variation tolerant pipeline design through a placement-aware multiple voltage island design style. *Design, Automation and Test in Europe*, 967-972.

Borkar, S., Karnik, T., & Vivek, D. (2004). Design and reliability challenges in nanometer technologies. In *Proceedings of the 41st Design Automation Conference* (p. 75).

Bowman, K. A., Duvall, S. G., & Meindl, J. D. (2002). Impact of die-to-die and within-die parameter fluctuations on the maximum clock frequency distribution for gigascale integration. *IEEE Journal of Solid-state Circuits*, *37*(2), 183–190. doi:10.1109/4.982424

Collins, H. A., & Nikel, R. E. (1999). *DDR-SDRAM, high-speed, source synchronous interfaces create design challenges.* Retrieved from http://www.edn.com/article/506435-DDR_SDRAM_high_speed_source_synchronous_interfaces_create_design_challenges.php

Dally, W. J., & Poulton, J. W. (1998). *Digital systems engineering*. Cambridge, UK: Cambridge University Press.

Diamos, G., Yalamanchili, S., & Duato, J. (2007). *STARS: A system for tuning and actively reconfiguring SoCs link.* Paper presented at the DAC Workshop on Diagnostic Services in NoCs.

Elrabaa, M. E. S. (2006). An all-digital clock frequency capturing circuitry for NRZ data communications. In *Proceedings of the IEEE International Conference on Electronics, Circuits and Systems* (pp. 106-109).

Ernst, D., Kim, N. S., Das, S., Pant, S., Rao, R., Pham, T., et al. (2003). Razor: A low-power pipeline based on circuit-level timing speculation. In *Proceedings of the 36th Annual IEEE/ACM International Symposium on Microarchitecture* (pp. 7-18).

Faiz-ul-Hassan. Cheng, B., Vanderbauwhede, W., & Rodriguez, F. (2009). Impact of device variability in the communication structures for future synchronous SoC designs. In *Proceedings of the International Symposium on System-on-Chip* (pp. 68-72).

Gilabert, F., Ludovici, D., Medardoni, S., Bertozzi, D., Benini, L., & Gaydadjiev, G. N. (2009). Designing regular network-on-chip topologies under technology, architecture and software constraints. In *Proceedings of the International Conference on Complex, Intelligent and Software Intensive Systems* (pp. 681-687).

Hernández, C., Roca, A., Silla, F., Flich, J., & Duato, J. (2010). Improving the performance of GALS-based NoCs in the presence of process variation. In *Proceedings of the Fourth ACM/IEEE International Symposium on Networks-on-Chip* (pp. 35-42).

Höppner, S., Walter, D., Eisenreich, H., & Schüffny, R. (2010). Efficient compensation of delay variations in high-speed network-on-chip data links. In *Proceedings of the International Symposium on System on Chip* (pp. 55-58).

ITRS. (2007). *International technology roadmap for semiconductors.* Retrieved from http://www.itrs.net/Links/2007ITRS/Home2007.htm

Jose, A. P., Patounakis, G., & Shepard, K. L. (2005). Near speed-of-light on-chip interconnects using pulsed current-mode signalling. In *Proceedings of the Digest of Technical Papers Symposium on VLSI Circuits* (pp. 108-111).

Kakoee, M. R., Loi, I., & Benini, L. (2010). A new physical routing approach for robust bundled signaling on NoC links. In *Proceedings of the 20th Symposium on the Great Lakes* (pp. 3-8).

Kenyon, C., Kornfeld, A., Kuhn, K., Liu, M., Maheshwari, A., Shih, W. et al. (2008). Managing process variation in Intel's 45nm CMOS technology. *Intel Technology Journal, 12*(2).

Kihara, M., Ono, S., & Eskelinen, P. (2003). *Digital clocks for synchronization and communications.* Norwood, MA: Artech House.

Loi, I., Angiolini, F., & Benini, L. (2008). Developing mesochronous synchronizers to enable 3D NoCs. *Design, Automation and Test in Europe*, 1414-1419.

Ludovici, D., Strano, A., Bertozzi, D., Benini, L., & Gaydadjiev, G. N. (2009). Comparing tightly and loosely coupled mesochronous synchronizers in a NoC switch architecture. In *Proceedings of the 3rd ACM/IEEE International Symposium on Networks-on-Chip* (pp. 244-249).

Ludovici, D., Strano, A., Gaydadjiev, G. N., Benini, L., & Bertozzi, D. (2010). Design space exploration of a mesochronous link for cost-effective and flexible GALS NOCs. *Design, Automation & Test in Europe*, 679-684.

Merdadoni, S., Lajolo, M., & Bertozzi, D. (2008). Variation tolerant NoC design by means of self-calibrating links. *Design, Automation and Test in Europe*, 1402-1407.

Mondal, M., Ragheb, T., Wu, X., Aziz, A., & Massoud, Y. (2007). Provisioning on-chip networks under buffered RC interconnect delay variations. In *Proceedings of the 8th International Symposium on Quality Electronic Design* (pp. 873-878).

Moore, S. W., & Taylor, G. S. (2000). Self calibrating clocks for globally asynchronous locally synchronous systems. In *Proceedings of the IEEE International Conference on Computer Design* (pp. 73-78).

Murali, S., Tamhankar, R., Angiolini, F., Pulling, A., Atienza, D., Benini, L., & De Micheli, G. (2006). Comparison of a timing-error tolerant scheme with a traditional re-transmission mechanism for networks on chips. In *Proceedings of the International Symposium on System-on-Chip* (pp. 1-4).

Nicopoulos, C., Srinivasan, S., & Yanamandra, A, Dongkook Park, Narayanan, V., Das, C. R., & Irwin, M. J. (2010). On the effects of process variation in network-on-chip architectures. *IEEE Transactions on Dependable and Secure Computing, 7*(3), 240–254. doi:10.1109/TDSC.2008.59

Orshansky, M., & Keutzer, K. (2002). A general probabilistic framework for worst case timing analysis. In *Proceedings of the 39th Design Automation Conference* (pp. 556- 561).

Paci, G., Bertozzi, D., & Benini, L. (2009). Effectiveness of adaptive supply voltage and body bias as post-silicon variability compensation techniques for full-swing and low-swing on-chip communication channels. *Design, Automation & Test in Europe*, 1404-1409.

Panades, I. M., Clermidy, F., Vivet, P., & Greiner, A. (2008). Physical implementation of the DSPIN network-on-chip in the FAUST Architecture. In *Proceedings of the International Symposium on Networks-on-Chip* (pp. 139-148).

Sarangi, S. R., Greskamp, B., Teodorescu, R., Nakano, J., Tiwari, A., & Torrellas, J. (2008). VARIUS: A model of process variation and resulting timing errors for microarchitects. *IEEE Transactions on Semiconductor Manufacturing*, *21*(1), 3–13. doi:10.1109/TSM.2007.913186

Stergiou, S., Angiolini, F., Carta, S., Raffo, L., Bertozzi, D., & De Micheli, G. (2005). Xpipes Lite: A synthesis oriented design library for networks on chips. In *Proceedings of the Conference on Design, Automation and Test in Europe* (Vol. 2, pp. 1188-1193).

Synopsys. (2007). *PrimeTime VX application note - Implementation methodology with variation-aware timing analysis, version 1.0.* Mountain View, CA: Synopsys.

Tran, A. T., Truong, D. N., & Baas, M. B. (2009). A low-cost high-speed source-synchronous interconnection technique for GALS chip multiprocessors. In *Proceedings of the IEEE International Symposium on Circuits and Systems* (pp. 996-999).

Worm, F., Ienne, P., Thiran, P., & De Micheli, G. (2005). A robust self-calibrating transmission scheme for on-chip networks. *IEEE Transactions on Very Large Scale Integration Systems*, *13*(1), 126–139. doi:10.1109/TVLSI.2004.834241

Yu, Z., & Baas, B. M. (2006). Implementing tile-based chip multiprocessors with GALS clocking style. In *Proceedings of the IEEE International Symposium on Circuits and Systems* (pp. 174-179).

This work was previously published in the International Journal of Embedded and Real-Time Communication Systems, Volume 2, Issue 4, edited by Seppo Virtanen, pp. 1-20, copyright 2011 by IGI Publishing (an imprint of IGI Global).

Chapter 12
Checkpointing SystemC-Based Virtual Platforms

Stefan Kraemer
RWTH Aachen University, Germany

Rainer Leupers
RWTH Aachen University, Germany

Dietmar Petras
Synopsys Inc., Germany

Thomas Philipp
Synopsys Inc., Germany

Andreas Hoffmann
Synopsys Inc., Germany

ABSTRACT

The ability to restore a virtual platform from a previously saved simulation state can considerably shorten the typical edit-compile-debug cycle for software developers and therefore enhance productivity. For SystemC based virtual platforms (VP), dedicated checkpoint/restore (C/R) solutions are required, taking into account the specific characteristics of such platforms. Apart from restoring the simulation process from a checkpoint image, the proposed checkpoint solution also takes care of re-attaching debuggers and interactive GUIs to the restored virtual platform. The checkpointing is handled automatically for most of the SystemC modules, only the usage of host OS resources requires user provision. A process checkpointing based C/R has been selected in order to minimize the adaption required for existing VPs at the expense of large checkpoint sizes. This drawback is overcome by introducing an online compression to the checkpoint process. A case study based on the SHAPES Virtual Platform is conducted to investigate the applicability of the proposed framework as well as the impact of checkpoint compression in a realistic system environment.

DOI: 10.4018/978-1-4666-2776-5.ch012

INTRODUCTION

Virtual platforms (VPs) are executable software models of hardware systems. The software running on top of the VP cannot tell the difference between the VP and the real hardware. Hence, VPs enable to simulate the complete system long before a first hardware prototype is ready. Compared to a hardware prototype, VPs exhibit a superior observability with deterministic execution and non-intrusive debugging capabilities, VPs form an ideal tool for fast and efficient software development.

Due to the growing complexity of the software being executed on modern systems-on-chip (SoCs) VPs have gained a lot of interest in industry and academia over the last years. Of course, the usability of a VP heavily depends on the achievable simulation speed. Only simulation environments that are sufficiently fast to simulate large workloads can be employed to efficiently develop software. It can be foreseen, that in future more complex multi-processor systems-on-chip (MPSoCs) will be deployed and, hence, there is a strong need to increase the simulation performance of VPs.

The *checkpoint/restore* (C/R) technique offers the possibility to reduce the number of simulation runs by storing the state of the simulation to a file, and to use this checkpoint image to restart the simulation in exactly the same state at a later point in time. For example, instead of booting the operating system (OS) each time an application is executed, a checkpoint permits to omit the time consuming boot process and to concentrate on the application execution itself. Although the C/R technique does not speed up the VP simulation itself, this technique can be used to reduce the time spent simulating.

For modeling VPs, SystemC (SystemC, 2005) has been established as the de-facto standard and it is supported by all major vendors of VP modeling environments. It allows to efficiently describe software as well as hardware aspects of VPs.[1]

The most important use cases for the C/R feature for VPs are listed:

- **Time Saving:** By restoring a checkpoint a certain simulation state can be reached without the need for lengthy simulation runs.
- **Move Around in Time:** The facility of creating multiple checkpoints at different points in time enables the user to move around in time.
- **Periodic Checkpointing:** Periodic checkpointing allows the user to quickly recover the state shortly before an error has occurred.
- **Simulation Transfer:** A developer of a VP can transfer a checkpoint image to another developer to request help for a specific problem or to demonstrate a certain behavior of the simulation.

How the C/R mechanism can be used to accelerate the edit-compile-debug cycle is clarified in the following example of a development task typically occurring when porting an OS to a new HW platform: Debug and analyze how an application running on an Android based smartphone interacts with the hardware after a touchscreen event occurred. This task requires a complete boot of the guest OS on top of the VP. Furthermore, the application has to be initialized before the actual debug task can be started.

For the actual debugging task the developer may also have to attach several debuggers to the VP. A typical collection of debug tools for the Synopsys Virtual Platform is shown in Figure 1. The Virtual Platform Analyzer (VPA) allows inspecting all details of the hardware platform, while it is possible to connect debuggers to the VPA for debugging the application software and the Linux kernel software.

Setting up the tools and driving the VP into the desired state for debugging the given problem can become a quite time consuming task. Figure

Figure 1. Debugging and analyzing a Android based smartphone virtual platform

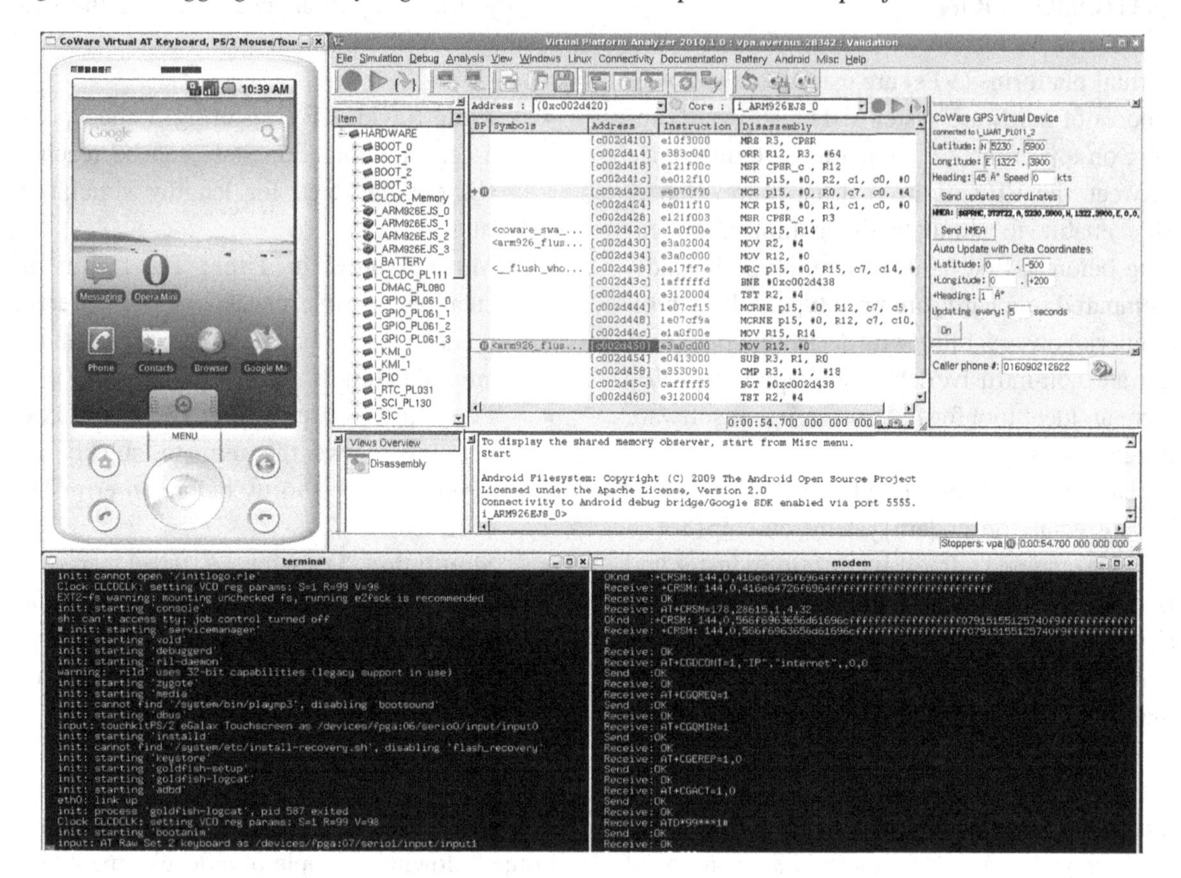

2 shows that this procedure can take several minutes. It requires to start the VP and the debugger tools, attach the debuggers to the platform, setup breakpoints and load the software images. Next, the VP has to be executed until the Linux OS boot is completed. Further user interaction may be required in order to start the desired application. For each debug cycle this procedure needs to be repeated. The C/R mechanism enables the developer to shorten the setup phase to several seconds, as shown in Figure 2. It is possible to create a checkpoint of the VP at any time, e.g., just at the beginning of the actual debug phase. When a checkpoint is restored, the VP is restarted in the exact state of the checkpoint creation time. Furthermore, all available debuggers are re-attached to the restored VP. Additionally all the external applications for touchscreen and console are also restarted if they are no longer running. The breakpoint settings in the VPA can be configured by a Tcl script that is automatically executed after each restoration of a checkpoint.

From the example introduced above it becomes clear that debugging complex VPs can benefit from the checkpointing functionality.

REQUIREMENTS

Depending on the exact use case, different requirements are imposed on the C/R capabilities. The most important requirements are briefly discussed in the following:

- **Reliability:** Reliability of the C/R functionality is of utmost importance. The reconstructed simulation is required to be in exactly the same state as the original simulation was before checkpointing.
- **Transparency:** In the context of C/R functionality transparency means that the checkpointing feature should be applicable without any modification of the simulation. Complete transparency of the C/R implementation is desirable but not required for SystemC simulations, because the re-compilation and the re-linking of the simulation is possible. Nevertheless, in order to minimize the effort for the user, the required adaptation should be as minimal as possible.
- **Performance:** The different use cases have different performance requirements on the C/R functionality. This includes the speed of checkpointing and restoring as well as the size of the created checkpoint image. For example, in case of periodic checkpointing the time required to checkpoint must be low in order to keep the impact on the simulation performance small. If a high number of images need to be stored at the same time, the size of the checkpoint also matters.
- **Support for External Applications:** Virtual platforms are often connected to external applications such as GUIs, consoles and debuggers in order to monitor and interact with the simulation (Figure 1). Therefore, the C/R functionality must be capable of re-establishing connections to external applications automatically in order to ensure seamless usage and debugging of the simulation.
- **Support for OS Resources:** SystemC simulations may contain dependencies on the OS. These resources must be handled separately by the C/R implementation as they are not part of the simulation process itself.

Due to the connection of external debuggers to processor models and frequent utilization of OS resources for logging data, the support of external applications and OS resources is of high impor-

Figure 2. The debug cycle is shortened by checkpointing and restoring a virtual platform

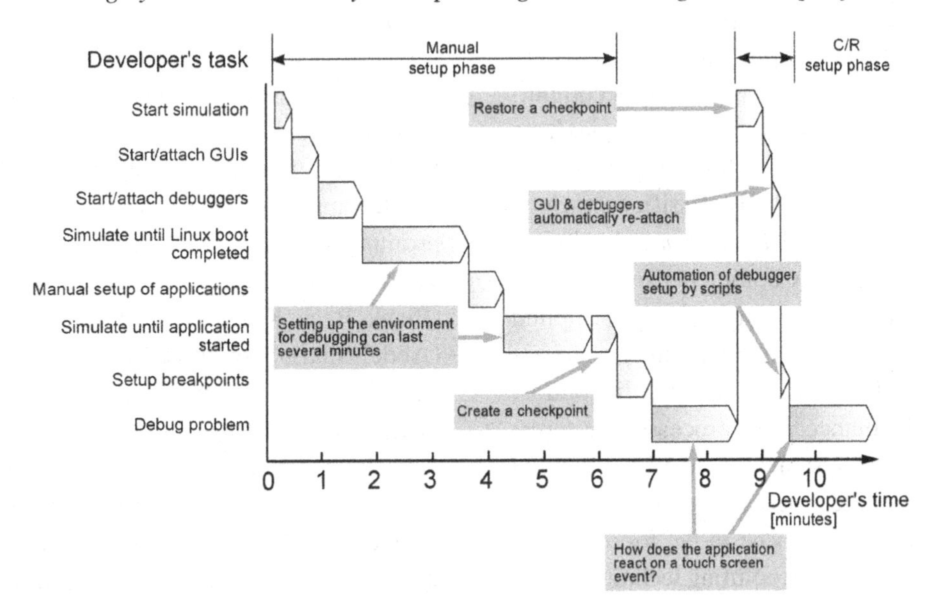

tance for SystemC based VPs. The transparency is ranked only second because VPs can easily be re-compiled and re-linked.

CHECKPOINTING CONCEPTS

Creating a checkpoint framework that supports SystemC modules is a complex task due to the representation of the internal state of the simulation. In SystemC the simulation state is described by global variables, local variables, heap values, and OS resources, e.g., file handles and sockets. The memory layout of those state variables heavily depends on external tools like compiler and linker. With respect to checkpointing these tools show a kind of black box behavior that cannot be influenced. Hence, extracting the necessary information to describe the simulation state is a very complex task.

Checkpointing is an important technique in the domain of high performance computing. Especially, with the trend towards large multiprocessor clusters based on commodity hardware checkpointing has gained a lot of attention. Checkpointing is primarily applied for dynamic load balancing, process migration and fault tolerance, in order to optimize the throughput, the flexibility and the reliability of the computing system. In the context of message passing parallel systems the global state is formed by the state of all computing nodes. Different checkpointing strategies have been developed in order to capture the global state. The techniques can be roughly classified into three categories: *uncoordinated*, *coordinated* and *communication induced*. A comprehensive analysis of the different checkpointing techniques applied in message passing based computing systems is presented by Elnozahy et al. (2002). In contrast to high performance multiprocessor systems, the simulation of SystemC based VPs usually is confined to a single process running on one processor. Therefore, we focus in the following on the different basic checkpointing techniques that are used for checkpointing individual processes.

Existing basic checkpointing approaches for SystemC simulations can be roughly grouped into three different types of approaches: *process checkpointing*, *model state serialization* and *operating system checkpointing*. The first approach, *process checkpointing*, stores the complete state of a process, i.e., the host processor registers, parts of the process address space, e.g., stack, heap, global data, and the state of allocated OS resources. This technique has originally been developed for process migration and fault tolerance and can be subdivided into two groups: *kernel space implementation* and *user space implementation*.

Kernel space implementations such as CRAK (Zhong & Nieh, 2001) and Berkeley Lab Checkpoint/Restart (BLCR) (Sankaran, Squyres, Barrett, & Lumsdaine, 2005) have the big advantage that they can directly access the complete state of all OS resources associated with the target process. This leads to a transparent checkpointing solution. However, additional effort must be spent to create the required kernel modules in order to access all OS states. Kernel space based checkpointing solutions can be realized either as kernel patch or as kernel module. Due to the continuously evolving Linux kernel, CRAK and BCLR are implemented as kernel module in order to minimize the dependencies on the Linux kernel. CRAK also provides support for migrating network sockets. However, this feature works only if it is possible to access the peer and modify the socket information after process migration. In contrast, the user space implementations usually require that the checkpointed application can be re-compiled. The most prominent implementations are Condor (Litzkow, Tannenbaum, Basney, & Livney, 1997), Libckpt (Plank, Beck, Kingsley, & Li, 1995) and WinCkp (Abdel-Shafi, Speight, & Bennett, 1999; Chung, Lee, Huang, Liang, & Wang, 1999). Condor and Libckpt are both implemented for UNIX whereas WinCkp is a process checkpoint implementation for Windows NT 4.0. Condor implements a load balancing system based on process checkpointing. It does not require any source code modifications of the application. The application needs only to be linked against the checkpoint library. Hence, it is required that all object files of the application are available. Libckpt is realized as library and provides a set of optimization features to reduce the memory consumption such as incremental checkpointing. Only pages that have been modified since the last checkpoint are stored. Furthermore, forked checkpointing is offered in order to reduce the impact of checkpointing on the application. After a checkpoint request an application copy is created by calling fork. The checkpointing is performed on the application copy while the original application continues running. Libckpt requires the modification of the application source code and linking the application against the checkpoint library. WinCkp is a library based checkpointing tool. The process control is realized by Win32 API calls which are commonly used by debuggers. Inter-process communication is not supported by this approach.

In the domain of virtual platform simulation, the concept process checkpointing is used by the Incisive Simulator from Cadence (Frazier, 2009) to provide the save and restore functionality. The developers of the C/C++ modules are responsible of storing the state of the modules. Callback functions are provided to mitigate the required development process. However, no support for external components such as debuggers is provided. Few details about the applied checkpointing technique are publically available.

The second type of checkpointing approaches for SystemC based simulations stores only the necessary information to describe the simulation state instead of the complete process state. The earliest mentioning of the utilization of checkpointing in a full-system simulator dates back to the mid nineties and is used inside the SimOS simulation environment (Rosenblum & Varadarajan, 1994) to switch between the fast and the detailed simulation mode. The switching between both simulation modes is realized by transferring the simulation state between both simulation modes. This kind of simulation state serialization is also used to realize the checkpointing functionality. The checkpointing capabilities of the IBM Mambo (Shafi, Bohrer, Phelan, Rusu, & Peterson, 2003) simulation system are based on the same principles as in the SimOS simulation system. Closely related to the previously presented checkpointing techniques is the approach utilized by Virtutech for checkpointing SystemC models (Montón, Engblom, & Burton, 2009) which are simulated together with the Virtutech Simics (Magnusson et al., 2002) simulator. The user needs to manually specify the data which represents the state of a SystemC module by utilizing a mechanism called attributes. Currently, this approach is limited to SystemC methods. It is not possible to checkpoint modules that make use of SystemC threads. This is due to the fact that SystemC threads, in contrast to SystemC methods, can be suspended during execution. The state of the SystemC thread is implicitly stored inside the user level thread implementation of the SystemC kernel. Hence, the state of a SystemC thread is part of the simulation environment and cannot be easily separated from it. The restriction to support only SystemC methods significantly limits the applicability of this approach, since most complex VPs rely on SystemC threads for modeling their behavior. In the context of computer architecture simulation the SimFlex (Hardavellas et al., 2004) environment shows an ambitious use of checkpointing for full system simulation. The approach of state serialization has the advantage of creating much smaller checkpoints. Furthermore, due to the lack of OS dependent information the checkpoints can be exchanged between OSs. However, the high flexibility comes at the price of a rigid programming model for all SystemC components to mark the information that needs to be checkpointed. Especially for complex SystemC blocks such as ISS based processor models with cache simulations, the effort marking all the necessary states can be quite huge and can only be performed by a developer with expert knowledge of this module. For example, the state of a processor module consists not only of the processor registers, the pipeline state and state of the branch predictor, but also of the state of all built-in peripherals, this includes the state of the built-in caches and the state of the memory management unit.

The third type of approach is the operating system checkpointing. It stores the state of the entire OS and all its applications, including the simulator and the debuggers. Hence, this approach is fully transparent and works for all types of simulation. Operating system checkpointing has become widely available during the last years especially by the usage of virtual machines like VMWare (VMWare Inc., 2010) and Xen. The drawbacks of this approach are the prohibitively large checkpoint images and the lack of controllability of the checkpointing process by the simulator.

Many modern software debuggers, such as gdb 7.x or totalview, provide reverse debugging as a feature to facilitate application debugging. This kind of lightweight checkpointing is realized using a trace based approach. The recorded traces are replayed in order to reach the desired state. However, the number of steps that can be reversed is usually limited. Furthermore, these debuggers focus only on processor cores. The peripheral components are not affected by the reverse debugging.

The evaluation of the different checkpointing techniques showed that there is currently no solution available which would support the complete set of uses cases. Model state serialization is capable of creating very small and easily exchangeable checkpoints at the expense of a high overhead during model creation. The VP developer needs to make sure that all variables representing the state of the platform have been identified and marked for serialization. This requires the capability to recompile the VP and thus this approach is not fully transparent.

The process checkpointing technique mitigates the effort required to apply checkpointing to VPs by storing the complete state of the VP simulation. Special care needs to be taken by the user to drive the simulation into a safe state prior to checkpointing. As for the model state serialization, process checkpointing requires the capability to recompile the VP. Hence, although less modifications are required than for the state serialization this approach is also not fully transparent. Inherently, process checkpointing produces comparatively large checkpoints that cannot be restored on different machines. Small changes in the memory layout will already invalidate the checkpoint.

Most solutions found in literature target one special use case or application type and support only a certain set of OS resources. Especially the support of inter-process communications to other user processes, such as external debuggers in the VP environment, is missing completely in all solutions.

NOVEL VP CHECKPOINT FRAMEWORK

Because of the high effort required to extract the simulation state from the simulation process in the model state serialization, we propose a checkpointing framework for VPs based on user level process checkpointing (Roman, 2002).

With the utilization of process checkpointing the complete simulation state is automatically captured without the need to adapt the platform source code. The sole exception is the handling of the inter-process communication channels and OS resources e.g., open files, sockets, threads or mutexes, which all form gateways from the simulation process into the host OS kernel and to other processes. In order to apply process checkpointing, the framework has to ensure that all the inter-process communication channels are closed during the checkpointing process. Afterwards the communication links need to be re-established before the simulation can be continued.

The central problem of handling OS resources and inter-process communication is solved by driving the simulation into a safe state where process checkpointing is applicable. Due to the usage of user level process checkpointing it is not possible to capture the state of OS resources like open files. Hence the state of these connections cannot be restored later on. The same applies to connections to external processes which are not controlled by the checkpointing framework. Therefore, a process is considered in a safe state, if no external dependencies exist. By closing all communication channels and releasing all OS resources it is always possible to drive the simulation in a safe state and later on successfully restore the process.

However, in case that the simulation communicates with an external partner the C/R framework is not able to fully control the communication link. Only the part of the communication channel implemented inside the simulation process is affected by the checkpointing framework. Hence, after restoring a checkpoint there is the risk of a potential mismatch between the restored communication state of the simulation platform and the state of the external simulation partner. For communication protocols used with external communication partners that are not stateless, e.g., TCP, this might lead to timeouts or communication errors inside the communication stack. For example, in case of the smartphone VP, the internet is accessed via the simulated Ethernet adapter. During the checkpointing process the network connection is closed, to drive the smartphone VP into a safe state. After re-loading this checkpoint into the VP, a previously ongoing data transaction will fail to resume, since the associated information stored inside the different layers of the local communication stack does not match anymore with the information inside the communication stack of the external partner. Although it is not possible to restore the ongoing transaction in the above mentioned example, it is of course possible to initiate a new data transaction after restoring the checkpoint. In case that restarting the transaction after a checkpoint is not desired, the user can temporally disable the checkpoint capability while a communication with the external partner is active. How to integrate user defined modules and temporally disable the C/R functionality is described in the next section.

Reaching this safe state is accomplished by providing a framework that releases all open OS resources and closes all communication channels before the actual initiation of the process checkpoint and re-establishing them afterwards. The C/R framework provides a set of callback functions to simplify the creation of C/R compliant user modules. The main components involved in this process are the *Virtual Platform Session Manager* and the *C/R engine*. The VP Session Manager is a separate process that is responsible for controlling the checkpointing procedure. It drives the communication between the debugger and GUI tools while they are detached from the simulation process. By introducing this background process it can be assured that the debuggers are not affected by the checkpointing procedure, since they remain attached to the VP Session Manager all the time. Third-party debuggers require a plug-in in order to connect to the VP Session Manager. The C/R engine is part of the simulation process itself and performs the process checkpointing once all external connections are closed.

The overall checkpointing procedure is illustrated in Figure 3. First, the debuggers and the GUIs are notified (1). If necessary the GUIs transfer data (2) to the checkpoint image. Then the debuggers and GUIs are detached from the simulation (3) to permit the release inter-process communication resources. In the next step, the user modules of the simulation are notified (4). After writing their user data (5) to the checkpoint image, the user modules release all OS resources (6). Now, the simulation process has no external dependencies and is in a safe state. Regular process checkpointing can be applied to store the process state (7). After the simulation process has been checkpointed the OS resources of the user modules are re-established (8) and the GUIs and debuggers are re-connected (9) to the simulation in order to continue simulation of the VP.

Figure 3. Checkpoint procedure

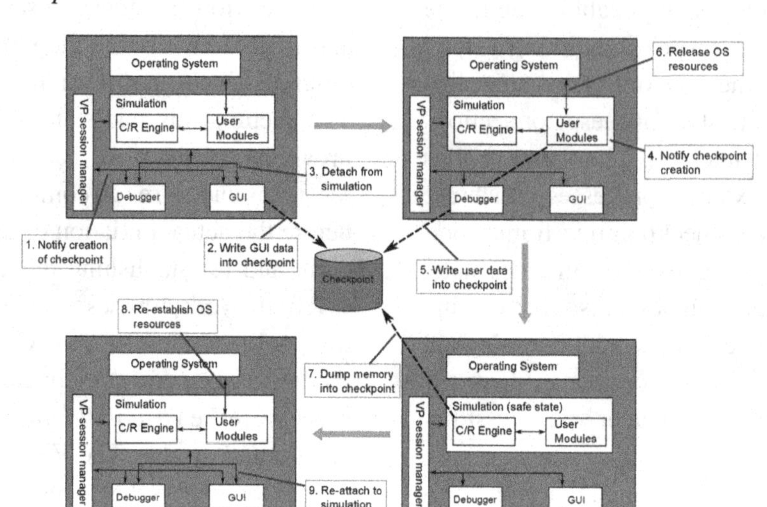

The restoration of a checkpoint follows the steps presented above in inverse order as shown in Figure 4. First a new simulation process is created (1). Then the stored data is used to restore the memory of this process (2). In the subsequent step the user modules are notified (3) and retrieve their user data from the checkpoint image (4). Furthermore, the connection to the OS resources is re-established (5). The GUIs and debuggers are notified (6) and the GUIs restore their state by loading the stored data from the checkpoint (7). Then the GUIs and debuggers are re-connected to the simulation (8). And finally, the debugger scripts are re-run to restore the breakpoint settings (9).

INTEGRATION OF USER-DEFINED MODULES

As described in the previous section, the framework for handling external communication channels and OS resources during the checkpoint and restore procedure allows the integration of user modules, debuggers and GUIs. This section focuses on the integration of user-defined SystemC modules, which is the most common task for model designers. For external applications such as GUIs and debuggers an API is also provided.

The user-defined modules are notified about checkpoint and restore events by inheriting from a special observer object provided by the C/R framework API. This observer object internally connects to the VP session manager. The C++ code example in Figure 5 illustrates how this observer is utilized for a SystemC module that continuously writes trace data into a file. The trace data that was produced before a checkpoint is created shall further be available when restoring the checkpoint. Therefore, the user module closes the file during the checkpoint procedure and copies its content into the checkpoint image. After the checkpoint was created, the file is re-opened. In case a checkpoint is restored, the module first restores the original content of the trace file by copying it back from the checkpoint image. Then, it re-opens the file to allow further dumping of data when the restored VP resumes the simulation.

The presented C/R framework does not handle temporary OS resources at all. Instead, the creation of a checkpoint is blocked for the short time when

Figure 4. Restore procedure

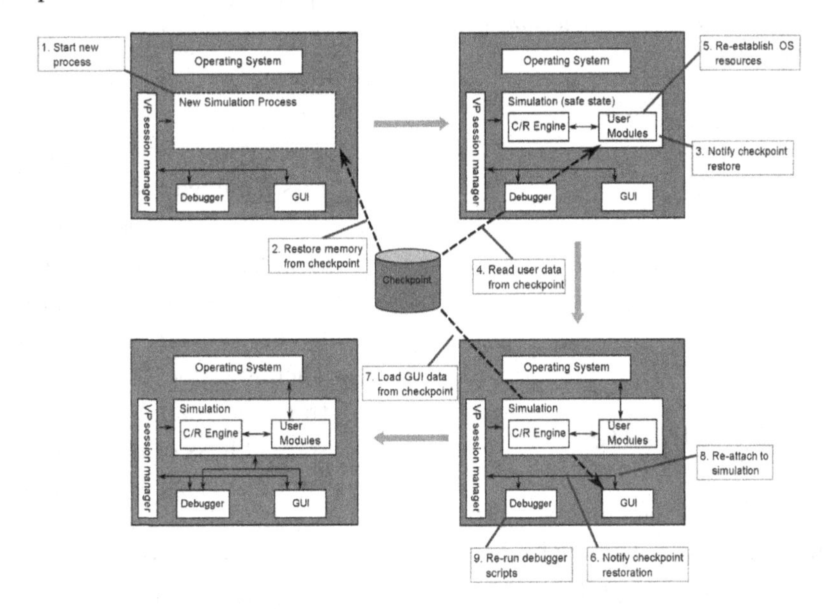

such resources are in use. Either the creation of a checkpoint is delayed until the appropriate resource is freed, or a message is printed to the GUI to inform the user why the creation of a checkpoint is not possible at the moment. Very often files or other OS resources are only used for a short period of time. Examples for a temporary file access are the reading of configuration information or loading of a binary image to the memory of a processor. For this kind of temporary OS resource utilization the C/R framework API provides a simple mechanism that blocks the creation of checkpoints while the simulation is inside such an unsafe section. This is achieved by temporarily instantiating a checkpoint blocker object, as illustrated by the code example in Figure 6. As long as the instance of the blocker is active checkpointing is not possible. The developer can pass a string to the constructor of the blocker object, which is used to inform the user at runtime, why the creation of checkpoints is temporarily disabled.

In case that not all OS resources had been released before the process checkpoint is created, the produced checkpoint image would be corrupted. As a result a restored simulation based on a corrupted checkpoint would fail when accessing a previously unreleased OS resource. Usually, a debugger will point the user to the source code location that uses the incorrectly handled OS resources. However, to ease the development of checkpointing enabled platforms the checkpointing environment supports the model developer by scanning the simulation process for open OS resources immediately before the process checkpoint is created. All open files, sockets and threads are reported to the user and the creation of a potentially corrupted checkpoint is aborted. This feedback helps the user to locate the incompatible code portion and correct it.

Figure 5. Adaptation of a user SystemC module: Writing user data into a checkpoint

```
1 #include <Base/SimBase/CheckpointRestoreObserver.h>
2
3 class MyModule : public nCheckpointRestore::Observer {
4 ...
5 private:
6 virtual std::string handleBeginOfCheckpointing(...) {
7 copyTraceFileToCheckpointDirectory();
8 closeTraceFile();
9 return "";
10 }
11
12 virtual void handleEndOfCheckpointing(...) {
13 reopenTraceFile();
14 }
15
16 virtual void handleEndOfRestore() {
17 copyTraceFileFromCheckpointDirectory();
18 reopenTraceFile();
19 }
20 ...
21 }
```

Figure 6. Adaptation of a user SystemC module: Temporarily blocking the creation of checkpointing while a file is opened

```
1 #include <Base/SimBase/CheckpointBlocker.h>
2
3 void MyModule::readConfiguration(...) {
4 ...
5 { // limit scope of variable config_file
6 nCheckpointRestore::copedCheckpointBlocker cp_blocker(
7 "MyModule reading config");
10 std::ifstream config_file(config_filename);
11 // read configuration
12 ...
13 }
14 // checkpointing is possible again
15 ...
16 }
```

PROCESS CHECKPOINTING

In this section the details of the employed process checkpointing are discussed on the basis of a Windows XP virtual memory layout as shown in Figure 7. The basic principles of the process checkpointing presented in the section are also valid for other OSs such as Linux.

Once the simulation process has been driven into a safe state, i.e., all external connections have been released; process checkpointing is applied to store the state of the simulation. The process checkpointing functionality is encapsulated into a separate process in order to guarantee that checkpointing does not interfere with the virtual memory space of the simulation process. In a first step, the virtual memory space of the simulation process is analyzed in order to identify the parts of the memory that need to be stored. The address space of a process is divided in two parts: The *user space*, ranging from 0x0000000 to 0x7FFFFFFF and the *kernel space* which ranges from 0x80000000 to 0xFFFFFFFF. The kernel space is not directly accessible by the user and is therefore excluded from checkpointing. The user space is flanked on both sides by not accessible and unused memory areas. These areas define the address frame for the identification process: It starts at 0x0010000 and ends at 0x7FFF0000. Next step is to divide this frame into different memory regions, where a region is defined as a range of memory pages which share the same attributes.

If an identified memory region needs to be part of the checkpoint depends on the following conditions:

- The environment string region and the process parameter region are not stored, since the C/R process assures that the simulation is started out of a consistent environment.
- Only the part of the main stack region used by the process stack needs to be stored. Therefore the stack pointer is extracted from the thread context and used as the start address of the stack region.
- All other stack regions found in the virtual memory space are fiber stacks. Thus, all regions identified as stack regions are stored.
- Regions with read-only access that never change their position or size, e.g., national language setting (NLS) and the code segment, are ignored.
- Static data regions with read-write access are stored.
- For the dynamic link libraries (DLL) all of their regions with read-write access are stored.

Figure 7. Simplified example of a Windows XP virtual memory layout

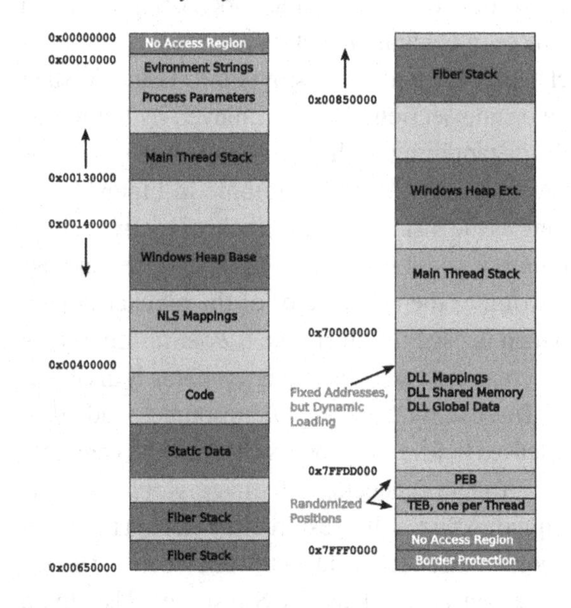

- Data regions are all regions with read-write protection which are not identified as one of the region types described above. This includes the heap regions as well as regions which have been created by the user. All these regions are stored.

During the restoration of a process checkpoint the stored memory region are restored from the hard disk. One of the following actions may be required before the data can be copied:

- Creation of new regions.
- Resizing of an existing region.
- Adjusting the protection attribute.

Process checkpointing relies on the fact that the memory layout of a process does not change between different execution runs. However, this assumption is not valid anymore with the introduction of *address space layout randomization* ASLR (Shacham, Page, & Pfaff, 2004). This feature of the loader randomizes the memory layout every time an application is loaded in order to reduce the threat of buffer overflow induced attacks. This randomization certainly aggravates the process checkpointing. In case of Windows XP the ASLR is restricted to the PEB and TEB regions and hence poses no problems for the C/R. However, for newer versions of Windows more regions are affected by the ASLR. Further research is required to analyze if process checkpointing in its current form can be utilized together with ASLR. In the context of this article ASLR has been deactivated.

PERFORMANCE EVALUATION

The time required for the creation and restoration of a checkpoint is proportional to the memory size of the simulation process. Depending on the file system and the used hard disk, the time for checkpointing can strongly vary. All measurements are conducted on an Intel Quad Core Q6700 based system with 2.66GHz and 16GB of RAM using Scientific Linux 5.0.

The smartphone platform shown in Figure 1 is based on an ARM926 processor with 128MB of RAM. For this platform the size of the checkpoint is approximately 569MB, the creation takes about 5s and the restore about 4s.

An inherent problem of process checkpointing based approaches is that checkpoint images are quite sensitive to recompilations of the virtual platform. Any change to the simulation executable of the virtual platform invalidates the checkpoint image. Furthermore, the portability of the checkpoints is limited. In general, it is possible to restore the checkpoint on a host with an identical installation. However, already a slight difference in the patch level of the OS may prohibit restoring a checkpoint.

Recompilations or other changes of the embedded software which is executed by the processor models of the VP do not corrupt the checkpoint image. After a checkpoint is restored, the VP always contains the original software, e.g., the OS, and behaves like the checkpointed platform. The user is now able to load the recompiled software image, e.g., an application, into the processor model and continue the simulation of the VP.

The comparatively large size of the checkpoints based on process checkpointing increases the checkpointing time as well as it limits the number of images that can be kept on a hard disk. For this reason the framework incorporates a stream-based compression utilizing the deflate algorithm (Deutsch, 1996) to compress the data before writing it to the hard disk. The deflate algorithm is a combination of the lossless Lempel-Ziv-Storer-Szymanski algorithm (Storer & Szymanski, 1982) and Huffman coding. This algorithm offers a set of parameters that enable the user to trade-off the compression rate for compression speed. This feature is very useful to determine the optimal working point for the compression in the context of checkpointing. The impact of the compression on the checkpoint time is twofold: First, the computational overhead introduced by the compression increases the checkpointing time. Second, less data needs to be stored, which leads to a faster checkpointing. Depending on which effect is stronger the checkpointing becomes faster or slower. Independent of the checkpointing time, compression clearly increases the number of checkpoints that can be stored simultaneously on the hard disk.

In order to analyze the implication of compressing the checkpoints while creating them, a scalable, complex VP is used for measurements.

The Virtual SHAPES Platform (VSP) has been developed in the context of the SHAPES (Paolucci, Jerraya, Leupers, Thiele, & Vicini, 2006) project. SHAPES is a tiled scalable software hardware architecture. The basic tile of this architecture comprises an ARM926, a VLIW mAgic DSP (Paolucci et al., 2001), a distributed network processor (DNP), 64MB of on-tile memories and a set of on-tile peripherals. All these components are modeled inside the VSP. Figure 8 shows the structure of the basic SHAPES tile. Before applying the checkpointing to the VSP, some modifications of the platform were required. More specific, the creation of log files by the different peripheral

Figure 8. Basic SHAPES tile

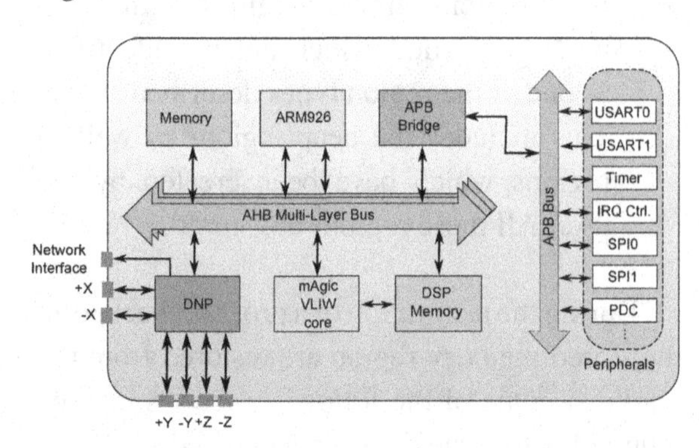

modules needed to be extended to support closing of the OS resources during the checkpointing. Updating all peripherals and succeeding validation required only two days for a developer who is already familiar with the SHAPES platform.

As test application a *Lattice Quantum Chromo-Dynamics* (LQCD) simulation from the domain of theoretical physics, specifically optimized for the SHAPES platform, is selected. Due to its excellent scalability the LQCD application is executed on different numbers of SHAPES tiles to evaluate the scaling of the checkpointing and the checkpoint compression. Figure 9 shows the time required for checkpointing and restoring compressed and non-compressed checkpoints.

Figure 9. Checkpoint/restore time for compressed and uncompressed checkpoints

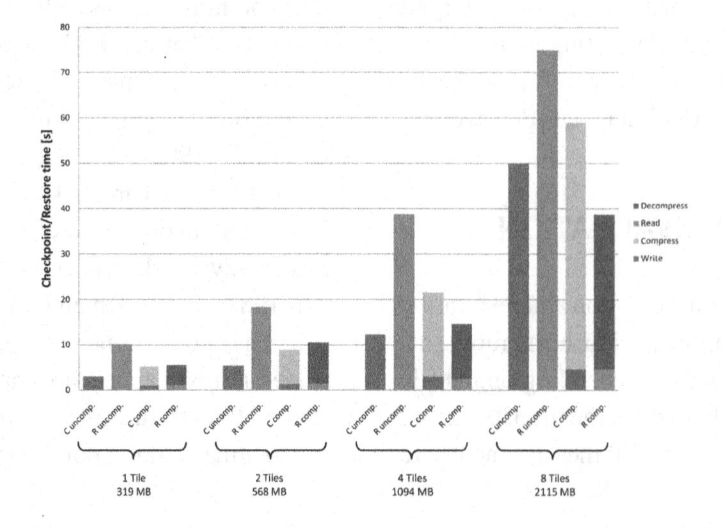

The simulation complexity has been varied by running the LQCD application on one, two, four and eight tiles, leading to uncompressed checkpoints ranging from 319MB to 2115MB. All experiments have been conducted using a local hard disk capable of storing approximately 42MB/s. On average the size of the compressed images is reduced by a factor of 10. However, due to the combination of a fast local hard disk and the caching effects of the underlying file system the compression overhead is not fully compensated by the smaller image size for the checkpoint process. In case of slow storage medium, e.g., laptop hard disk, the picture is completely different. Here, the compression is capable to reduce the checkpoint size and the checkpoint time. In contrast to the checkpointing the restore process significantly benefits from the compression.

CONCLUSION

A checkpoint/restore feature considerably helps the software developers to shorten their typical edit-compile-debug cycles. In order to evaluate the applicability of this approach to SystemC-based VPs it has been integrated into the Synopsys Virtual Platform environment. Compared to other checkpointing approaches this framework is specifically tuned for SystemC by supporting the checkpointing of SystemC simulations with debuggers and GUIs connected to it. Complete transparency has not been reached by the proposed approach; however the changes required are only minimal and usually consist of only a few lines of code. In order to mitigate the large checkpoint size, the checkpoints can be compressed before being stored. On average a compression of 10x is reached. Depending on the underlying file system and on the speed of the storage medium the checkpointing time may benefit from the compression. Due to the utilization of process checkpointing the presented approach is sensitive to changes of the VP. The C/R feature is most useful for engineers developing software on top of a VP.

ACKNOWLEDGMENT

This work was partly funded by the European project SHAPES. For more information visit http://apegate.roma1.infn.it/SHAPES. This work was presented in part at the International Symposium on System-on-Chip (SoC) 2009, Tampere, Finland, October 5-7, 2009 (Kraemer, Leupers, Petras, & Philipp, 2009).

REFERENCES

Abdel-Shafi, H., Speight, E., & Bennett, J. K. (1999). Efficient user-level thread migration and checkpointing on Windows NT clusters. In *Proceedings of the 3rd USENIX Windows NT Symposium*, Seattle, WA.

Chung, P. E., Lee, W.-J., Huang, Y., Liang, D., & Wang, C.-Y. (1999). Winckp: A transparent checkpointing and rollback recovery tool for Windows NT applications. In *Proceedings of the International Symposium on Fault-Tolerant Computing* (pp. 220-223). Washington, DC: IEEE Computer Society.

Deutsch, P. (1996). *DEFLATE compressed data format specification version 1.3*. Retrieved from http://tools.ietf.org/pdf/rfc1951.pdf

Elnozahy, E. N., Alvisi, L., Wang, Y., & Johnson, D. (2002). A survey of rollback-recovery protocols in message-passing systems. *ACM Computing Surveys, 34*(3), 375–408. doi:10.1145/568522.568525

Frazier, G. (2009). *SystemC save and restore part 2 - Advanced usage.* Retrieved from http://www.cadence.com/community/blogs/sd/archive/2009/03/09/systemc-save-and-restore-part-2-advanced-usage.aspx

Hardavellas, N., Somogyi, S., Wenisch, T. F., Wunderlich, R. E., Chen, S., & Kim, J. (2004). SimFlex: A fast, accurate, flexible full-system simulation framework for performance evaluation of server architecture. *ACM Special Interest Group on Measurement and Evaluation. Performance Evaluation Review*, *31*(4), 31–34. doi:10.1145/1054907.1054914

Institute of Electrical and Electronics Engineers, Inc. (IEEE). (2005). *IEEE 1666: Standard, open SystemC language reference manual.* Washington, DC: Institute of Electrical and Electronics Engineers, Inc.

Kraemer, S., Leupers, R., Petras, D., & Philipp, T. (2009). A checkpoint/restore framework for SystemC-based virtual platforms. In *Proceedings of the International Conference on System-on-Chip* (pp. 161-167). Washington, DC: IEEE Computer Society.

Litzkow, M., Tannenbaum, T., Basney, J., & Livney, M. (1997). *Checkpoint and migration of UNIX processes in the condor distributed processing system* (Tech. Rep. No. 1346). Madison, WI: University of Wisconsin Madison.

Magnusson, P. S., Christensson, M., Eskilson, J., Forsgren, D., Hållberg, G., & Högberg, J. (2002). Simics: A full system simulation platform. *IEEE Computer*, *35*(2), 50–58.

Montón, M., Engblom, J., & Burton, M. (2009). Checkpoint and restore for SystemC models. In *Proceedings of the Forum on Specification & Design Languages*, Sophia Antipolis, France.

Paolucci, P. S., Jerraya, A. A., Leupers, R., Thiele, L., & Vicini, P. (2006). SHAPES: A tiled scalable software hardware architecture platform for embedded systems. In *Proceedings of the International Conference on Hardware/Software Codesign and System Synthesis* (pp. 167-172).

Paolucci, P. S., Kajfasz, P., Bonnot, P., Candaele, B., Maufroid, D., & Pastorelli, E. (2001). mAgic-FPU and MADE: A customizable VLIW core and the modular VLIW processor architecture description environment. *Computer Physics Communications*, *139*(1), 132–143. doi:10.1016/S0010-4655(01)00235-1

Plank, J. S., Beck, M., Kingsley, G., & Li, K. (1995). Libckpt: Transparent checkpointing under Unix. In *Proceedings of the USENIX Technical Conference*, New Orleans, LA (pp. 213-224).

Roman, E. (2002). *A survey of checkpoint/restart implementations* (Tech. Rep. No. LBNL-54942). Berkeley, CA: Lawrence Berkeley National Laboratory.

Rosenblum, M., & Varadarajan, M. (1994). *SimOS: A fast operating system simulation environment* (Tech. Rep. No. CSL-TR-94-631). Stanford, CA: Stanford University.

Sankaran, S., Squyres, M. J., Barrett, B., & Lumsdaine, A. (2005). The LAM/MPI checkpoint/RestartFramework: System-initiated checkpointing. *International Journal of High Performance Computing Applications*, *4*(19), 479–493. doi:10.1177/1094342005056139

Shacham, H., Page, M., & Pfaff, B. (2004). On the effectiveness of address-space randomization. In *Proceedings of the 11th ACM Conference on Computer and Communications Security* (pp. 298-307).

Shafi, H., Bohrer, P. J., Phelan, J., Rusu, C. A., & Peterson, J. L. (2003). Design and validation of a performance and power simulator for PowerPC systems. *IBM Journal of Research and Development, 47*(5), 641–651. doi:10.1147/rd.475.0641

Storer, J. A., & Szymanski, T. G. (1982). Data compression via textual substitution. *Journal of the ACM, 29*(4), 928–951. doi:10.1145/322344.322346

VMWare Inc. (2010). *VMWare server.* Retrieved from http://www.vmware.com/products/server/

Zhong, H., & Nieh, J. (2001). CRAK: Linux checkpoint/restart as a kernel module *(Tech. Rep. No. CUCS-014-01). New York, NY: Columbia University.*

This work was previously published in the International Journal of Embedded and Real-Time Communication Systems, Volume 2, Issue 4, edited by Seppo Virtanen, pp. 21-37, copyright 2011 by IGI Publishing (an imprint of IGI Global).

Compilation of References

Aarts, E. L., & Lenstra, J. K. (2003). *Local search in combinatorial optimization.* Princeton, NJ: Princeton University Press.

Abdel-Shafi, H., Speight, E., & Bennett, J. K. (1999). Efficient user-level thread migration and checkpointing on Windows NT clusters. In *Proceedings of the 3rd USENIX Windows NT Symposium*, Seattle, WA.

Ahonen, T., Sigüenza-Tortosa, D. A., Bin, H., & Nurmi, J. (2004). *Topology optimization for application-specific networks-on-chip.* In *Proceedings of the 2004 international workshop on System Level Interconnect Prediction (SLIP '04)* (pp. 53-60). New York: ACM.

Aiken, A. (1999). Introduction to set constraint-based program analysis. *Science of Computer Programming, 35*(2-3), 79–111. doi:10.1016/S0167-6423(99)00007-6

Aizenbud-Reshef, N., Nolan, B., Rubin, J., & Shaham-Gafni, Y. (2006). Model traceability. *IBM Systems Journal, 45*(3), 515–526. doi:10.1147/sj.453.0515

Akyildiz, I., Weilian, S., Sankarasubramaniam, Y., & Cayirci, E. (2002). A survey on sensor networks. *IEEE Communications Magazine, 40*(8), 102–114. doi:10.1109/MCOM.2002.1024422

Al Faruque, M., Ebi, T., & Henkel, J. (2008, November). ROAdNoC: Runtime Observability for an Adaptive Network on Chip Architecture. In *Proceedings of the IEEE/ACM International Conference on Computer-Aided Design (ICCAD 2008)* (pp. 543-548).

Alliance, Z. (2010). *ZigBee specification.* Retrieved from http://www.zigbee.org/Standards/ZigBeeSmartEnergy/Specification.aspx

Alur, R. (1999). Timed automata. In N. Halbwachs & D. Peled (Eds.), *Proceedings of the 11th International Conference on Computer Aided Verification* (LNCS 1633, pp. 8-22).

Alur, R., & Henzinger, T. A. (1992). Logics and models of real time: A survey. In J. W. de Bakker, C. Huizing, W. P. de Roever, & G. Rozenberg (Eds.), *Proceedings of the Workshop on Real-Time: Theory in Practice* (LNCS 600, pp. 74-106).

Alur, R., Courcoubetis, C., & Dill, D. (1990). Model-checking for real-time systems. In *Proceedings of the 5th IEEE Annual Symposium on Logic in Computer Science*, Philadelphia, PA (pp. 414-425). Washington, DC: IEEE Computer Society.

Alur, R., Peled, D., & Penczek, W. (1995). Model-checking of causality properties. In *Proceedings of the 10th IEEE Symposium on Logic in Computer Science*, San Diego, CA (pp. 90-100). Washington, DC: IEEE Computer Society.

ARINC 664. (2003). *Aircraft Data Network, Part 7: Deterministic Networks* (Computer software manual).

Asanovic, K., Bodik, R., Catanzaro, B. C., Gebis, J. J., Husbands, P., Keutzer, K., et al. (2006). *The landscape of parallel computing research: A view from Berkeley* (Tech. Rep. No. UCB/EECS-2006-183). Berkeley, CA: University of California.

Asenov, A., Kaya, S., & Davies, J. H. (2002). Intrinsic threshold voltage fluctuations in decanano MOSFETs due to local oxide thickness variations. *IEEE Transactions on Electron Devices, 49*(1), 112–119. doi:10.1109/16.974757

Atallah, M. J., & Blanton, M. (2009). *Algorithms and Theory of Computation Handbook* (*Vol. 2*). Boca Raton, FL: CRC Press LLC.

Atmel Corporation. (2011). DIOPSIS, D940. Retrieved from http://www.atmel.com/dyn/products/devices.asp?family_id=680

Aydal, E. G., Paige, R. F., Utting, M., & Woodcock, J. (2009). Putting formal specifications under the magnifying glass: Model-based testing for validation. In *Proceedings of the 2nd International Conference on Software Testing Verification and Validation*, Denver, CO (pp. 131-140). Washington, DC: IEEE Computer Society.

Bacherini, S., Fantechi, A., Tempestini, M., & Zingoni, N. (2006). A story about formal methods adoption by a railway signaling manufacturer. In J. Misra, T. Nipkow, & E. Sekerinski (Eds.), *Proceedings of the 14th International Symposium on Formal Methods*, Hamilton, ON, Canada (LNCS 4085, pp. 179-189).

Baier, C., & Katoen, J. P. (2008). *Principles of model checking*. Cambridge, MA: MIT Press.

Banerjee, A., Mullins, R., & Moore, S. (2007, May). A power and energy exploration of network-on-chip architectures. In *Proceedings of the First International Symposium on Networks-on-Chip* (pp. 163-172). Washington, DC: IEEE Computer Society.

Baresel, A., Conrad, M., Sadeghipour, S., & Wegener, J. (2003). The interplay between model coverage and code coverage. In *Proceedings of the 11th European International Conference on Software Testing, Analysis and Review*, Amsterdam, Netherlands.

Barkah, D., Ermedahl, A., Gustafsson, J., Lisper, B., & Sandberg, C. (2008). Evaluation of automatic flow analysis for WCET calculation on industrial real-time system code. In *Proceedings of the 20th Euromicro Conference on Real-Time Systems*, Prague, Czech Republic (pp. 331-340). Washington, DC: IEEE Computer Society.

Barr, M., & Massa, A. (2006). *Programming Embedded Systems: With C and GNU Development Tools, Second edition*. New York: O'Reilly Media, Inc.

Basili, V. R., Caldiera, G., & Rombach, H. D. (1994). The goal question metric approach. In *Encyclopedia of software engineering*. New York, NY: John Wiley & Sons.

Bauer, T., Böhr, F., Landmann, D., Beletski, T., Eschbach, R., & Poore, J. H. (2007). From requirements to statistical testing of embedded systems. In *Proceedings of the Software Engineering for Automotive Systems Workshops*, Minneapolis, MN.

Bauer, T., Eschbach, R., Groessl, M., Hussain, T., Streitferdt, D., & Kantz, F. (2009). Combining combinatorial and model-based test approaches for highly configurable safety-critical systems. In *Proceedings of the 2nd Workshop on Model-based Testing in Practice at the 5th European Conference on Model-Driven Architecture Foundations and Applications*, Enschede, The Netherlands.

Bauer, T., Stallbaum, H., Metzger, A., & Eschbach, R. (2008). Risikobasierte Ableitung und Priorisierung von Testfällen für den modellbasierten Systemtest. In *Proceedings of the Software Engineering Conference*, Munich, Germany.

Beer, S., Ginosar, R., Priel, M., Dobkin, R., & Kolodny, A. (2010). The devolution of synchronizers. In *Proceedings of the IEEE Symposium on Asynchronous Circuits and Systems* (pp. 94-103).

Benini, L., & De Micheli, G. (2002). Networks on chips: A new SoC paradigm. *IEEE Computer*, *35*(1), 70–78.

Bertozzi, D., Benini, L., & De Micheli, G. (2005, June). Error Control Schemes for On-chip Communication Links: The Energy-reliability Tradeoff. *IEEE Transactions on Computer-Aided Design of Integrated Circuits and Systems*, *24*(6), 818–831. doi:10.1109/TCAD.2005.847907

Bik, A. J. C., & Wijshoff, H. A. G. (1995). Implementation of Fourier-Motzkin elimination. In *Proceedings of the 1st Annual Conference of the Advanced School for Computing and Imaging* (pp. 377-386).

Black, P. E. (2006, May). *Manhattan distance*. Retrieved from http://www.itl.nist.gov/div897/sqg/dads/

Black, D., & Donovan, J. (2004). *SystemC: From the Ground Up*. New York: Springer. doi:10.1007/0-387-30864-4

Blanchet, B., Cousot, P., Cousot, R., Feret, R., Mauborgne, R., Miné, A., et al. (2003). A static analyzer for large safety-critical software. In *Proceedings of the ACM SIGPLAN Conference on Programming Language Design and Implementation*, San Diego, CA (pp. 196-207). New York, NY: ACM Press.

Bonesi, S., Bertozzi, D., Benini, L., & Macii, E. (2008). Process variation tolerant pipeline design through a placement-aware multiple voltage island design style. *Design, Automation and Test in Europe*, 967-972.

Borkar, S., Karnik, T., & Vivek, D. (2004). Design and reliability challenges in nanometer technologies. In *Proceedings of the 41st Design Automation Conference* (p. 75).

Bose, P., Morin, P., Stojmenovic, I., & Urrutia, J. (1999). Routing with guaranteed delivery in ad-hoc wireless networks. In *Proceedings of the 3rd International Workshop on Discrete Algorithms and Methods for Mobile Computing and Communications*, Seattle, WA (pp. 48-55).

Bowman, K. A., Duvall, S. G., & Meindl, J. D. (2002). Impact of die-to-die and within-die parameter fluctuations on the maximum clock frequency distribution for gigascale integration. *IEEE Journal of Solid-state Circuits*, *37*(2), 183–190. doi:10.1109/4.982424

Bozga, M., Fernandez, J. C., Ghirvu, L., Jard, C., Jéron, T., & Kerbrat, A. (2000). Verification and test generation for the Sscop protocol. *Science of Computer Programming*, *36*(1), 27–52. doi:10.1016/S0167-6423(99)00017-9

Braberman, V., Kicillof, N., & Olivero, A. (2005). A scenario-matching approach to the description and model checking of real-time properties. *IEEE Transactions on Software Engineering*, *31*, 1028–1041. doi:10.1109/TSE.2005.131

Brat, G., & Klemm, R. (2003). Static analysis of the mars exploration rover flight software. In *Proceedings of the 1st International Space Mission Challenges for Information Technology*, Pasadena, CA (pp. 321-326). Washington, DC: IEEE Computer Society.

Broy, M., & Jonsson, B. Katoen, Philipps, J., Leucker, M., & Pretschner, A. (eds.). (2005). *Model-based testing of reactive systems* (LNCS 3472, pp. 281-291). Berlin, Germany: Springer-Verlag.

Bykhteev, A. (2008). Methods and facilities for systems-on-chip design. *ChipInfo microchip manual.* Retrieved from http://www.chipinfo.ru/literature/chipnews/200304/1.html

Cai, X., & Giannakis, G. (2003, November). A two-dimensional channel simulation model for shadowing processes. *IEEE Transactions on Vehicular Technology*, *52*(6), 1558–1567. doi:10.1109/TVT.2003.819627

Castrillon, J., Zhang, D., Kempf, T., Vanthournout, B., Leupers, R., & Ascheid, G. (2009). Task management in MPSoCs: An ASIP approach. In Proceedings of the International Conference on Computer-Aided Design (pp. 587-594). New York, NY: ACM Press.

Cerpa, A., Wong, J. L., Kuang, L., Miodrag, P., & Estrin, D. (2005). Statistical model of lossy links in wireless sensor networks. In *Proceedings of the 4th International Symposium on Information Processing in Sensor Networks*, Los Angeles, CA (pp. 81-88).

Chardaire, P., & Lutton, J. L. (1993). In Vidal, R. V. (Ed.), *Applied Simulated Annealing* (pp. 173–199). New York: Springer Verlag.

Chen, B. B., Hao, S., Zhang, M., Chan, M. C., & Ananda, A. I. (2009). DEAL: Discover and exploit asymmetric links in dense wireless sensor networks. In *Proceedings of the 6th Annual IEEE Communications Society Conference on Sensor, Mesh and Ad Hoc Communications and Networks*, Rome, Italy (pp. 297-305).

Chen, G., Li, F., & Kandemir, M. (2007). Compiler-directed application mapping for Noc based chip multiprocessors. *SIGPLAN Notes*, *42*(7), 155–157. doi:10.1145/1273444.1254796

Chung, P. E., Lee, W.-J., Huang, Y., Liang, D., & Wang, C.-Y. (1999). Winckp: A transparent checkpointing and rollback recovery tool for Windows NT applications. In *Proceedings of the International Symposium on Fault-Tolerant Computing* (pp. 220-223). Washington, DC: IEEE Computer Society.

Ciordas, C., Goossens, K., Basten, T., Radulescu, A., & Boon, A. (2006, October). Transaction Monitoring in Networks on Chip: The On-chip Run-time Perspective. In *Proceedings of the International Symposium on Industrial Embedded Systems (IES '06).*

Ciordas, C., Goossens, K., Radulescu, A., & Basten, T. (2006, May). NoC Monitoring: Impact on the Design Flow. In *Proceedings of the IEEE International Symposium on Circuits and Systems (ISCAS 2006)* (pp. 1981-1984).

Ciordas, C., Basten, A., Radulescu, A., Goossens, K., & Meerbergen, J. V. (2005, October). An Event-based Monitoring Service for Networks on Chip. *ACM Transactions on Design Automation of Electronic Systems*, *10*(4), 702–723. doi:10.1145/1109118.1109126

Clarke, E. M., Grumberg, O., & Peled, D. A. (2000). *Model checking*. Cambridge, MA: MIT Press.

Clarke, E., Biere, A., Raimi, R., & Zhu, Y. (2001). Bounded model checking using satisfiability solving. *Formal Methods in System Design*, *19*(1), 7–34. doi:10.1023/A:1011276507260

Cohen, M. B., Dwyer, M. B., & Shi, J. (2006). Coverage and adequacy in software product line testing. In *Proceedings of the ISSTA Workshop on Role of Software Architecture for Testing and Analysis* (pp. 53-63).

Collins, H. A., & Nikel, R. E. (1999). *DDR-SDRAM, high-speed, source synchronous interfaces create design challenges*. Retrieved from http://www.edn.com/article/506435-DDR_SDRAM_high_speed_source_synchronous_interfaces_create_design_challenges.php

Comité Européen de Normalisation en Électronique et en Électrotechnique. (1999). *EN 50126: Railway applications - the specification and demonstration of dependability, reliability, availability, maintainability and safety (RAMS)* (Tech. Rep. No. TX9X). Brussels, Belgium: CENELEC.

Comité Européen de Normalisation en Électronique et en Électrotechnique. (2001). *EN 50128: Railway applications - communications, signalling and processing systems - software for railway control and protection systems*. Brussels, Belgium: CENELEC.

Conrad, M. (2009). Testing-based translation validation of generated code in the context of IEC 61508. *Formal Methods in System Design*, *35*(3), 389–401. doi:10.1007/s10703-009-0082-0

Cormen, T. H., Leiserson, C. E., Rivest, R. L., & Stein, C. (2001). *Introduction to Algorithms* (2nd ed.). Cambridge, MA: MIT Press.

Corporation, I. B. M. (2010). CoreConnect bus architecture. Retrieved from https://www-01.ibm.com/chips/techlib/techlib.nsf/productfamilies/CoreConnect_Bus_Architecture

Cousot, P., & Cousot, R. (1977). Abstract interpretation: A unified lattice model for static analysis of programs by construction or approximation of fixpoints. In *Proceedings of the 4th ACM SIGACT-SIGPLAN Symposium on Principles of Programming Languages*, Los Angeles, CA (pp.238-353). New York, NY: ACM Press.

Cousot, P., & Cousot, R. (1979). Systematic design of program analysis frameworks. In *Proceedings of the 6th Annual ACM SIGPLAN-SIGACT Symposium on Principles of Programming Languages*, San Antonio, TX (pp. 269-282). New York, NY: ACM Press.

Cousot, P., & Cousot, R. (1992). Abstract interpretation frameworks. *Journal of Logic and Computation*, *2*(4), 511–547. doi:10.1093/logcom/2.4.511

Crossbow Technologies. (2003). *Mote in-network programming user reference*. Retrieved from http://www.tinyos.net/tinyos-1.x/doc/Xnp.pdf

Dalal, S. R., Jain, A., Karunanithi, N., Leaton, J. M., Lott, C. M., Patton, G. C., et al. (1999). Model-based testing in practice. In *Proceedings of the 21st International Conference on Software Engineering*, Los Angeles, CA (pp. 285-294). New York, NY: ACM Press.

Dally, W. J., & Towles, B. (2001). Route packets, not wires: On-chip interconnection networks. In *Proceedings of the 38th Conference on Design Automation* (pp. 684-689). New York, NY: ACM Press.

Dally, W. J., & Poulton, J. W. (1998). *Digital systems engineering*. Cambridge, UK: Cambridge University Press.

Dally, W. J., & Towles, B. (2004). *Principles and Practices of Interconnection Networks*. San Francisco: Morgan Kaufmann.

Davies, J., & Woodcock, J. (1996). *Using Z: Specification, refinement and proof*. Upper Saddle River, NJ: Prentice Hall.

De Couto, D., Aguayo, D., Bicket, J., & Morris, R. (2003). A high-throughput path metric for multi-hop wireless routing. In *Proceedings of the 9th Annual International Conference on Mobile Computing and Networking*, San Diego, CA (pp. 134-146).

De Micheli, G., & Benini, L. (2006). *Networks on Chips*. San Francisco: Morgan Kaufmann.

Decotignie, J.-D. (2005, June). Ethernet-based real-time and industrial communications. In *Proceedings of the IEEE*, Neuchatel, Switzerland (Vol. 93, p. 1102-1117). Washington, DC: IEEE.

Delmas, D., & Souyris, J. (2007). Astrée: From research to industry. In H. R. Nielson & G. Filé (Eds.), *Proceedings of the 14th International Static Analysis Symposium*, Lyngby, Denmark (LNCS 4634, pp. 437-451).

Deng, J., Han, R., & Mishra, S. (2006). Secure code distribution in dynamically programmable wireless sensor networks. In *Proceedings of the Fifth International Conference on Information Processing in Sensor Networks* (pp. 292-300).

Deutsch, A. (2004). *Static verification of dynamic properties*. Retrieved from http://www.sigada.org/conf/sigada2003/SIGAda2003-CDROM

Deutsch, P. (1996). *DEFLATE compressed data format specification version 1.3*. Retrieved from http://tools.ietf.org/pdf/rfc1951.pdf

Diamos, G., Yalamanchili, S., & Duato, J. (2007). *STARS: A system for tuning and actively reconfiguring SoCs link*. Paper presented at the DAC Workshop on Diagnostic Services in NoCs.

Duato, J., Yalamanchili, S., & Ni, L. (2003). *Interconnection Networks: an Engineering Approach*. San Francisco: Morgan Kaufmann.

Du, D., & Hu, X. (2008). *Steiner Tree Problems in Computer Communication Networks*. New York: World Scientific. doi:10.1142/9789812791450

Dunkels, A., Finne, N., Eriksson, J., & Voigt, T. (2006). Run-time dynamic linking for reprogramming wireless sensor networks. In *Proceedings of the Fourth ACM Conference on Embedded Networked Sensor Systems* (pp. 15-28).

Dunkels, A., Gronvall, B., & Voigt, T. (2004). Contiki - a lightweight and flexible operating system for tiny networked sensors. In *Proceedings of the 29th Annual IEEE International Conference on Local Computer Networks* (pp. 455-462).

Duros, E., & Dabbous, W. (1996). *Handling of unidirectional links with OSPF*. Retrieved from http://tools.ietf.org/html/draft-ietf-ospf-unidirectional-link-00

Dwyer, M. B., Avrunin, G. S., & Corbett, J. C. (1999). Patterns in property specifications for finite-state verification. In *Proceedings of the 21st International Conference on Software Engineering* (pp. 411-420). New York, NY: ACM Press.

Eggers, S. J., & Katz, R. H. (1988). Evaluating the performance of four snooping cache coherency protocols (Tech. Rep. No. UCB/CSD-88-478). Berkeley, CA: University of California.

Egyed, A., & Gruenbacher, P. (2002). Automating requirements traceability: Beyond the record & replay paradigm. In *Proceedings of the 17th IEEE International Conference on Automated Software Engineering* (pp. 163-171). Washington, DC: IEEE Computer Society.

El-Far, I. K., & Whittaker, J. A. (2002). Model-based software testing. In Marciniak, J. J. (Ed.), *Encyclopedia of software engineering* (*Vol. 1*, pp. 825–837). New York, NY: John Wiley & Sons.

Elnozahy, E. N., Alvisi, L., Wang, Y., & Johnson, D. (2002). A survey of rollback-recovery protocols in message-passing systems. *ACM Computing Surveys*, *34*(3), 375–408. doi:10.1145/568522.568525

Elrabaa, M. E. S. (2006). An all-digital clock frequency capturing circuitry for NRZ data communications. In *Proceedings of the IEEE International Conference on Electronics, Circuits and Systems* (pp. 106-109).

Ernst, D., Kim, N. S., Das, S., Pant, S., Rao, R., Pham, T., et al. (2003). Razor: A low-power pipeline based on circuit-level timing speculation. In *Proceedings of the 36th Annual IEEE/ACM International Symposium on Microarchitecture* (pp. 7-18).

Faiz-ul-Hassan. Cheng, B., Vanderbauwhede, W., & Rodriguez, F. (2009). Impact of device variability in the communication structures for future synchronous SoC designs. In *Proceedings of the International Symposium on System-on-Chip* (pp. 68-72).

Felser, M. (2005, June). Real-Time Ethernet – Industry Prospective. *Proceedings of the IEEE, 93*(6), 1118–1129. doi:10.1109/JPROC.2005.849720

Ferdinand, C., Heckmann, R., & Franzel, B. (2007). Static memory and timing analysis of embedded systems code. In *Proceedings of the 3rd European Symposium on Verification and Validation of Software Systems*, Eindhoven, The Netherlands (pp. 153-163).

Ferrari, A., Fantechi, A., Tempestini, M., & Zingoni, N. (2009). Modeling guidelines for code generation in the railway signaling context. In *Proceedings of the 1st NASA Formal Methods Symposium*, Moffet Field, CA (pp. 166-170).

Ferrari, A., Grasso, D., Magnani, G., Fantechi, A., & Tempestini, M. (2010, September). The Metrô Rio ATP case study. In S. Kowalewski & M. Roveri (Eds.), *Proceedings of the 15th International Workshop on Formal Methods for Industrial Critical Systems*, Antwerp, Belgium (LNCS 6371, pp. 1-16).

Fiorin, L., Palermo, G., & Silvano, C. (2009, April). MPSoCs Run-time Monitoring Through Networks-on-Chip. In *Proceedings of the Design, Automation & Test in Europe Conference and Exhibition (DATE '09)* (pp. 558-561).

Fonseca, P. (2008). SDL, a graphical language useful to describe social simulation models. In *Proceedings of the 2nd Workshop on Social Simulation and Artificial Societies Analysis (SSASA'08)*. Retrieved from http://ceur-ws.org/Vol-442/p4_Fonseca.pdf

Forkel, I., Schinnenburg, M., & Ang, M. (2004, September). Generation of Two-Dimensional Correlated Shadowing for Mobile Radio Network Simulation. In *Proceedings of the 7th International Symposium on Wireless Personal Multimedia Communications (WPMC)*, Abano Terme (Padova), Italy (p. 5).

Fosdick, L. D., & Osterweil, L. J. (1976). Data flow analysis in software reliability. *ACM Computing Surveys, 8*(3), 305–330. doi:10.1145/356674.356676

Fraser, G., & Wotawa, F. (2008). Using model-checkers to generate and analyze property relevant test-cases. *Software Quality Journal, 16*, 161–183. doi:10.1007/s11219-007-9031-6

Frazier, G. (2009). *SystemC save and restore part 2 - Advanced usage.* Retrieved from http://www.cadence.com/community/blogs/sd/archive/2009/03/09/systemc-save-and-restore-part-2-advanced-usage.aspx

Ganesan, D., Estrin, D., Woo, A., Culler, D., Krishnamachari, B., & Wicker, S. (2002). *Complex behavior at scale: An experimental study of low-power wireless sensor networks* (Tech. Rep. No. CSD-TR 02-0013). Los Angeles, CA: University of California at Los Angeles.

Garey, M. R., Graham, R. L., & Johnson, D. (1977, June). The complexity of computing Steiner minimal trees. *SIAM Journal on Applied Mathematics, 31*(4), 835–859. doi:10.1137/0132072

Gaston, C., & Seifert, D. (2005). Evaluating coverage based testing. In M. Broy, B. Jonsson, J.-P. Katoen, M. Leucker, & A. Pretschner (Eds.), *Proceedings of the International Conference on Model-Based Testing of Reactive Systems* (LNCS 3472, pp. 293-322).

Gilabert, F., Ludovici, D., Medardoni, S., Bertozzi, D., Benini, L., & Gaydadjiev, G. N. (2009). Designing regular network-on-chip topologies under technology, architecture and software constraints. In *Proceedings of the International Conference on Complex, Intelligent and Software Intensive Systems* (pp. 681-687).

Gill, A. (1962). *Introduction to the theory of finite-state machines.* New York, NY: McGraw-Hill.

Gillet, M. (2008). Hardware/software co-simulation for conformance testing of embedded networks. In *Proceedings of the 6th Seminar of Finnish-Russian University Cooperation in Telecommunications (FRUCT) Program.* Retrieved October 31, 2008, from http://fruct.org/index.php?option=com_content&view=article&id=68&Itemid=73

Gipper, J. (2007). SystemC the SoC system-level modeling language. *Embedded computing Design.* Retrieved from www.embedded-computing.com/pdfs/OSP2.May07.pdf

Glover, F., & Laguna, M. (1997). *Tabu Search.* Dordrecht, The Netherlands: Kluwer Academic Publisher.

Gonzales, T. F. (Ed.). (2007). *Handbook of Approximation Algorithms and Metaheuristics*. Boca Raton, FL: Chapman & Hall CRC.

Goossens, K., Dielissen, J., & Radulescu, A. (2005, September-October). Æthereal Network on Chip: Concepts, Architectures and Implementations. *IEEE Design & Test of Computers*, *22*(5), 414–421. doi:10.1109/MDT.2005.99

Goyal, P., Guo, X., & Vin, H. M. (1996). A hierarchical CPU scheduler for multimedia operating systems. In Proceedings of the Symposium on Operating Systems Design and Implementation (pp. 107-121). New York, NY: ACM Press.

Grasso, D., Fantechi, A., Ferrari, A., Becheri, C., & Bacherini, S. (2010). Model based testing and abstract interpretation in the railway signaling context. In *Proceedings of the Third International Conference on Software Testing, Verification and Validation, Paris, France* (pp. 103-106). Washington, DC: IEEE Computer Society.

Grimpel, E., Timmermann, B., Fandrey, T., Biniasch, R., & Oppenheimer, F. (2002). SystemC Object-Oriented Extensions and Synthesis Features. In *Proceedings of the European Electronic Chips & Systems design Initiative*. Retrieved from www.ecsiassociation.org/ecsi/projects/odette/files/fdl2002.pdf

Gulavani, B. S., & Rajamani, S. K. (2006). Counterexample driven refinement for abstract interpretation. In H. Hermanns & J. Palsberg (Eds.), *Proceedings of the 12th International Conference on Tools and Algorithms for the Construction and Analysis of Systems*, Wien, Austria (LNCS 3920, pp. 474-488).

Hardavellas, N., Somogyi, S., Wenisch, T. F., Wunderlich, R. E., Chen, S., & Kim, J. (2004). SimFlex: A fast, accurate, flexible full-system simulation framework for performance evaluation of server architecture. *ACM Special Interest Group on Measurement and Evaluation. Performance Evaluation Review*, *31*(4), 31–34. doi:10.1145/1054907.1054914

Harel, D., & Marelly, R. (2002). Playing with time: On the specification and execution of time-enriched LSCs. In *Proceedings of the 10th IEEE International Symposium on Modeling, Analysis, and Simulation of Computer and Telecommunications Systems* (pp. 193-202). Washington, DC: IEEE Computer Society.

Harel, D., & Thiagarajan, P. S. (2003). *Message sequence charts*. Retrieved from http://www.comp.nus.edu.sg/~thiagu/public_papers/surveymsc.pdf

Harel, D., Lachover, H., Naamad, A., Pnueli, A., Politi, M., Sherman, R., et al. (1988). Statemate: A working environment for the development of complex reactive systems. In *Proceedings of the 10th International Conference on Software Engineering*, Raffles City, Singapore (pp. 396-496). Washington, DC: IEEE Computer Society.

Harmatos, J., Jüttner, A., & Szentesi, A. (1999, September). Cost-based UMTS Transport Network Topology Optimisation. In *Proceedings of the International Conference on Computer Communication (ICCC'99)*, Tokyo, Japan.

Haroud, M., & Blazevic, L. (2006). HW accelerated Ultra Wide Band MAC protocol using SDL and SystemC. In *Proceedings of the Fourth IEEE International Conference on Pervasive Computing and Communications Workshops (PERCOMW'06)*. Retrieved from http://fmv.jku.at/papers/HaroudBlazevicBiere-RAWCON04.pdf

Heath, S. (2003). *Embedded Systems Design* (2nd ed.). New York: Newnes.

Hernández, C., Roca, A., Silla, F., Flich, J., & Duato, J. (2010). Improving the performance of GALS-based NoCs in the presence of process variation. In *Proceedings of the Fourth ACM/IEEE International Symposium on Networks-on-Chip* (pp. 35-42).

Hessel, A., & Pettersson, P. (2006). Model-based testing of a WAP gateway: An industrial study. In L. Brim & M. Leucker (Eds.), *Proceedings of the 11th International Workshop on Formal Methods for Industrial Critical Systems*, Bonn, Germany (LNCS 4346, pp. 116-131).

Hierons, R. M., Bowen, J. P., & Harman, M. (Eds.). (2008). *Formal methods and testing* (LNCS 4949, pp. 1-38). Berlin, Germany: Springer-Verlag.

Hierons, R. M., Bogdanov, K., Bowen, J. P., Cleaveland, R., Derrick, J., & Dick, J. (2009). Using formal specifications to support testing. *ACM Computing Surveys*, *41*(2), 1–9. doi:10.1145/1459352.1459354

Hill, J. W., Szewczyk, R., Woo, A., Hollar, S., Culler, D., & Pister, K. (2000). System architecture directions for networked sensors. *SIGPLAN Notes*, *35*(11), 93–104. doi:10.1145/356989.356998

Hoare, C. A. R. (1985). *Communicating sequential processes*. Upper Saddle River, NJ: Prentice Hall.

Hollis, S. J., & Jackson, C. (2009, September). When does network-on-chip bypassing make sense? In *Proceedings of the 22nd IEEE International SOCC Conference* (pp. 143-146). Washington, DC: IEEE Computer Society.

Höppner, S., Walter, D., Eisenreich, H., & Schüffny, R. (2010). Efficient compensation of delay variations in high-speed network-on-chip data links. In *Proceedings of the International Symposium on System on Chip* (pp. 55-58).

Howden, W. E. (1975). Methodology for the generation of program test data. *IEEE Transactions on Computers*, *24*(5), 554–560. doi:10.1109/T-C.1975.224259

Hui, J. W. (2005). *Deluge 2.0 - TinyOS network programming*. Retrieved from http://www.cs.berkeley.edu/~jwhui/deluge/deluge-manual.pdf

Hui, J. W., & Culler, D. (2004). The dynamic behavior of a data dissemination protocol for network programming at scale. In *Proceedings of the 2nd International Conference on Embedded Networked Sensor Systems* (pp. 81-94).

Hu, J., & Marculescu, R. (2005). Energy- and performance-aware mapping for regular Noc architectures. *IEEE Transactions on Computer-Aided Design of Integrated Circuits and Systems*, *24*(4).

IBM. (2009). *SDL Suite and TTCN Suite Help*. IBM Rational SDL and TTCN Suite.

IEEE Standards Association. (2008). *Part 15.4: Wireless medium access control (MAC) and physical layer (PHY) specifications for low-rate wireless personal area networks (WPANs)*. Retrieved from http://standards.ieee.org/getieee802/download/802.15.4a-2007.pdf

Ienne, P., & Leupers, R. (Eds.). (2006). *Customizable embedded processors: Design technologies and applications (systems on silicon)*. San Francisco, CA: Morgan Kaufmann.

Institute of Electrical and Electronics Engineers, Inc. (IEEE). (2005). *IEEE 1666: Standard, open SystemC language reference manual*. Washington, DC: Institute of Electrical and Electronics Engineers, Inc.

International Telecommunication Union. (2002). *Recommendation Z.100. Specification and Description Language (SDL)*. Geneva, Switzerland: ITU.

ITRS. (2007). *International technology roadmap for semiconductors*. Retrieved from http://www.itrs.net/Links/2007ITRS/Home2007.htm

ITU. (1999). *Series Z: Languages and general software aspects for telecommunication systems: Formal Description Techniques (FDT) - Message Sequence Chart*. Retrieved from http://www.itu.int/ITU-T/studygroups/com17/languages/Z120.pdf

ITU-T. (2000). *Specification and Description Language (SDL)*. Retrieved from http://www.itu.int

Jackson, C., & Hollis, S. J. (2011). A deadlock-free routing algorithm for dynamically reconfigurable networks-on-chip. *Microprocessors and Microsystems*, *35*(2), 139–151. doi:10.1016/j.micpro.2010.09.004

Jantsch, A. (2004). *Modeling Embedded Systems and SoCs*. San Francisco: Morgan Kaufmann Publishers.

Jantsch, A., & Tenhunen, H. (Eds.). (2003). *Networks on Chip*. Boston, MA: Kluwer Academic.

JiST. (2005). *Java in simulation time/scalable wireless ad hoc network simulator*. Retrieved from http://jist.ece.cornell.edu/

Johnson, B. W. (1989). *Design and Analysis of Fault-Tolerant Digital Systems*. Boston: Addison-Wesley.

Johnson, D. B., & Maltz, D. A. (1996). Dynamic source routing in ad hoc wireless networks. *Mobile Computing*, *5*, 153–181. doi:10.1007/978-0-585-29603-6_5

Jose, A. P., Patounakis, G., & Shepard, K. L. (2005). Near speed-of-light on-chip interconnects using pulsed current-mode signalling. In *Proceedings of the Digest of Technical Papers Symposium on VLSI Circuits* (pp. 108-111).

Jozawa, T., Huang, L., Sakai, E., Takeuchi, S., & Kasslin, M. (2006). Heterogeneous Co-simulation with SDL and SystemC for Protocol Modeling. In *Proceedings of the IEEE Radio and Wireless Symposium, 2006*, 603–606. Retrieved from http://research.nokia.com/node/5789. doi:10.1109/RWS.2006.1615229

Jung, Y., Kim, J., Shin, J., & Yi, K. (2005). Taming false alarms from a domain-unaware c analyzer by a Bayesian statistical post analysis. In C. Hankin & I. Siveroni (Eds.), *Proceedings of the 12th International Static Analysis Symposium*, London, UK (LNCS 3672, pp. 203-217).

Juntunen, J., Kuorilehto, M., Kohvakka, M., Kaseva, V., Hännikäinen, M., & Hämäläinen, T. (2006). WSN API: Application programming interface for wireless sensor networks. In *Proceedings of the IEEE 17th International Symposium on Personal, Indoor and Mobile Radio Communications* (pp. 1-5).

Kakoee, M. R., Loi, I., & Benini, L. (2010). A new physical routing approach for robust bundled signaling on NoC links. In *Proceedings of the 20th Symposium on the Great Lakes* (pp. 3-8).

Kamal, R. (2008). *Embedded systems: architecture, programming and design* (2nd ed.). New Delhi, India: Tata McGraw-hill Publishing Company Limited.

Kamsties, E., Reuys, A., Pohl, K., & Reis, S. (2004). Testing variabilities in use case models. In F. van der Linden (Ed.), *Proceedings of the 5th International Workshop on Software Product-Family Engineering* (LNCS 3014, pp.6-18).

Kantz, F., Ruschival, T., Nenninger, P., & Streitferdt, D. (2009). Testing with large parameter sets for the development of embedded systems in the automation domain. In *Proceedings of the 2nd International Workshop on Component-Based Design of Resource-Constrained Systems at the 33rd Annual IEEE International Computers, Software and Applications Conference*, Seattle, WA (pp. 504-509).

Karabegov, A., & Ter-Mikaelyan, T. (1993). *Introduction to the SDL language*. Moscow, Russia: Radio and communication.

Karp, B., & Kung, H. T. (2000). Greedy perimeter stateless routing for wireless networks. In *Proceedings of the 6th Annual International Conference on Mobile Computing and Networking*, Boston, MA.

Kenyon, C., Kornfeld, A., Kuhn, K., Liu, M., Maheshwari, A., Shih, W. et al. (2008). Managing process variation in Intel's 45nm CMOS technology. *Intel Technology Journal, 12*(2).

Kihara, M., Ono, S., & Eskelinen, P. (2003). *Digital clocks for synchronization and communications*. Norwood, MA: Artech House.

Kim, D., Toh, C. K., & Chou, Y. (2000). RODA: A new dynamic routing protocol using dual paths to support asymmetric links in mobile ad hoc networks. In *Proceedings of the 9th International Conference on Computer Communications and Networks*, Las Vegas, NV (pp. 4-8).

Kim, Y. J., Govidan, R., Karp, B., & Shenker, S. (2004). Practical and robust geographic routing in wireless networks. In *Proceedings of the 2nd International Conference on Embedded Networked Sensor Systems*, Baltimore, MD (pp. 295-296).

Kim, Y., Lee, J., Han, H., & Choea, K. M. (2010). Filtering false alarms of buffer overflow analysis using SMT solvers. *Information and Software Technology, 52*(2), 210–219. doi:10.1016/j.infsof.2009.10.004

Kinniment, D. J. (2007). *Synchronization and arbitration in digital systems*. Chichester, UK: John Wiley & Sons. doi:10.1002/9780470517147

Kirkpatrick, S., Gelatt, C. D. Jr, & Vecchi, M. P. (1983). Optimization by simulated annealing. *Science, 220*, 671–680. doi:10.1126/science.220.4598.671

Kloos, J., & Eschbach, R. (2009). Generating system models for a highly configurable train control system using a domain-specific language: A case study. In *Proceedings of the 5th Workshop on Advances in Model Based Testing*, Denver, CO.

Kohout, P., Ganesh, B., & Jacob, B. (2003). Hardware support for real-time operating systems. In Proceedings of the IEEE/ACM/IFIP International Conference on Hardware/Software Codesign and System Synthesis (pp. 45-51). New York, NY: ACM Press.

Kraemer, S., Leupers, R., Petras, D., & Philipp, T. (2009). A checkpoint/restore framework for SystemC-based virtual platforms. In *Proceedings of the International Conference on System-on-Chip* (pp. 161-167). Washington, DC: IEEE Computer Society.

Kremenek, T., & Engler, D. R. (2003). Z-Ranking: Using statistical analysis to counter the impact of static analysis approximations. In R. Cousot (Ed.), *Proceedings of the 10th International Static Analysis Symposium*, San Diego, CA (LNCS 2694, pp. 295-315).

Krstic, M., Grass, E., Gurkaynak, F. K., & Vivet, P. (2007). Globally asynchronous, locally synchronous circuits: Overview and outlook. *IEEE Design & Test of Computers*, *24*, 430–441. doi:10.1109/MDT.2007.164

Kruskal, J. B. (1956, February). On the Shortest Spanning Subtree of a Graph and the Traveling Salesman Problem. *Proceedings of the American Mathematical Society*, *7*(1), 48–50. doi:10.1090/S0002-9939-1956-0078686-7

Kuhn, D., Wallace, D., & Gallo, A.M., J. (2004). Software fault interactions and implications for software testing. *IEEE Transactions on Software Engineering*, *30*(6), 418–421. doi:10.1109/TSE.2004.24

Kulkarni, S., & Wang, L. (2005). MNP: Multihop network reprogramming service for sensor networks. In *Proceedings of the 25th IEEE International Conference on Distributed Computing Systems* (pp. 7-16).

Kumar, A., Peh, L.-S., Kundu, P., & Jha, N. K. (2007). Express virtual channels: Towards the ideal interconnection fabric. *SIGARCH Computer Architecture News*, *35*(2), 150–161. doi:10.1145/1273440.1250681

Kuorilehto, M., Alho, T., Hännikäinen, M., & Hämäläinen, T. D. (2007). SensorOS: A new operating system for time critical WSN applications. In *Proceedings of the 7th International Conference on Embedded Computer Systems: Architectures, Modeling, and Simulation* (pp. 431-442).

Kuorilehto, M., Kohvakka, M., Suhonen, J., Hämäläinen, P., Hännikäinen, M., & Hämäläinen, T. D. (2007). *Ultra-low energy wireless sensor networks in practice: Theory, realization and deployment*. New York, NY: John Wiley & Sons. doi:10.1002/9780470516805

Langendoen, K., Baggio, A., & Visser, O. (2006). Murphy loves potatoes: Experiences from a pilot sensor network deployment in precision agriculture. In *Proceedings of the 20th International Parallel and Distributed Processing Symposium* (p. 8).

Larsen, K. G., Mikucionis, M., Nielsen, B., & Skou, A. (2005). Testing real-time embedded software using UPPAAL-TRON - an industrial case study. In *Proceedings of the 5th ACM International Conference on Embedded Software*, Jersey City, NJ (pp. 299-306). New York, NY: ACM Press.

Lee, K., Lee, S.-J., & Yoo, H.-J. (2006). Low-power network-on-chip for high-performance soc design. *IEEE Transactions on Very Large Scale Integration Systems*, *14*, 148–160. doi:10.1109/TVLSI.2005.863753

Lehtonen, T. (2009) *On Fault Tolerance Methods for Networks-on-Chip*. Unpublished doctoral dissertation, Turku Centre for Computer Science (TUCS), Turku, Finland.

Levis, P., Patel, N., Culler, D., & Shenker, S. (2004). Trickle: A self-regulating algorithm for code propagation and maintenance in wireless sensor networks. In *Proceedings of the 1st Conference on Networked Systems Design and Implementation* (p. 2).

Levis, P., & Culler, D. (2002). Maté: A tiny virtual machine for sensor networks. *SIGOPS Operating Systems Review*, *36*(5), 85–95. doi:10.1145/635508.605407

Limberg, T., Winter, M., Bimberg, M., Klemm, R., Tavares, M., Ahlendorf, H., et al. (2009). A heterogeneous MPSoC with hardware supported dynamic task scheduling for software defined radio. Paper presented at the DAC/ISSCC Student Design Contest, San Francisco, CA.

Lippett, M. (2004). An IP core based approach to the on-chip management of heterogeneous SoCs. In Proceedings of the IP Based SoC Design Forum & Exhibition.

Litzkow, M., Tannenbaum, T., Basney, J., & Livney, M. (1997). *Checkpoint and migration of UNIX processes in the condor distributed processing system* (Tech. Rep. No. 1346). Madison, WI: University of Wisconsin Madison.

Liu, R. P., Rosberg, Z., Collings, I. B., Wilson, C., Dong, A., & Jha, S. (2008). Overcoming radio link asymmetry in wireless sensor networks. In *Proceedings of the International Symposium on Personal, Indoor and Mobile Radio Communications*, Cannes, France (pp. 1-5).

Loghi, M., Poncino, M., & Benini, L. (2006). Cache coherence tradeoffs in shared-memory MPSoCs. *ACM Transactions on Embedded Computing Systems*, *5*(2), 383–407. .doi:10.1145/1151074.1151081

Loi, I., Angiolini, F., & Benini, L. (2008). Developing mesochronous synchronizers to enable 3D NoCs. *Design, Automation and Test in Europe*, 1414-1419.

Ltd, A. R. M. (2010). AMBA on-chip connectivity. Retrieved from http://www.arm.com/products/system-ip/interconnect/

Ludovici, D., Strano, A., Bertozzi, D., Benini, L., & Gaydadjiev, G. N. (2009). Comparing tightly and loosely coupled mesochronous synchronizers in a NoC switch architecture. In *Proceedings of the 3rd ACM/IEEE International Symposium on Networks-on-Chip* (pp. 244-249).

Ludovici, D., Strano, A., Gaydadjiev, G. N., Benini, L., & Bertozzi, D. (2010). Design space exploration of a mesochronous link for cost-effective and flexible GALS NOCs. *Design, Automation & Test in Europe*, 679-684.

MacDonald, J. T., & Roberson, D. A. (2007). Spectrum occupancy estimation in wireless channels with asymmetric transmitter powers. In *Proceedings of the 2nd International Conference on Cognitive Radio Oriented Wireless Networks and Communications*, Orlando, FL (pp. 245-249).

Magnusson, P. S., Christensson, M., Eskilson, J., Forsgren, D., Hållberg, G., & Högberg, J. (2002). Simics: A full system simulation platform. *IEEE Computer, 35*(2), 50–58.

Marescaux, T., Rångevall, A., Nollet, V., Bartic, A., & Corporaal, H. (2005, September). Distributed Congestion Control for Packet Switched Networks on Chip. In *Proceedings of the International Conference on Parallel Computing: Current & Future Issues of High-End Computing (ParCo 2005)* (pp. 761-768).

Marina, M. K., & Das, S. K. (2002). Routing performance in the presence of unidirectional links in multihop wireless networks. In *Proceedings of the 3rd International Symposium on Mobile Ad Hoc Networking and Computing*, Lausanne, Switzerland (pp. 12-23).

Matlab Automotive Advisory Board. (2007). *Control algorithm modeling guidelines using Matlab, Simulink and Stateflow, version 2.0.* Retrieved from http://www.mathworks.com/automotive/standards/maab.html;jsessionid=WTtvNL6XcCkNpFBMJQsPdQ6tnhcNsPQXJx9Jxffh1rLDnTBKKh7w!892165066

McGregor, J. D. (2001). *Testing a software product line* (Tech. Rep. CMU/SEI-2001-TR-022). Pittsburgh, USA, Carnegie Mellon University

Meenderinck, C., Azevedo, A., Alvarez, M., Juurlink, B., & Ramirez, A. (2008). Parallel scalability of H.264. In Proceedings of the Workshop on Programmability Issues for Multi-Core Computers.

Merdadoni, S., Lajolo, M., & Bertozzi, D. (2008). Variation tolerant NoC design by means of self-calibrating links. *Design, Automation and Test in Europe*, 1402-1407.

Michelogiannakis, G., & Dally, W. J. (2009). Router designs for elastic buffer on-chip networks. In *Proceedings of the Conference on High Performance Computing Networking, Storage and Analysis* (pp. 1-10). New York, NY: ACM Press.

Microchip Technology. (2008). *PIC18F8722 product page*. Retrieved from http://www.microchip.com/

Microchip Technology. (2009). *MPLAB C compiler for PIC18MCUs*. Retrieved from http://www.microchip.com/

Miller, C., & Poellabauer, C. (2008). PALER: A reliable transport protocol for code distribution in large sensor networks. In *Proceedings of the 5th Annual IEEE Communications Society Conference on Sensor, Mesh and Ad Hoc Communications and Network* (pp. 206-214).

Miller, K., Morell, L., Noonan, R., Park, S., Nicol, D., & Murril, B. (1992). Estimating the probability of failure when testing reveals no failures. *IEEE Transactions on Software Engineering, 18*(1), 33–43. doi:10.1109/32.120314

Miller, S. P., Whalen, M. W., & Cofer, D. D. (2010). Software model checking takes off. *Communications of the ACM, 53*(2), 58–64. doi:10.1145/1646353.1646372

Misra, R., & Mandal, C. R. (2005). Performance comparison of AODV/DSR on-demand routing protocols for ad-hoc networks in constrained situation. In *Proceedings of the IEEE International Conference on Personal Wireless Communications*, New Delhi, India (pp. 86-89).

Mohagheghi, P., & Dehlen, V. (2010). Where is the proof? A review of experiences from applying MDE in industry. In I. Schieferdecker & A. Hartman (Eds.), *Proceedings of the International Conference on Model Based Testing of Reactive Systems* (LNCS 5095, pp. 432-443).

Mondal, M., Ragheb, T., Wu, X., Aziz, A., & Massoud, Y. (2007). Provisioning on-chip networks under buffered RC interconnect delay variations. In *Proceedings of the 8th International Symposium on Quality Electronic Design* (pp. 873-878).

Montón, M., Engblom, J., & Burton, M. (2009). Checkpoint and restore for SystemC models. In *Proceedings of the Forum on Specification & Design Languages*, Sophia Antipolis, France.

Moore, S. W., & Taylor, G. S. (2000). Self calibrating clocks for globally asynchronous locally synchronous systems. In *Proceedings of the IEEE International Conference on Computer Design* (pp. 73-78).

Moore, S., & Greenfield, D. (2008). The next resource war: Computation vs. communication. In *Proceedings of the International Workshop on System Level Interconnect Prediction* (pp. 81-86). New York, NY: ACM Press.

Mouhoub, R., & Hammami, O. (2006, October). NoC Monitoring Hardware Support for Fast NoC Design Space Exploration and Potential NoC Partial Dynamic Reconfiguration. In *Proceedings of the International Symposium on Industrial Embedded Systems (IES '06)*.

Mtt L., Suhonen, J., Laukkarinen, T., Hmlinen, T., & Hnnikinen, M. (2010). Program image dissemination protocol for low-energy multihop wireless sensor networks. In *Proceedings of the International Symposium on System on Chip* (pp. 133-138).

Mukherjee, B., Banerjee, D., Ramamurthy, S., & Mukherjee, A. (1996). Some principles for designing a wide-area WDM optical network. [TON]. *IEEE/ACM Transactions on Networking*, *4*(5), 684–696. doi:10.1109/90.541317

Mukhtar, H., Kim, B. W., Kim, B. S., & Joo, S.-S. (2009). An efficient remote code update mechanism for wireless sensor networks. In *Proceedings of the IEEE Military Communications Conference* (pp. 1-7).

Mullins, R. (2006, Nov.). Minimising dynamic power consumption in on-chip networks. In *Proceedings of the International Symposium on System-on-Chip* (pp. 1-4).

Murali, S., Tamhankar, R., Angiolini, F., Pulling, A., Atienza, D., Benini, L., & De Micheli, G. (2006). Comparison of a timing-error tolerant scheme with a traditional re-transmission mechanism for networks on chips. In *Proceedings of the International Symposium on System-on-Chip* (pp. 1-4).

Murali, S., Theocharides, T., Vijaykrishnan, N., Irwin, M., Benini, L., & De Micheli, G. (2005, September-October). Analysis of Error Recovery Schemes for Networks on Chips. *IEEE Design & Test of Computers*, *22*(5), 434–442. doi:10.1109/MDT.2005.104

Murtaza, Z., Khan, S. A., Rafique, A., Bajwa, K. B., & Zaman, U. (2006). Silicon real time operating system for embedded DSPs. In Proceedings of the International Conference on Emerging Technologies (pp. 188-191). Washington, DC: IEEE Computer Society.

Musa, J. D. (1993). Operational profiles in software-reliability engineering. *IEEE Software*, *10*(2), 14–32. doi:10.1109/52.199724

Muscholl, A., Peled, D., & Su, Z. (2005). Deciding properties of message sequence charts. In M. Nivat (Ed.), *Proceedings of the 1st International Conference on Foundations of Software Science and Computation Structure* (LNCS 1378, pp. 226-242).

Nácul, A. C., Regazzoni, F., & Lajolo, M. (2007). Hardware scheduling support in SMP architectures. In Proceedings of the Conference on Design, Automation and Test in Europe (pp. 642-647).

Nakano, T., Utama, A., Itabashi, M., Shiomi, A., & Imai, M. (1995). Hardware implementation of a real-time operating system. In Proceedings of the TRON Project International Symposium (pp. 34-42). Washington, DC: IEEE Computer Society.

Naslavsky, L., Alspaugh, T. A., Richardson, D. J., & Ziv, H. (2005). Using scenarios to support traceability. In *Proceedings of the 3rd International Workshop on Traceability in Emerging Forms of Software Engineering* (pp. 25-30). New York, NY: ACM Press.

Naslavsky, L., Ziv, H., & Richardson, D. J. (2007). Towards traceability of model-based testing artefacts. In *Proceedings of the 3rd International Workshop on Advances in Model-Based Testing* (pp. 105-114). New York, NY: ACM Press.

Nemydrov, V., & Martin, G. (2004). *Systems-on-chip. Design and evaluation problems*. Moscow, Russia: Technosphera.

Nicopoulos, C., Srinivasan, S., & Yanamandra, A, Dongkook Park, Narayanan, V., Das, C. R., & Irwin, M. J. (2010). On the effects of process variation in network-on-chip architectures. *IEEE Transactions on Dependable and Secure Computing*, *7*(3), 240–254. doi:10.1109/TDSC.2008.59

Nilsson, E., Millberg, M., Öberg, J., & Jantsch, A. (2003, March). Load Distribution with the Proximity Congestion Awareness in a Network on Chip. In *Proceedings of the Design Automation and Test Europe (DATE)* (pp. 1126-1127).

Nollet, V., Marescaux, T., & Verkest, D. (2004, June). Operating-system Controlled Network on Chip. In *Proceedings of the Design Automation Conference (DAC '04)* (pp. 256-259).

Nordic Semiconductors. (2007). *nRF24L01 product specification*. Retrieved from http://www.nordicsemi.com/

Nordström, S., & Asplund, L. (2007). Configurable hardware/software support for single processor real-time kernels. In Proceedings of the International Symposium on System-on-Chip (pp. 1-4).

Nurmi, J., Tenhunen, H., Isoaho, J., & Jantsch, A. (Eds.). (2004). *Interconnect-Centric Design for Advanced SoC and NoC*. Dordrecht, The Netherlands: Kluwer Academic Publishers.

Object Management Group (OMG). (2000). *Object constraint language specification: OMG unified modeling language specification, version* 1.3. Retrieved from http://www.omg.org/spec/UML/1.3

Object Management Group. (2010). *OMG systems modeling language (OMG SysML™), Version 1.2*. Retrieved from http://www.omg.org/spec/SysML/1.2

Object Management Group. (2010). *OMG unified modeling language™(OMG UML), Infrastructure, Version 2.3*. Retrieved from http://www.omg.org/spec/UML/2.3/Infrastructure

Ogras, U. Y., & Marculescu, R. (2006). It's a small world after all: Noc performance optimization via long-range link insertion. *IEEE Transactions on Very Large Scale Integration Systems*, *14*(7), 693–706. doi:10.1109/TVLSI.2006.878263

Olenev, V. (2009). Different approaches for the stacks of protocols SystemC modeling analysis. In *Proceedings of the Saint-Petersburg University of Aerospace Instrumentation scientific conference* (pp. 112-113). Saint-Petersburg, Russia: Saint-Petersburg University of Aerospace Instrumentation (SUAI).

Olenev, V., Onishenko, L., & Eganyan, A. (2008). Connections in SystemC Models of Large Systems. In *Proceedings of the Saint-Petersburg University of Aerospace Instrumentation scientific student's conference* (pp. 98-99). Saint-Petersburg, Russia: Saint-Petersburg University of Aerospace Instrumentation (SUAI).

Olenev, V., Rabin, A., Stepanov, A., & Lavrovskaya, I. (2009). SystemC and SDL Co-Modeling Methods. In *Proceedings of the 6th Seminar of Finnish-Russian University Cooperation in Telecommunications (FRUCT) Program* (pp. 136-140). Saint-Petersburg, Russia: Saint-Petersburg University of Aerospace Instrumentation (SUAI).

Olimpiew, E. M., & Gomaa, H. (2005). Model-based testing for applications derived from software product lines. In *Proceedings of the 1st International Workshop on Advances in Model-Based Testing* (pp. 1-7).

Open SystemC Initiative (OSCI). (2005). *IEEE 1666™-2005 Standard for SystemC*. Retrieved from http://www.systemc.org

Orshansky, M., & Keutzer, K. (2002). A general probabilistic framework for worst case timing analysis. In *Proceedings of the 39th Design Automation Conference* (pp. 556- 561).

Paci, G., Bertozzi, D., & Benini, L. (2009). Effectiveness of adaptive supply voltage and body bias as post-silicon variability compensation techniques for full-swing and low-swing on-chip communication channels. *Design, Automation & Test in Europe*, 1404-1409.

Panades, I. M., Clermidy, F., Vivet, P., & Greiner, A. (2008). Physical implementation of the DSPIN network-on-chip in the FAUST Architecture. In *Proceedings of the International Symposium on Networks-on-Chip* (pp. 139-148).

Pan, Z., & Wells, B. E. (2008). Hardware supported task scheduling on dynamically reconfigurable SoC architectures. *IEEE Transactions on Very Large Scale Integration Systems*, *16*(11), 1465–1474. .doi:10.1109/TVLSI.2008.2000974

Paolucci, P. S., Jerraya, A. A., Leupers, R., Thiele, L., & Vicini, P. (2006). SHAPES: A tiled scalable software hardware architecture platform for embedded systems. In *Proceedings of the International Conference on Hardware/Software Codesign and System Synthesis* (pp. 167-172).

Paolucci, P. S., Kajfasz, P., Bonnot, P., Candaele, B., Maufroid, D., & Pastorelli, E. (2001). mAgic-FPU and MADE: A customizable VLIW core and the modular VLIW processor architecture description environment. *Computer Physics Communications*, *139*(1), 132–143. doi:10.1016/S0010-4655(01)00235-1

Park, S., Hong, D.-s., & Chae, S.-I. (2008). A hardware operating system kernel for multi-processor systems. *IEICE Electronics Express*, *5*(9), 296–302. .doi:10.1587/elex.5.296

Pasareanu, C. S., Schumann, J., Mehlitz, P., Lowry, M., Karsai, G., Nine, H., et al. (2009). Model based analysis and test generation for flight software. In *Proceedings of the 3rd IEEE International Conference on Space Mission Challenges for Information Technology*, Pasadena, CA (pp. 83-90). Washington, DC: IEEE Computer Society.

Patzold, M., & Nguyen, V. (2004, September). A spatial simulation model for shadow fading processes in mobile radio channels. In *Proceedings of the 15th IEEE International Symposium on Personal, Indoor and Mobile Radio Communications (PIMRC 2004)* (Vol. 3, pp. 1832-1838).

Peleska, J. (2009). The automation problems and their reduction to bounded model checking. In Drechsler, R. (Ed.), *Model-based testing of embedded control systems in the railway, avionic and automotive domain*. [Proceedings of the TuZ, 21 Workshop für Testmethoden und Zuverlässigkeit von Schaltungen und Systemen]

Perkins, C., Belding-Rower, E. M., & Das, S. (2003). *Ad-hoc on demand vector (AODV) routing*. Retrieved from http://www.ietf.org/rfc/rfc3561.txt

Peterson, J. L. (1981). *Petri net theory and the modeling of systems*. Upper Saddle River, NJ: Prentice Hall.

Pfaller, C. (2008). Requirements-based test case specification by using information from model construction. In *Proceedings of the 3rd International Workshop on Automation of Software Test* (pp. 7-16). New York, NY: ACM Press.

Pióro, M., & Mehdi, D. (2004). *Routing, Flow, and Capacity Design in Communication and Computer Networks*. San Francisco: Morgan Kaufmann Publishers.

Plank, J. S., Beck, M., Kingsley, G., & Li, K. (1995). Libckpt: Transparent checkpointing under Unix. In *Proceedings of the USENIX Technical Conference*, New Orleans, LA (pp. 213-224).

Pnueli, A. (1977). The temporal logic of programs. In *Proceedings of the 18th IEEE Annual Symposium on Foundations of Computer Science* (pp. 46-57). Washington, DC: IEEE Computer Society.

Post, H., & Sinz, C. Kaiser, A., & Gorges, T. (2008). Reducing false positives by combining abstract interpretation and bounded model checking. In *Proceedings of the 23rd IEEE/ACM International Conference on Automated Software Engineering*, L'Aquila, Italy (pp. 188-197). Washington, DC: IEEE Computer Society.

Prenninger, W., El-Ramly, M., & Horstmann, M. (2005). Case studies. In M. Broy, B. Jonsson, J. P. Katoen, M. Leucker, & A. Pretschner (Eds.), *Proceedings of the International Conference on Model Based Testing of Reactive Systems* (LNCS 3472, pp. 439-461).

Pretschner, A., & Phillips, J. (2005). Methodological issues in model-based testing. In M. Broy, B. Jonsson, J. P. Katoen, M. Leucker, & A. Pretschner (Eds.), *Proceedings of the International Conference on Model Based Testing of Reactive Systems* (LNCS 3472, pp. 281-291).

Pretschner, A., Prenninger, W., Wagner, S., Kühnel, C., Baumgartner, M., Sostawa, B., et al. (2005). One evaluation of model-based testing and its automation. In *Proceedings of the 27th International Conference on Software Engineering*, St. Louis, MO (pp. 392-401). New York, NY: ACM Press.

Pretschner, A., Slotosch, O., Aiglstorfer, A., & Kriebel, S. (2004). Model-based testing for real. *Software Tools for Technology Transfer*, *5*(2-3), 140–157. doi:10.1007/s10009-003-0128-3

Prim, R. (1957). Shortest connection networks and some generalizations. *The Bell System Technical Journal*, *36*, 1389–1401.

Prowell, S. (2005). Using Markov chain usage models to test complex systems. In *Proceedings of the 38th Annual Hawaii International Conference on System Sciences* (p. 318).

Prowell, S., & Poore, J. H. (2003). Foundations of sequence-based software specification. *IEEE Transactions on Software Engineering*, *29*(5), 417–429. doi:10.1109/TSE.2003.1199071

Prowell, S., Trammell, C., Linger, R., & Poore, J. (1999). *Cleanroom software engineering: Technology and process*. Reading, MA: Addison-Wesley.

Rabaey, J. M. (1996). *Digital Integrated Circuits: A Design Perspective*. Upper Saddle River, NJ: Prentice Hall, Inc.

Rahmani, M., Hillebrand, J., Hintermaier, W., Bogenberger, R., & Steinbach, E. (2007, May). A Novel Network Architecture for In-Vehicle Audio and Video Communication. In *Proceedings of the 2nd IEEE/IFIP International Workshop on Broadband Convergence Networks (BcN '07)* (pp. 1-12).

Rahmani, M., Steffen, R., Tappayuthpijarn, K., Steinbach, E., & Giordano, G. (2008, February). Performance Analysis of Different Network Topologies for In-vehicle Audio and Video Communication. In *Proceedings of the 4th International Telecommunication Networking Workshop on QoS in Multiservice IP Networks* (pp. 179-184).

Rahmani, M., Tappayuthpijarn, K., Krebs, B., Steinbach, E., & Bogenberger, R. (2009). Traffic

Rantala, V., Lehtonen, T., & Plosila, J. (2006, August). *Network on Chip Routing Algorithms* (Tech. Rep. No. 779). Turku, Finland: Turku Centre for Computer Science (TUCS).

Regehr, J., Reid, A., & Webb, K. (2005). Eliminating stack overflow by abstract interpretation. *ACM Transactions on Embedded Computing Systems*, *4*(4), 751–778. doi:10.1145/1113830.1113833

Reijers, N., & Langendoen, K. (2003). Efficient code distribution in wireless sensor networks. In *Proceedings of the 2nd ACM International Conference on Wireless Sensor Networks and Applications* (pp. 60-67).

Reijers, N., Halkes, G., & Langendoen, K. (2004). Link layer measurements in sensor networks. In *Proceedings of the 1st IEEE International Conference on Mobile Ad-hoc and Sensor Systems*, Fort Lauderdale, FL (pp. 224-234).

Reinhold, M. (2003). *Praxistauglichkeit von Vorgehensmodellen: Specification of large IT-systems – integration of requirements engineering and UML based on V-Model'97*. North Rhine-Westphalia, Germany: Shaker Verlag.

Reuys, A., Kamsties, E., Pohl, K., & Reis, S. (2005). Model-based system testing of software product families. In O. Pastor & J. Falcão e Cunha (Eds.), *Proceedings of the 17th International Conference on Advanced Information Systems Engineering* (LNCS 3520, pp. 519-534).

Rival, X. (2005). Understanding the origin of alarms in Astrée. In C. Hankin & I. Siveroni (Eds), *Proceedings of the 12th International Static Analysis Symposium*, London, UK (LNCS 3672, pp. 303-319).

Roman, E. (2002). *A survey of checkpoint/restart implementations* (Tech. Rep. No. LBNL-54942). Berkeley, CA: Lawrence Berkeley National Laboratory.

Roosta, T., Menzo, M., & Sastry, S. (2005). Probabilistic geographic routing in ad hoc and sensor networks. In *Proceedings of the Wireless Networks and Emerging Technologies*, Banff, AB, Canada.

Rosenblum, M., & Varadarajan, M. (1994). *SimOS: A fast operating system simulation environment* (Tech. Rep. No. CSL-TR-94-631). Stanford, CA: Stanford University.

Ruschival, T., Nenninger, P., Kantz, F., & Streitferdt, D. (2009). Test case mutation in hybrid state space for reduction of no-fault-found test results in the industrial automation domain. In *Proceedings of the 2nd International Workshop on Industrial Experience in Embedded Systems Design at the 33rd Annual IEEE International Computers, Software and Applications Conference*, Seattle, WA (pp. 528-533).

Ryser, J., Berner, S., & Glinz, M. (1998). *On the state of the art in requirements-based validation and test of software* (Tech. Rep. No. 12). Zurich, Switzerland: University of Zurich.

Sang, L., Arora, A., & Zhang, H. (2007). On exploiting asymmetric wireless links via one-way estimation. In *Proceedings of the 8th International Symposium on Mobile Ad Hoc Networking and Computing*, Montreal, QC, Canada (pp. 11-21).

Sankaran, S., Squyres, M. J., Barrett, B., & Lumsdaine, A. (2005). The LAM/MPI checkpoint/ RestartFramework: System-initiated checkpointing. *International Journal of High Performance Computing Applications*, *4*(19), 479–493. doi:10.1177/1094342005056139

Sarangi, S. R., Greskamp, B., Teodorescu, R., Nakano, J., Tiwari, A., & Torrellas, J. (2008). VARIUS: A model of process variation and resulting timing errors for microarchitects. *IEEE Transactions on Semiconductor Manufacturing*, *21*(1), 3–13. doi:10.1109/TSM.2007.913186

Schloer, R. (2001). *Symbolic timing diagrams: A visual formalism for model verification.* Unpublished doctoral dissertation, University of Oldenburg, Germany.

Schmid, H. (2008). *Hardware-in-the-loop Technologie: Quo Vadis?* [Tagungsband Simulation und Test in der Funktions- und Softwareentwicklung für die Automobilelektronik]. (pp. 195–202). Stuttgart, Germany: Verlag.

Schwiebert, L., & Jayasimha, D. N. (1993). Optimal fully adaptive wormhole routing for meshes. In *Proceedings of the ACM/IEEE Conference on Supercomputing* (pp. 782-791). New York, NY: ACM Press.

Seidel, H. (2006). *A task-level programmable processor.* Duisburg, Germany: WiKu-Verlag.

Shacham, H., Page, M., & Pfaff, B. (2004). On the effectiveness of address-space randomization. In *Proceedings of the 11th ACM Conference on Computer and Communications Security* (pp. 298-307).

Shafi, H., Bohrer, P. J., Phelan, J., Rusu, C. A., & Peterson, J. L. (2003). Design and validation of a performance and power simulator for PowerPC systems. *IBM Journal of Research and Development*, *47*(5), 641–651. doi:10.1147/rd.475.0641

Shalf, J., Kamil, S., Oliker, L., & Skinner, D. (2005). Analyzing ultra-scale application communication requirements for a reconfigurable hybrid interconnect. In *Proceedings of the ACM/IEEE Conference on Supercomputing* (p. 17). Washington, DC: IEEE Computer Society.

Shaping for Resource-Efficient In-Vehicle Communication. *IEEE Transactions on Industrial Informatics.*

Skiena, S. (2009). *GeoSteiner: Software for Computing Steiner Trees.*

Sloss, A., Symes, D., & Wright, C. (2004). *ARM system developer's guide: Designing and optimizing system software.* San Francisco, CA: Morgan Kaufmann.

Smith, M. H., Holzmann, G. J., & Etessami, K. (2001). Events and constraints: A graphical editor for capturing logic requirements of programs. In *Proceedings of the 5th IEEE International Symposium on Requirements Engineering* (pp. 14-22). Washington, DC: IEEE Computer Society.

Sommer, J., & Doumith, E. A. (2008, July). Topology Optimization of In-vehicle Multimedia Communication Systems. In *Proceedings of the First Annual International Symposium on Vehicular Computing Systems (ISVCS)*, Dublin, Ireland.

Sommer, J., Gunreben, S., Feller, F., Köhn, M., Mifdaoui, A., Saß, D., et al. (2010). Ethernet – a survey on its fields of application. *IEEE Communications Surveys & Tutorials, 12*(2).

Son, D., Helmy, A., & Krishnamachari, B. (2004). The effect of mobility-induced location errors on geographic routing in ad hoc networks: Analysis and improvement using mobility prediction. *IEEE Transactions on Mobile Computing*, 233–245. doi:10.1109/TMC.2004.28

Soteriou, V., & Peh, L.-S. (2004, October). Design-space exploration of power- aware on/off interconnection networks. In *Proceedings of the 22nd International Conference on Computer Design* (pp. 510-517). Washington, DC: IEEE Computer Society.

Soteriou, V., Wang, H., & Peh, L.-S. (2006). A statistical traffic model for on-chip interconnection networks. In *Proceedings of the 14th IEEE International Symposium on Modeling, Analysis, And Simulation* (pp. 104-116). Washington, DC: IEEE Computer Society.

Sparso, J., & Furber, S. (2001). *Principles of asynchronous circuit design - a systems perspective*. Dordrecht, The Netherlands: Kluwer Academic.

Srinivasan, K., & Levis, P. (2006). RSSI is under appreciated. In *Proceedings of the 3rd Workshop on Embedded Networked Sensors*, Cambridge, MA.

Srinivasan, K., Dutta, P., Tavakoli, A., & Levis, P. (2006). Understanding the causes of packet delivery success and failure in dense wireless sensor networks. In *Proceedings of the 4th International Conference on Embedded Networked Sensor Systems*, Boulder, CO (pp. 419-420).

Stargate. (n. d.). *Resource links*. Retrieved from http://platformx.sourceforge.net/Links/resource.html

Stathopoulos, T., Heidemann, J., Estrin, D., & SENSING, C. U. (2003). *A remote code update mechanism for wireless sensor networks*. Retrieved from http://www.isi.edu/~johnh/PAPERS/Stathopoulos03b.html

Stensgaard, M. B., & Sparsø, J. (2008). Renoc: A network-on-chip architecture with reconfigurable topology. In *Proceedings of the Second ACM/IEEE International Symposium on Networks-on-Chip* (pp. 55–64). Washington, DC: IEEE Computer Society.

Stepanov, A. (2009). Comparison of SDL and SystemC Languages applicability for the protocol stack modeling. In *Proceedings of the Saint-Petersburg University of Aerospace Instrumentation scientific student's conference* (pp. 76-80). Saint-Petersburg, Russia: Saint-Petersburg University of Aerospace Instrumentation (SUAI).

Stergiou, S., Angiolini, F., Carta, S., Raffo, L., Bertozzi, D., & De Micheli, G. (2005). Xpipes Lite: A synthesis oriented design library for networks on chips. In *Proceedings of the Conference on Design, Automation and Test in Europe* (Vol. 2, pp. 1188-1193).

Stine, J. E., Castellanos, I., Wood, M., Henson, J., Love, F., Davis, W. R., et al. (2007). Freepdk: An open-source variation-aware design kit. In *Proceedings of the IEEE International Conference on Microelectronic Systems Education* (pp. 173-174). Washington, DC: IEEE Computer Society.

Storer, J. A., & Szymanski, T. G. (1982). Data compression via textual substitution. *Journal of the ACM, 29*(4), 928–951. doi:10.1145/322344.322346

Streitferdt, D., Nenninger, P., Bilich, C., Kantz, F., Bauer, T., & Eschbach, R. (2008). Model-based testing in the automation domain, safety enabled. In *Proceedings of the 1st Workshop on Model-based Testing in Practice*, Berlin, Germany (pp. 83-89).

Suhonen, J., Kuorilehto, M., Hännikäinen, M., & Hämäläinen, T. (2006). Cost-aware dynamic routing protocol for wireless sensor networks - design and prototype experiments. In *Proceedings of the IEEE 17th International Symposium on Personal, Indoor and Mobile Radio Communications* (pp. 1-5).

Suvorova, E. (2007). A Methodology and the Tool for Testing SpaceWire Routing Switches. In *Proceedings of the first International SpaceWire Conference*. Retrieved September 19, 2007, from http://spacewire.computing.dundee.ac.uk/proceedings/Papers/Test and Verification 2/suvorova2.pdf

Suvorova, E., & Sheynin, Y. (2003). *Digital systems design on VHDL language*. Saint-Petersburg, Russia: BHV-Petersburg.

Swan, S. (2003). *A Tutorial Introduction to the SystemC TLM Standard.* Retrieved July 7, 2008, from http://www-ti.informatik.uni-tuebingen.de/~systemc/Documents/Presentation-13-OSCI_2_swan.pdf

Synopsys, Inc. (2010). Synopsys platform architect. Retrieved from http://www.synopsys.com/Tools/SLD/VirtualPrototyping/Pages/PlatformArchitect.aspx

Synopsys, Inc. (2010). Synopsys processor designer. Retrieved from http://www.synopsys.com/Tools/SLD/ProcessorDev/Pages/default.aspx

Synopsys. (2007). *PrimeTime VX application note - Implementation methodology with variation-aware timing analysis, version 1.0.* Mountain View, CA: Synopsys.

Tagel, M., Ellervee, P., & Jervan, G. (2009, October). Scheduling Framework for Real-time Dependable NoC-based Systems. In *Proceedings of the International Symposium on System-on-Chip (SOC 2009)* (pp. 95-99).

Tan, L., Sokolsky, O., & Lee, I. (2004). Specification-based testing with linear temporal logic. In *Proceedings of the IEEE International Conference on Information Reuse and Integration* (pp. 483-498). Washington, DC: IEEE Computer Society.

Texas Instruments, Inc. (2010). OMAP. Retrieved from http://focus.ti.com/docs/prod/folders/print/omap3530.html

Thesing, S., Souyris, J., Heckmann, R., Randimbivololona, F., Langenbach, M., Wilhelm, R., et al. (2003). An abstract interpretation-based timing validation of hard real-time avionics software systems. In *Proceedings of the International Conference on Dependable Systems and Networks*, San Francisco, CA (pp. 625-632). Washington, DC: IEEE Computer Society.

Tianxu, Z., & Xuchao, D. (2003, November). Reliability Estimation Model of IC's Interconnect Based on Uniform Distribution of Defects on a Chip. In *Proceedings of the 18th IEEE International Symposium on Defect and Fault Tolerance in VLSI Systems* (pp. 11-17).

Tomasevic, M., & Milutinovic, V. (1994). Hardware approaches to cache coherence in shared-memory multiprocessors, part 1. *IEEE Micro*, *14*(5), 52–59. .doi:10.1109/MM.1994.363067

Towles, B., Dally, W. J., & Boyd, S. (2003). Throughput-centric routing algorithm design. In *Proceedings of the Fifteenth Annual ACM Symposium on Parallel Algorithms and Architectures* (pp. 200-209). New York, NY: ACM Press.

Tran, A. T., Truong, D. N., & Baas, M. B. (2009). A low-cost high-speed source-synchronous interconnection technique for GALS chip multiprocessors. In *Proceedings of the IEEE International Symposium on Circuits and Systems* (pp. 996-999).

Utting, M., & Legeard, B. (2007). *Practical model-based testing*. San Francisco, CA: Morgan Kaufmann.

van den Brand, J., Ciordas, C., Goossens, K., & Basten, T. (2007, April). Congestion Controlled Best-effort Communication for Networks-on-Chip. In *Proceedings of the Design, Automation & Test in Europe Conference and Exhibition (DATE '07)* (pp. 1-6).

Venet, A., & Brat, G. (2004). Precise and efficient static array bound checking for large embedded C programs. In *Proceedings of the ACM SIGPLAN Conference on Programming Language Design and Implementation*, Washington, DC (pp. 231-242). New York, NY: ACM Press.

Vidal, R. V. (Ed.). (1993). *Applied Simulated Annealing (Vol. 1)*. New York: Springer Verlag.

VMWare Inc. (2010). *VMWare server.* Retrieved from http://www.vmware.com/products/server/

Vouk, M. A. (1990). Back-to-back testing. *Information and Software Technology*, *32*(1), 34–45. doi:10.1016/0950-5849(90)90044-R

Wakileh, J., & Pahwa, A. (1996, November). Distribution system design optimization for cold load pickup. *IEEE Transactions on Power Systems*, *11*(4), 1879–1884. doi:10.1109/59.544658

Wang, G., Ji, Y., & Turgut, D. (2004). A routing protocol for power constrained networks with asymmetric links. In *Proceedings of the International Workshop on Performance Evaluation of Wireless Ad Hoc, Sensor, and Ubiquitous Networks*, Venice, Italy (pp. 69-76).

Wang, Q., Zhu, Y., & Cheng, L. (2006). Reprogramming wireless sensor networks: Challenges and approaches. *IEEE Network*, *20*(3), 48–55. doi:10.1109/MNET.2006.1637932

Weiser, M. (1981). Program slicing. In *Proceedings of the 5th International Conference on Software Engineering*, San Diego, CA (pp. 439-449). Washington, DC: IEEE Computer Society.

Wong, H.-S. P., Frank, D. J., Solomon, P. M., Wann, C. H. J., & Welser, J. J. (2005). Nanoscale CMOS. In Ionescu, A. M., & Banerjee, K. (Eds.), *Emerging Nanoelectronics: Life with and after CMOS* (*Vol. 1*, pp. 46–83). Norwell, MA: Kluwer Academic Publishers.

Woo, A., Tong, T., & Culler, D. (2003). Taming the underlying challenges of reliable multi-hop routing in wireless networks. In *Proceedings of the 1st International Conference on Embedded Networked Sensor Systems*, Los Angeles, CA (pp. 14-27).

Worm, F., Ienne, P., Thiran, P., & De Micheli, G. (2005). A robust self-calibrating transmission scheme for on-chip networks. *IEEE Transactions on Very Large Scale Integration Systems*, *13*(1), 126–139. doi:10.1109/TVLSI.2004.834241

Ye, T., Benini, L., & De Micheli, G. (2002). Analysis of power consumption on switch fabrics in network routers. In *Proceedings of the 39th Design Automation Conference* (pp. 524-529). New York, NY: ACM Press.

Yu, Z., & Baas, B. M. (2006). Implementing tile-based chip multiprocessors with GALS clocking style. In *Proceedings of the IEEE International Symposium on Circuits and Systems* (pp. 174-179).

Zamalloa, M. Z., & Krishnamachari, B. (2007). An analysis of unreliability and asymmetry in low-power wireless links. *ACM Transactions on Sensor Networks*, *3*(2).

Zander, J., & Schieferdecker, I. (2010). Model-based testing of embedded systems exemplified for the automotive domain. In Gomes, L., & Fernandes, J. M. (Eds.), *Behavioral modeling for embedded systems and technologies: Applications for design and implementation* (pp. 377–412). Hershey, PA: IGI Global. doi:10.4018/978-1-60566-750-8.ch015

Zhang, C., Yu, Q., Huang, X., & Yang, C. (2008). An RC4-based lightweight security protocol for resource-constrained communications. In *Proceedings of the 11th IEEE International Conference on Computational Science and Engineering Workshops* (pp. 133-140).

Zhao, Y. J., & Govidan, R. (2003). Understanding packet delivery performance in dense wireless sensor network. In *Proceedings of the 1st International Conference on Embedded Networked Sensor Systems*, Los Angeles, CA (pp. 1-13).

Zhong, H., & Nieh, J. (2001). CRAK: Linux checkpoint/restart as a kernel module *(Tech. Rep. No. CUCS-014-01). New York, NY: Columbia University.*

Zhou, G., He, T., Krishnamurthy, S., & Stankovic, J. A. (2004). Impact of radio irregularity on wireless sensor networks. In *Proceedings of the 2nd International Conference on Mobile Systems, Applications, and Services*, Boston, MA (pp. 125-138).

Zimmermann, F., Kloos, J., Eschbach, R., & Bauer, T. (2009). Risk-based statistical testing: A refinement-based approach to the reliability analysis of safety-critical systems. In *Proceedings of the European Workshop on Dependable Systems*, Toulouse, France.

Zisman, A., Spanoudakis, G., Perez-Miana, E., & Krause, P. (2003). Tracing software requirements artefacts. In *Proceedings of the International Conference on Software Engineering Research and Practise*, *1*, 448–455.

Zuse Institute Berlin (ZIB). (2009). *SCIP: Solving Constraint Integer Programs.*

About the Contributors

Seppo Virtanen received his BSc in Applied Physics, MSc in Electronics and Information Technology (1998), and DSc (Tech.) in Communication Systems (2004) from the University of Turku (Finland). Since 2009 he has been Adjunct Professor of Embedded Communication Systems in University of Turku. He is Editor-in-Chief of *International Journal of Embedded and Real-Time Communication Systems* (IJERTCS) and Senior Member of the IEEE. His current research interests include the design and methodological aspects of reliable and secure systems.

* * *

Gerd Ascheid received his Diploma and Ph.D. degrees in electrical engineering (communications eng.) from RWTH Aachen University, Aachen, Germany, in 1977 and 1984. In 1988 he started as a co-founder of CADIS GmbH. The company has successfully brought the system simulation tool COSSAP to the market. In 1994 CADIS GmbH was acquired by Synopsys, a California-based EDA market leader where his last position was Senior Director (executive management), wireless and broadband communications service line, Synopsys Professional Services. Since April 2003 he heads the Chair for Integrated Signal Processing Systems (ISS) of the Institute for Communication Technologies and Embedded Systems (ICE), RWTH Aachen University. He has published several scientific papers in reputed journals and conferences. He is also the chairman of the cluster of excellence in "Ultra-high speed Mobile Information and Communication (UMIC)" at RWTH Aachen University.

Sergey Balandin is Principle Scientist of Nokia Research Center and Adjunct Professor of Tampere University of Technology. Sergey Balandin holds M.Sc. in Computer Science from St. Petersburg Electrotechnical University "LETI", the main research subject: "Efficient pattern recognition in highly dynamic environment"; M.Sc. in Telecommunications from Lappeenranta University of Technology, main subject "Efficient IP routing in backbone networks"; and Ph.D. in Telecommunications and Control Theory from Nokia PhD school and St.Petersburg Electrotechnical University "LETI". Sergey Balandin has over 11 years of industrial experience. His main research interests are: smart spaces, embedded networks, routing and resource management, load balancing in fixed and wireless networks, peer-to-peer mobile and sensor networks. In 2006-2008 Sergey Balandin was Nokia representative in MIPI UniPro standardization. Currently he is Nokia's University cooperation representative in Russian and CIS and the general chair of Finnish-Russian University Cooperation in Telecommunications (FRUCT) program.

Davide Bertozzi got his PhD in Electrical Engineering from University of Bologna (Italy) in 2003. Since 2005 he is Assistant Professor at University of Ferrara (Italy), where he leads the research activities on Multi-Processor Systems-on-Chip and in Networks-on-Chip in particular. He has been visiting researcher at international academic institutions (Stanford University) and large semiconductor companies (NEC America Labs, USA; NXP Semiconductors, Holland; STMicroelectronics, Italy; Samsung Electronics, South Korea). Bertozzi was Program Chair of the Int. Symposium on Networks-on-Chip (2008), of the NoC track at the Design Automation and Test in Europe Conference (2010,2011) and guest editor of special issues on NoCs of IET CDT and Hindawi Journals (2007 and 2009). Bertozzi is a member of the Hipeac-2 NoE (Interconnect Cluster) and actively involved in STREP projects funded by the EU (Galaxy project, NaNoC project).

Jeronimo Castrillon received the electronics engineering degree from the Pontificia Bolivariana University in Colombia in 2004 and a master degree from the ALaRI Institute in Switzerland in 2006. Thereafter he joined the Chair for Software for Systems on Silicon (SSS) of the Institute for Communication Technologies and Embedded Systems (ICE), RWTH Aachen University, Germany, where he is working as a full time researcher while pursuing his PhD. His research interests lie on multi-processor systems-on-chip (MPSoC) programming: automatic code partitioning, code generation from abstract programming models, compile time mapping and scheduling as well as hardware/software support for run-time systems.

Elias A. Doumith is Assistant Professor in the Department of Networks and Computer Science at TELECOM ParisTech, France. He received his M.Sc. degree (2003) and Ph.D. degree (2007) from the Ecole Nationale Supérieure des Télécommunications, France. Between 2007 and 2009, he worked as Junior Research Engineer at the Institute of Communication Networks and Computer Engineering (IKR) at the University of Stuttgart, Germany. His domain of interest covers network planning and traffic engineering for networks, ranging from access networks to core networks including embedded networks. His current research works deal with cloud computing, radio over fiber, monitoring in optical networks, and scalable optical network design.

Alessandro Fantechi is full professor at the Faculty of Engineering of the University of Florence, where he teaches at the Computer Engineering and mechanical Engineering curricula. Prof. Fantechi is an active researcher in the areas of formal description techniques, temporal logics, formal verification, and their applications to the development of safety-critical systems, with a particular experience on railway signalling systems. His current research is focused on the application of model-checking based formal verification methods to safety-critical systems, on service formal description languages, and on formal aspects of software product lines. He has written over eighty papers for international journals and conferences, and has acted as the project team leader within several national and international research projects.

Alessio Ferrari graduated from the University of Florence with a MSc degree in Computer Engineering in 2007. In 2008, he started an Industrial Ph.D. programme in Computer Engineering, in collaboration with the Railway Signalling division of the General Electric Transportation Systems (GETS) company, where he has been System Engineer until September 2010. His role during the three years of collaboration consisted in defining and implementing strategies for the introduction of formal/semi-formal modelling and code generation within the development process of the company. The results of the research have been used for completely restructuring the software process of the company and have been successfully applied in the development of operating products. Alessio Ferrari is currently completing his academic studies and he is expected to discuss his Ph.D. dissertation concerning its research experience in GETS in March 2011.

Michel Gillet is an architect at Nokia Devices in Helsinki, Finland. Gillet has a MSc in computer science from University of Liege, Belgium, 1999, with the title "Ethernet ISDN gateway for VoIP". From 1999 till 2008, he was in the Nokia Research Center with the following research activities: high-speed serial links, signal integrity, electromagnetic simulations, sensor integration and embedded networks. In 2008, he joined Nokia Devices to drive the productization of embedded network technologies. Since its creation, he is the main Nokia representative in the MIPI UniPro standardization. His research interests include on-chip, off-chip and die-to-die embedded networks, system architecture, system modeling and distributed systems.

Daniele Grasso graduated with a MSc in Computer Science from the University of Florence in 2009. In 2009 started a research grant in formal verification with General Electric Transportation Systems (GETS) on behalf of the University of Florence. In 2010, he started an industrial Ph.D. programme in Computer Engineering in collaboration with GETS and in the same year he was hired by GETS as Verification Engineer in the Delta Project. First as Ph.D. student, and then in the Verification Team, he has been working in the definition and implementation of a new verification process that fulfils the development of innovative products in the railway signalling context through the integration of formal methods in the verification phases of the development process.

Vanessa Grosch received the diploma degree in Mathematics and Economics from Ulm University, Ulm, Germany, in 2004. She then worked as a software engineer in Great Britain, followed by the position as research assistant at Hohenheim University, Stuttgart, Germany. The main research area included software reliability. In 2007, she started as Ph.D. student at Ulm University (in cooperation with Daimler, Inc.) with model-based testing as her research interest. She is presently a test engineer at Daimler, Inc., Stuttgart, Germany.

Timo D. Hämäläinen received the M.Sc. degree in electrical engineering from Tampere University of Technology (TUT), Finland, in 1993, and PhD degree in electrical engineering from TUT, Finland, in 1997. He is a Professor at Department of Computer Systems (DCS) at TUT since 2001. He is author of over 60 journal and 200 conference publications and holds several patents. His research interests include wireless sensor networks, parallel system-on-chip architectures and design tools.

Marko Hännikäinen received his M.Sc. in information technology in 1998, and PhD in information technology in 2002 from the Tampere University of Technology (TUT). He was nominated an adjunct professor in 2005, and since 2007 he has been a Professor at the Department of Computer Systems, TUT. He has authored over 120 publications and holds several patents. His research interests include wireless sensor networks, new wireless applications concepts, and model-based system design.

Carles Hernández received the MS in Telecomunications Engineering and Computer Engineering from the Technical University of Valencia, Spain, in 2006 and 2008, respectively. Currently, he is a PhD candidate at the Technical University of Valencia. His research areas include Network-on-Chip architectures to address process variation as well as system level solutions to minimize the effects of variability on NoC performance.

Andreas Hoffmann is currently a R&D Director at Synopsys, responsible for Virtual Prototyping. In this role, he heads the development of Synopsys flagship products in the virtual platforms space. He received his MSc and PhD in electrical engineering and information technology from Ruhr University of Bochum, Germany and RWTH Aachen University, Germany, respectively. He has published more than 50 technical papers and articles and holds numerous patents in the field of embedded processor design and virtual platforms for HW and SW development.

Simon J. Hollis received the BA, MA and PhD. degrees from the University of Cambridge in 2003 and 2007. He is currently a lecturer in the Department of Computer Science at the University of Bristol. He has designed security-hardened embedded microprocessors and their interconnection networks, including the RasP Network-on-Chip. His current interests include massively parallel systems, on-chip interconnection networks, energy efficiency and asynchronous circuit design.

Chris Jackson earned his M.Eng. in 2007 in Computer systems and Software Engineering from the University of York, UK. Since 2008 he has been working in the Department of Computer Science, Bristol where he is currently pursuing his Ph.D. He has authored and co-authored papers on Network-on-Chip architecture and analysis. His current research interests include on-chip interconnection networks, routing algorithms, software simulation, synthetic traffic generation and network topology.

Florian Kantz is a Scientist at the ABB Corporate Research Center in Ladenburg, Germany. He joined ABB in 2008 with a background in automated testing of networked embedded systems. Currently he focuses on technologies for the integration of embedded devices in automation systems. His background is in electronic engineering and information technology. He received his Dipl.-Ing. from the Technical University of Darmstadt in 2002.

Torsten Kempf received the Diploma degree in electrical engineering from RWTH Aachen University, Aachen, Germany, in 2003. In 2004 he joined the Chair for Integrated Signal Processing Systems (ISS) of the Institute for Communication Technologies and Embedded Systems (ICE) where he is currently chief engineer. His research interests include multi-processor systems-on-chip (MPSoCs) and software defined radios (SDRs).

Stefan Kraemer is pursuing his PhD in electrical engineering and information technology at RWTH Aachen University, Germany. His research interests include efficient and fast processor and system simulation, code instrumentation and code optimization. He obtained his diploma in electrical engineering and information technology from RWTH Aachen University, Germany.

Teemu Laukkarinen received the MSc degree in computer science from Tampere University of Technology (TUT) in 2010. He is currently pursuing towards PhD in the Department of Computer Systems at TUT. His research interests include operating systems, applications and high level abstractions in wireless sensor networks.

Irina Lavrovskaya is a M. Sc student of the Saint-Petersburg State University of Airspace Instrumentation (SUAI). She holds B. Sc. in Computer Science from Saint-Petersburg State University of Airspace Instrumentation, the main research subject: "Investigation of SDL models validation methods on the example of distributed interrupts mechanism of SpaceWire standard". Irina Lavrovskaya has over 3 years of industrial experience and her main research interests are: embedded systems, modelling, SDL and SystemC modelling languages. Currently Irina Lavrovskaya is an engineer in a SUAI-Nokia Embedded Computing for Mobile Communications Lab.

Teijo Lehtonen received the M.Sc. degree in electronics and communication technology and the Ph.D. degree in electronics from the University of Turku, Turku, Finland, in 2004 and 2009, respectively. In 2008, he interned at the Embedded Integrated Systems-on-Chip (EdISon) Research Group, University of Rochester, Rochester, NY. He is currently working as a Senior Researcher at the University of Turku, Department of Information Technology. His research interests include the design of fault-tolerant nanoscale VLSI systems, networks-on-chip, reconfigurable FPGAs in space applications, and agile methods in the design of digital circuits.

Rainer Leupers received the M.Sc. and PhD degrees in Computer Science with honors from the Technical University of Dortmund, Germany, in 1992 and 1997. From 1997 to 2001 he was the chief engineer at the Embedded Systems chair at TU Dortmund. He joined RWTH Aachen University as a professor in 2002, where he currently heads the Chair for Software for Systems on Silicon (SSS) of the Institute for Communication Technologies and Embedded Systems (ICE). He published numerous books and technical papers, and served as a program committee member and topic chair of leading international conferences, including DAC, DATE, and ICCAD. He has been a co-founder of LISATek, an EDA tool provider for embedded processor design, acquired by Synopsys Inc. His research and teaching activities comprise software development tools, processor architectures and electronic design automation for embedded systems, with emphasis on multiprocessor system-on-chip design tools.

Pasi Liljeberg received his M.Sc. and Ph.D. degrees in electronics and information technology from the University of Turku, Finland, in 2000 and 2005, respectively. Since January 2010 he has been working in the Computer Systems laboratory, University of Turku as a senior lecturer. During the period 2007-2009 he has worked as an Academy of Finland postdoctoral researcher. His current research interests include network-on-chip intelligent communication architectures, fault tolerant design aspects, 3D multiprocessor system architectures, globally-asynchronous locally-synchronous platforms for nanoscale NoC and formal approaches in embedded system development. He has established and is leading a research group focusing on reliable and fault tolerant self-timed communication platforms for chip multiprocessor systems.

Lasse Määttä received the M.Sc. degree in computer science from Tampere University of Technology (TUT) in 2010. He is currently working as a researcher in the DACI research group in the Department of Computer Systems at TUT. His research interests include software design for low-power wireless sensor networks.

Gianluca Magnani obtained his MSc from University of Florence in 2009, discussing the thesis "Static analysis of Stateflow Models", realized in collaboration with General Electric Transportation Systems (GETS). In late 2009, he started a research grant concerning model-driven development and automatic code generation applied to the development process of GETS, and in early 2010, he started an Industrial Ph.D. programme in Computer Engineering in collaboration with the Safety and Validation office of GETS. During this time, he studied the application of formal verification techniques in the development of safety-critical systems. Today, he is hired as Software Engineer in the context of Delta Project, working on innovative signalling products according to CENELEC norms.

Pramita Mitra received her B.S. degree from Jadavpur University, India in 2006 and M.S. degree from University of Notre Dame, US in 2009, both in Computer Science and Engineering. She is currently a PhD student in the Computer Science and Engineering Department at the University of Notre Dame. Her area of research is efficient group communication and service sharing in mobile ad-hoc networks. Her research has been recognized with the Best Poster Award at the ACM Student Research Competition held at the 2009 Richard Tapia Celebration of Diversity in Computing Conference. She is a student member of ACM, IEEE and USENIX.

Philipp Nenninger is a Principal Scientist at the ABB Corporate Research Center in Ladenburg, Germany where he currently focuses on the low-power optimization of embedded devices, with focus on wireless sensors and energy harvesting as well as the development process of these devices. His background is in electronic engineering and industrial information technology. He received his Dipl.-Ing. and Dr.-Ing. in 2003 and 2007, respectively, both from the University of Karlsruhe (now KIT).

Valentin Olenev is Senior Researcher of the Saint-Petersburg State University of Airspace Instrumentation (SUAI). He holds M. Sc. in Computer Science from Saint-Petersburg State University of Airspace Instrumentation, the main research subject: "Research and development of a system for Wi-Fi networks security increase". Valentin Olenev has over 7 years of industrial experience and his main research interests are: embedded systems, modelling, SDL and SystemC modeling languages, models architecture, Petri Nets, SpaceWire, onboard systems. Currently Valentin Olenev is a project leader in a SUAI-Nokia Embedded Computing for Mobile Communications Lab. Alexey Rabin is Assistant Professor of St.-Petersburg State University of Aerospace Instrumentation.

Dietmar Petras is currently a Senior Manager at Synopsys. Within the department for System Level Solutions he is responsible for all aspects of the SystemC kernel and debug & analysis infrastructure. He has 20 years experience in system level simulation of multi-core platforms and communication systems. He obtained his diploma and PhD in electrical engineering from RWTH Aachen University, Germany. His research interests include system level simulation, software engineering processes and communication networks. He holds 10 patents and has published more than 40 papers and articles in the field of virtual platforms and mobile communications.

Thomas Philipp is a Senior R&D Engineer at Synopsys. Within the department for System-Level Solutions he is working in a team responsible for the SystemC kernel and debug & analysis infrastructure. His research interests include system-level simulation and checkpoint/restore methods. Thomas obtained his diploma in electrical engineering and information technology from RWTH Aachen University, Germany.

Juha Plosila received his M.Sc. and Ph.D. degrees in electronics and communication technology from the University of Turku, Turku, Finland, in 1993 and 1999, respectively. He is an Adjunct Professor in digital systems design at the University of Turku, Department of Information Technology, and holds currently a five year position of Academy Research Fellow at the Academy of Finland (2006-2011). He directs, at the University of Turku, an active research group focusing on modeling, design, and verification of network-on-chip (NoC) and 3D integrated systems at different abstraction levels. Plosila's research interests include formal methods for NoC and embedded system design, reliable on-chip communication techniques, on-chip thermal monitoring and control methods, dynamically reconfigurable systems, and application mapping on NoC based systems.

Christian Poellabauer received his Dipl. Ing. degree from the Vienna University of Technology, Austria and Ph.D. degree from the Georgia Institute of Technology, Atlanta, GA, both in Computer Science. He is currently an Associate Professor in the department of Computer Science and Engineering at the University of Notre Dame and Associate Director of the Wireless Institute at the University of Notre Dame. His research interests are in the areas of real-time systems, wireless ad-hoc and sensor networks, mobile computing, energy-efficient systems, and vehicular networks. He has published more than 80 papers in these areas and he has co-authored a textbook on Wireless Sensor Networks. His research has received funding through the National Science Foundation (including a CAREER award in 2006), Army Research Office, Office of Naval Research, IBM, Intel, Toyota ITC, and Motorola Labs. He is a senior member of ACM and IEEE.

Alexey Rabin is graduated engineer in Computer Science from St.-Petersburg State University of Aerospace Instrumentation, the main research subject: "Implementation of the switches in systems with arithmetical code division of channels", and holds Ph.D. in System Analysis, Control and Information Processing from St.-Petersburg State University of Aerospace Instrumentation, the main research subject: "Use of orthogonal coding for increase of noise immunity of information transmission systems". Alexey Rabin has over 8 years of industrial experience. His main research interests are: information theory, noise immunity coding, data transmission systems and networks, data transmission protocols, embedded systems modelling, embedded computing and digital signal processing.

Ville Rantala received the B.Sc. (Tech.) degree in communication systems and the M.Sc. (Tech.) degree in computer systems from the University of Turku, Turku, Finland, in 2007 and 2008, respectively, where he is currently working toward the Ph.D. degree in the Department of Information Technology. He is currently with the Turku Centre for Computer Science (TUCS), Turku, Finland. His current research interests include design of intelligent and fault-tolerant network-on-chip communication architectures, routing algorithms as well as monitoring and diagnostic services.

Andreas Reifert received his degree as Dipl.-Inform. in Computer Science from the Julius-Maximilians-Universität Würzburg, Germany, in 2003. He then moved to the Institute of Communication Networks and Computer Systems (IKR) at the University of Stuttgart as a member of the research staff. There, he worked in various national and international projects focusing on peer-to-peer technologies and service architectures for future telecommunication infrastructures. Mr. Reifert is currently working towards his PhD in Electrical Engineering. He concentrates on service component placement in wide-area communication infrastructures, specializing in planning aspects and resource provisoning.

Thomas Ruschival works as Scientist at ABB Corporate Research Center, Germany. His focus is on integration of wireless communication in industrial control systems and device management systems. In the context of industrial automation, he is also involved in improvement of development processes for application engineering projects. Thomas Ruschival graduated as Dipl.-Ing. in Electrical Engineering at University of Stuttgart in 2008.

Federico Silla received the MS and PhD degrees in Computer Engineering from the Technical University of Valencia, Spain, in 1995 and 1999, respectively. He is currently an associate professor at the Department of Computer Engineering (DISCA) at that university, and external contributor of the Advanced Computer Architecture research group at the Department of Computer Engineering at the University of Heidelberg. He is also member of the Advanced Technology Group of the HyperTransport Consortium, whose main result to date has been the development and standardization of an extension to HyperTransport (High Node Count HyperTransport Specification 1.0). His research addresses high performance on-chip and off-chip interconnection networks as well as distributed systems. He has published numerous papers in peer-reviewed conferences and journals, as well as several book chapters. He has been member of the Program Committee in several of the most prestigious conferences in his area, including IPDPS, HiPC, SC, etc.

Jörg Sommer received his Diploma degree in information technology from the Baden-Württemberg Cooperative State University, Heidenheim, Germany, and his Master degree in computer science from the University of Ulm, Germany, in 2002 and 2004 respectively. Since 2005, he has been with the Institute of Communication Networks and Computer Engineering (IKR) at the University of Stuttgart, Germany. His research interests include performance evaluation of communication networks as well as topology design and optimization problems. He is a member of the German Gesellschaft für Informatik (Computer Science Society).

Alexander Stepanov is a M. Sc student of the Saint-Petersburg State University of Airspace Instrumentation (SUAI). He holds B. Sc. in Computer Science from Saint-Petersburg State University of Airspace Instrumentation, the main research subject: "Data transmission protocol physical adapter layer research by modeling in SDL language". Alexander Stepanov has 3 years of industrial experience and his main research interests are: embedded systems, modelling, SDL and SystemC modelling languages. Currently Alexander Stepanov is an engineer in a SUAI-Nokia Embedded Computing for Mobile Communications Lab.

Alessandro Strano received the bachelor's degree and the master degree from the University of Ferrara, Italy, in Electronic and Telecomunication Engeneering in 2005 and 2009 respectively. Now he is pursuing PhD under the supervision of Dr.D.Bertozzi and, from 2010, he is involved as technical leader in the European NANOC project. His current research interests include Globally-Asynchronous-Locally-Synchronous Network on Chip with primary focus on synchronization issues and fault-tollerant Network on Chip.

Detlef Streitferdt is the head of the Software Architectures and Product Lines group at the Ilmenau University of Technology since August 2010. The research fields are the efficient development of software architectures and product lines, their analysis and their assessment as well as software development processes and model-driven development. Before that he was Principal Scientist at the ABB AG Corporate Research Center in Ladenburg, Germany, where he started 2005. He was working in the field of software development for embedded systems. Detlef studied Computer Science at the University of Stuttgart, spent a year of his studies at the University of Waterloo in Canada and graduated 1999. He received his doctoral degree from the Technical University of Ilmenau in the field of requirements engineering for product-line software development in 2004.

Jukka Suhonen received the M.Sc. degree in computer science from the Tampere University of Technology (TUT), Finland in 2004. He is currently pursuing his PhD in the Department of Computer Systems at TUT. His research interests include wireless networking, network protocol and algorithm design, and Quality of Service issues in wireless networks.

Matteo Tempestini, MSc degree in Computer Science (2005), has realized the project of thesis titled "*Stateflow System Modelling of a Railway Signalling System*" in collaboration with the Verification and Validation office of General Electric Transportation Systems (GETS) in Florence. Involved since 2006 in the software testing process of GETS, he has written the publications "*A Story About Formal Methods Adoption by a Railway Signalling Manufacturer*" (2006) and "*Formal Specifications of Railway Signalling Systems using Stateflow*" (2007). An expert in safety-critical software process, especially in CENELEC EN 50128 applications, he has studied the possibilities to use formal methods and model-based design in safety-critical systems. Today he is responsible for the Safety and Validation Laboratory in GETS and manages validation-testing activities for embedded signalling systems.

Bart Vanthournout received the electrical engineering degree from the Katholieke Universiteit Leuven, Heverlee, Belgium, in 1990. Subsequently, he joined the VLSI Systems Design Group of IMEC, Leuven, Belgium, and was involved in developing scheduling and assignment techniques as part of a behavioral synthesis environment. From 1994 to 1996 he was a member of the Hardware and Software Systems group, developing a synthesis and simulation environment for Hardware and Software codesign. Since 1997 he continued this work at CoWare, of which he is one of the cofounders. At CoWare he held several positions managing applications and engineering groups. He joined Synopsys in 2010 after the acquisition of CoWare. Most recently he is responsible for product definition and research activities as product architect. Currently areas of interest are transactional modeling (TLM) and MPSOC design methodologies.

Diandian Zhang received the Diploma degree in electrical engineering from the Chair for Integrated Signal Processing Systems (ISS) of the Institute for Communication Technologies and Embedded Systems (ICE), RWTH Aachen University, Germany, in 2006. He is currently a Ph.D. candidate in electrical engineering at RWTH Aachen University. His current research interests are development of application-specific instruction-set processors (ASIPs), communication architecture exploration and high-level power estimation in multi-processor systems-on-chip (MPSoCs) in the embedded domain.

Han Zhang received his Diploma degree in electrical engineering from the Chair for Integrated Signal Processing Systems (ISS) of the Institute for Communication Technologies and Embedded Systems (ICE), RWTH Aachen University, Germany, in 2009. His research interests include the development of communication architectures in multi-processor systems-on-chip (MPSoCs). Now he is with AAEON Technology GmbH.

Index

F

G

H

I

K

L

M

N

O

P

R

S

T

U

V

W